Functional Biodegradable Nanocomposites

Functional Biodegradable Nanocomposites

Editors

Daniel López
Coro Echeverría
Águeda Sonseca

MDPI • Basel • Beijing • Wuhan • Barcelona • Belgrade • Manchester • Tokyo • Cluj • Tianjin

Editors
Daniel López
CSIC—Instituto de Ciencia y
Tecnología de Polímeros
(ICTP)
Spain

Coro Echeverría
Instituto de Ciencia y
Tecnología de Polímeros
(ICTP-CSIC) &
Interdisciplinary Platform for
Sustainable Plastics towards a
Circular Economy,
SusPlast-CSIC
Spain

Águeda Sonseca
Universitat Politècnica de
València
Spain

Editorial Office
MDPI
St. Alban-Anlage 66
4052 Basel, Switzerland

This is a reprint of articles from the Special Issue published online in the open access journal *Nanomaterials* (ISSN 2079-4991) (available at: https://www.mdpi.com/journal/nanomaterials/special_issues/func_biodeg_nano).

For citation purposes, cite each article independently as indicated on the article page online and as indicated below:

LastName, A.A.; LastName, B.B.; LastName, C.C. Article Title. *Journal Name* **Year**, *Volume Number*, Page Range.

ISBN 978-3-0365-5697-0 (Hbk)
ISBN 978-3-0365-5698-7 (PDF)

© 2022 by the authors. Articles in this book are Open Access and distributed under the Creative Commons Attribution (CC BY) license, which allows users to download, copy and build upon published articles, as long as the author and publisher are properly credited, which ensures maximum dissemination and a wider impact of our publications.
The book as a whole is distributed by MDPI under the terms and conditions of the Creative Commons license CC BY-NC-ND.

Contents

Preface to "Functional Biodegradable Nanocomposites" vii

Agueda Sonseca, Coro Echeverría and Daniel López
Functional Biodegradable Nanocomposites
Reprinted from: *Nanomaterials* **2022**, *12*, 2500, doi:10.3390/nano12142500 1

Tiphaine Messin, Nadège Follain, Quentin Lozay, Alain Guinault, Nicolas Delpouve, Jérémie Soulestin, Cyrille Sollogoub and Stéphane Marais
Biodegradable PLA/PBSA Multinanolayer Nanocomposites: Effect of Nanoclays Incorporation in Multinanolayered Structure on Mechanical and Water Barrier Properties
Reprinted from: *Nanomaterials* **2020**, *10*, 2561, doi:10.3390/nano10122561 5

Agueda Sonseca, Salim Madani, Gema Rodríguez, Víctor Hevilla, Coro Echeverría, Marta Fernández-García, Alexandra Muñoz-Bonilla, Noureddine Charef and Daniel López
Multifunctional PLA Blends Containing Chitosan Mediated Silver Nanoparticles: Thermal, Mechanical, Antibacterial, and Degradation Properties
Reprinted from: *Nanomaterials* **2020**, *10*, 22, doi:10.3390/nano10010022 31

Agueda Sonseca, Salim Madani, Alexandra Muñoz-Bonilla, Marta Fernández-García, Laura Peponi, Adrián Leonés, Gema Rodríguez, Coro Echeverría and Daniel López
Biodegradable and Antimicrobial PLA–OLA Blends Containing Chitosan-Mediated Silver Nanoparticles with Shape Memory Properties for Potential Medical Applications
Reprinted from: *Nanomaterials* **2020**, *10*, 1065, doi:10.3390/nano10061065 49

G. M. Nazmul Islam, Stewart Collie, Mohammad Qasim and M. Azam Ali
Highly Stretchable and Flexible Melt Spun Thermoplastic Conductive Yarns for Smart Textiles
Reprinted from: *Nanomaterials* **2020**, *10*, 2324, doi:10.3390/nano10122324 65

Coro Echeverría and Carmen Mijangos
A Way to Predict Gold Nanoparticles/Polymer Hybrid Microgel Agglomeration Based on Rheological Studies
Reprinted from: *Nanomaterials* **2019**, *9*, 1499, doi:10.3390/nano9101499 89

Shi Su and Peter M. Kang
Systemic Review of Biodegradable Nanomaterials in Nanomedicine
Reprinted from: *Nanomaterials* **2020**, *10*, 656, doi:10.3390/nano10040656 103

Haichao Liu, Ranran Jian, Hongbo Chen, Xiaolong Tian, Changlong Sun, Jing Zhu, Zhaogang Yang, Jingyao Sun and Chuansheng Wang
Application of Biodegradable and Biocompatible Nanocomposites in Electronics: Current Status and Future Directions
Reprinted from: *Nanomaterials* **2019**, *9*, 950, doi:10.3390/nano9070950 125

Madison Bardot and Michael D. Schulz
Biodegradable Poly(Lactic Acid) Nanocomposites for Fused Deposition Modeling 3D Printing
Reprinted from: *Nanomaterials* **2020**, *10*, 2567, doi:10.3390/nano10122567 157

Preface to "Functional Biodegradable Nanocomposites"

This Special Issue presents several examples of the latest advances in functional biodegradable nanocomposites for different properties and applications, namely packaging, electronic, conductive, and biomedical applications. It is aimed at scientists in all application areas interested in developing environmentally friendly materials while improving their properties and functionalities. Authors contributing to this Special Issue are acknowledged. We trust that readers will find this content of interest.

Daniel López, Coro Echeverría, and Águeda Sonseca
Editors

Editorial

Functional Biodegradable Nanocomposites

Agueda Sonseca [1,*,†], Coro Echeverría [1,2,*] and Daniel López [1,2,*]

1. MacroEng Group, Instituto de Ciencia y Tecnología de Polímeros (ICTP-CSIC), C/Juan de la Cierva 3, 28006 Madrid, Spain
2. Interdisciplinary Plataform for "Sustainable Plastics towards a Circular Economy" (SUSPLAST-CSIC), Madrid, Spain
* Correspondence: agsonol@upvnet.upv.es (A.S.); cecheverria@ictp.csic.es (C.E.); daniel.l.g@csic.es (D.L.)
† Current address: Instituto de Tecnología de Materiales, Universitat Politècnica de València (UPV), Camino de Vera s/n, 46022 Valencia, Spain.

Over 367 million tons of plastics are produced annually worldwide, and the growth of plastic pollution has become a global concern [1]. Environmental issues related to the persistence of plastic waste have urged the development of more sustainable biodegradable alternatives. Consequently, in 2018, the European Commission adopted a circular economy plan for the management of plastics based on innovative research on polymers derived from natural resources [2]. Thus, biodegradable polymers have been effectively developed over the last few years as promising alternatives to mostly non-degradable commodity polymers, meeting the demands of a broad range of fields, including the medical, packaging, agricultural, personal care, and automotive industries [3].

Individually, biodegradable polymers do not possess physical properties or mechanical strengths comparable to their non-degradable counterparts, limiting their application. Significant research efforts have been made for the development of biodegradable polymeric formulations with mechanical and physical properties comparable to those of non-biodegradable ones [4]. As a result, biodegradable nanocomposites entered the research scene, offering the possibility of new, enhanced properties and fields of application [5].

One of the reviews in this Special Issue focuses on the application of biodegradable and biocompatible nanocomposites in electronics, highlighting the need for degradable functional systems based on nanocomposites to deal with the problem of electronic waste [6]. Nanoparticles have also found applications in nanomedicine, providing unique properties and great advantages thanks to their small size that is favorable from a therapeutic point of view. However, their safety has been questioned many times. In this context, biodegradable nanomaterials, degradable under biological conditions, hold great promise in the biomedical field, and the latest advances are reviewed in this Special Issue by Su et al. [7].

The properties of nanocomposites depend not only on the properties of individual materials, but also on their interfacial interactions and morphology, which are significantly affected by processing methods. In this context, Echeverría et al. present a detailed rheological study that investigates how gold nanoparticles (AuNP) affect the properties of a hybrid poly(acrylamide-co-acrylic acid) P(AAm-co-AAc) microgel matrix. The knowledge presented through this work facilitates the prediction of system behavior, consequently allowing the preparation of reproducible systems, for instance, as injectable systems [8]. Bardot et al. review the development of nanocomposites based on polylactic acid (PLA), a biodegradable biopolymer obtained from agricultural products, by means of fused deposition modelling (3D printing). They demonstrate the possibility of obtaining biodegradable systems without compromising mechanical robustness, which is key in industrial applications [9].

As evidenced in the review described above, polylactic acid (PLA) represents a promising alternative to mostly non-degradable commodity polymers; moreover, the modulation

Citation: Sonseca, A.; Echeverría, C.; López, D. Functional Biodegradable Nanocomposites. *Nanomaterials* **2022**, *12*, 2500. https://doi.org/10.3390/nano12142500

Received: 29 June 2022
Accepted: 30 June 2022
Published: 21 July 2022

Publisher's Note: MDPI stays neutral with regard to jurisdictional claims in published maps and institutional affiliations.

Copyright: © 2022 by the authors. Licensee MDPI, Basel, Switzerland. This article is an open access article distributed under the terms and conditions of the Creative Commons Attribution (CC BY) license (https://creativecommons.org/licenses/by/4.0/).

of its mechanical performance can be controlled with nanocomposites formation and specific processing methods. Messin et al. developed multi-nanolayered nanocomposites via the coextrusion of polylactic acid and poly(butylene succinate-co-butylene adipate) filled with nanoclays in order to obtain enhanced water barrier properties [10]. Sonseca et al. developed plasticized PLA nanocomposites with potential application for use as antibacterial food packaging degradable materials, incorporating silver nanoparticles obtained from a green synthesis procedure. The same materials were demonstrated to be useful as shape memory nanocomposites for potential medical application, thanks to the synergistic effect of lactic acid oligomer (OLA) and silver nanoparticles. The incorporation of OLA as a plasticizer located the glass transition of the system near to the physiological one, while the silver nanoparticles fastened the recovery process and imparted antimicrobial activity [11,12]. Nazmul et al. produced scalable environmentally friendly smart interactive textiles by means of melt spun thermoplastic conductive yarns based on PLA, polypropylene (PP), and their mixtures (PLA/PP) [13].

In summary, this Special Issue presents several examples of the latest advances in functional biodegradable nanocomposites for different applications. We would like to thank all authors for contributing to this collection, and we hope readers will find the content interesting, enjoyable, and useful.

Funding: This research was funded by Spanish Ministry of Science and Innovation (AEIMICINN/FEDER); Projects MAT2016-78437-R, MAT2017-88123-P and PCIN-2017-036 and by the Valencian Autonomous Government, Generalitat Valenciana, GVA (GV/2021/182).

Acknowledgments: A. S. acknowledges her "APOSTD/2018/228" and "PAID-10-19" postdoctoral contracts from the Education, Research, Culture and Sport Council from the Government of Valencia and from the Polytechnic University of Valencia, respectively. C.E. acknowledges IJCI-2015-26432 contract from MICINN.

Conflicts of Interest: The authors declare no conflict of interest.

References

1. Plastics Europe Market Research Group (PEMRG) and Conversion Market & Strategy GmbH. Plastics Europe Plastics the Fact 2021. 2021. Available online: https://plasticseurope.org/knowledge-hub/plastics-the-facts-2021/ (accessed on 20 June 2022).
2. European Commission, Directorate-General for Research and Innovation. *A Circular Economy for Plastics: Research and Innovation for Systemic Change*; Publications Office of the European Union: Luxembourg City, Luxembourg, 2020; Available online: https://data.europa.eu/doi/10.2777/192216 (accessed on 20 June 2022).
3. Rai, P.; Mehrotra, S.; Priya, S.; Gnansounou, E.; Sharma, S.K. Recent Advances in the Sustainable Design and Applications of Biodegradable Polymers. *Bioresour. Technol.* **2021**, *325*, 124739. [CrossRef] [PubMed]
4. Abioye, A.A.; Fasanmi, O.O.; Rotimi, D.O.; Abioye, O.P.; Obuekwe, C.C.; Afolalu, S.A.; Okokpujie, I.P. Review of the Development of Biodegradable Plastic from Synthetic Polymers and Selected Synthesized Nanoparticle Starches. *J. Phys. Conf. Ser.* **2019**, *1378*, 42064. [CrossRef]
5. Hwang, S.Y.; Yoo, E.S.; Im, S.S. The Synthesis of Copolymers, Blends and Composites Based on Poly(Butylene Succinate). *Polym. J.* **2012**, *44*, 1179–1190. [CrossRef]
6. Liu, H.; Jian, R.; Chen, H.; Tian, X.; Sun, C.; Zhu, J.; Yang, Z.; Sun, J.; Wang, C. Application of Biodegradable and Biocompatible Nanocomposites in Electronics: Current Status and Future Directions. *Nanomaterials* **2019**, *9*, 950. [CrossRef] [PubMed]
7. Su, S.; Kang, P.M. Systemic Review of Biodegradable Nanomaterials in Nanomedicine. *Nanomaterials* **2020**, *10*, 656. [CrossRef] [PubMed]
8. Echeverría, C.; Mijangos, C. A Way to Predict Gold Nanoparticles/Polymer Hybrid Microgel Agglomeration Based on Rheological Studies. *Nanomaterials* **2019**, *9*, 1499. [CrossRef] [PubMed]
9. Bardot, M.; Schulz, M.D. Biodegradable Poly(Lactic Acid) Nanocomposites for Fused Deposition Modeling 3d Printing. *Nanomaterials* **2020**, *10*, 2567. [CrossRef] [PubMed]
10. Messin, T.; Follain, N.; Lozay, Q.; Guinault, A.; Delpouve, N.; Soulestin, J.; Sollogoub, C.; Marais, S. Biodegradable Pla/Pbsa Multinanolayer Nanocomposites: Effect of Nanoclays Incorporation in Multinanolayered Structure on Mechanical and Water Barrier Properties. *Nanomaterials* **2020**, *10*, 2561. [CrossRef] [PubMed]
11. Sonseca, A.; Madani, S.; Rodríguez, G.; Hevilla, V.; Echeverría, C.; Fernández-García, M.; Muñoz-Bonilla, A.; Charef, N.; López, D. Multifunctional PLA Blends Containing Chitosan Mediated Silver Nanoparticles: Thermal, Mechanical, Antibacterial, and Degradation Properties. *Nanomaterials* **2020**, *10*, 22. [CrossRef]

2. Sonseca, A.; Madani, S.; Muñoz-Bonilla, A.; Fernández-García, M.; Peponi, L.; Leonés, A.; Rodríguez, G.; Echeverría, C.; López, D. Biodegradable and Antimicrobial Pla–Ola Blends Containing Chitosan-Mediated Silver Nanoparticles with Shape Memory Properties for Potential Medical Applications. *Nanomaterials* **2020**, *10*, 1065. [CrossRef]
3. Nazmul Islam, G.M.; Collie, S.; Qasim, M.; Azam Ali, M. Highly Stretchable and Flexible Melt Spun Thermoplastic Conductive Yarns for Smart Textiles. *Nanomaterials* **2020**, *10*, 2324. [CrossRef]

Article

Biodegradable PLA/PBSA Multinanolayer Nanocomposites: Effect of Nanoclays Incorporation in Multinanolayered Structure on Mechanical and Water Barrier Properties

Tiphaine Messin [1], Nadège Follain [1,*], Quentin Lozay [1], Alain Guinault [2], Nicolas Delpouve [3], Jérémie Soulestin [4], Cyrille Sollogoub [2] and Stéphane Marais [1,*]

1. Normandie Univ, UNIROUEN, INSA Rouen, CNRS, PBS, 76000 Rouen, France; tiphaine.messin@gmail.com (T.M.); Lozay.quentin@yahoo.fr (Q.L.)
2. Laboratoire PIMM, Arts et Métiers Institute of Technology, CNRS, Cnam, Hesam Université, 151, Bd de l'Hôpital, 75013 Paris, France; alain.guinault@lecnam.net (A.G.); cyrille.sollogoub@cnam.fr (C.S.)
3. Normandie Univ, UNIROUEN Normandie, INSA Rouen, CNRS, Groupe de Physique des Matériaux, 76000 Rouen, France; nicolas.delpouve1@univ-rouen.fr
4. Département Technologie des Polymères et Composites & Ingénierie Mécanique (TPCIM), Institut Mines Telecom Lille Douai (IMT Lille Douai), 59508 Douai, France; jeremie.soulestin@imt-lille-douai.fr
* Correspondence: nadege.follain@univ-rouen.fr (N.F.); stephane.marais@univ-rouen.fr (S.M.)

Received: 26 November 2020; Accepted: 15 December 2020; Published: 20 December 2020

Abstract: Biodegradable PLA/PBSA multinanolayer nanocomposites were obtained from semi-crystalline poly(butylene succinate-*co*-butylene adipate) (PBSA) nanolayers filled with nanoclays and confined against amorphous poly(lactic acid) (PLA) nanolayers in a continuous manner by applying an innovative coextrusion technology. The cloisite 30B (C30B) filler incorporation in nanolayers was considered to be an improvement of barrier properties of the multilayer films additional to the confinement effect resulting to forced assembly during the multilayer coextrusion process. 2049-layer films of ~300 µm thick were processed containing loaded PBSA nanolayers of ~200 nm, which presented certain homogeneity and were mostly continuous for the 80/20 wt% PLA/PBSA composition. The nanocomposite PBSA films (monolayer) were also processed for comparison. The presence of exfoliated and intercalated clay structure and some aggregates were observed within the PBSA nanolayers depending on the C30B content. A greater reduction of macromolecular chain segment mobility was measured due to combined effects of confinement effect and clays constraints. The absence of both polymer and clays interdiffusions was highlighted since the PLA glass transition was unchanged. Besides, a larger increase in local chain rigidification was evidenced through RAF values due to geometrical constraints initiated by close nanoclay contact without changing the crystallinity of PBSA. Tortuosity effects into the filled PBSA layers adding to confinement effects induced by PLA layers have caused a significant improvement of water barrier properties through a reduction of water permeability, water vapor solubility and water vapor diffusivity. The obtaining barrier properties were successfully correlated to microstructure, thermal properties and mobility of PBSA amorphous phase.

Keywords: biodegradable polymers; coextrusion; multilayer film; barrier properties; montmorillonite fillers

---

1. Introduction

Over the past thirty years, numerous works have been carried out in the field of polymer materials with the aim of reducing permeability as much as possible and approaching properties of other materials like glass or metal. The strategies developed to achieve this aim were the deposition of a thin layer by cold plasma [1,2]; the mixtures of highly barrier polymers, like EVOH for example with standard polymers used in packaging field like PE [3], PET [4] or PP [5]; the incorporation of fillers like montmorillonite [6,7] or graphene [8,9]; and even the development of a multilayer structure, containing 5–7 layers, each one having a specific characteristic. During the last 15 years, an innovative coextrusion process has been developed at the laboratory scale, allowing production of multinanolayer films containing up to 2000 or 4000 layers of two different polymers [10]. The most important results concerning the improvement of the barrier properties thanks to this multinanolayer structure, were obtained by confining PEO with EAA [11] and PS [12] or PCL with PS [13] and PMMA [14]. The confinement effect, induced by the forced assembly multinanolayer co-extrusion process, led to a spectacular decrease in the oxygen permeability coefficient of more than 20 times and was explained by the induced morphology of the crystals creating more tortuosity into the film. In fact, in multilayered structures by nanoscale confinement, the polymer layers crystallized as single crystals in in-place orientation parallel to the layers. More recently, the confinement of PLLA by PC and PS amorphous polymer layers have been performed by means of the multilayer coextrusion process [15], and the authors have highlighted the impact on the oxygen barrier properties of the rigid amorphous fraction (RAF) generated by postannealing of the multilayer films. Annealing PLLA/PS films allowed for obtaining a high crystallinity degree without RAF and this has induced an oxygen barrier improvement by a factor of 10 with 20 nm thick PLLA layers [15]. Concerning the water barrier properties for multilayer films, nowadays, only few works refer to this and the water barrier improvement was mainly obtained afterheat treatment or after a biaxial drawing [16] utilized to promote polymer crystallization. Such post treatments have improved water barrier properties. Interesting results were put forward with PC/MXD6 multilayer films [17] without annealing treatment. The authors have reported improvement of water barrier properties, only due to the nanoscale confinement effect induced by the alternating polymer multilayering, even if water-crystallization effect for MXD6 was evidenced causing an increase in MXD6 crystallinity. Again, in a previous work, a PLA/PBSA [18] multilayer film was elaborated and a reduction in permeability coefficient to different molecules was observed, by 100 times in the case of carbon dioxide. In order to further improve the barrier performances of this material, it has been considered the possibility to couple the multilayer structure with one of the approaches mentioned above, namely by incorporating fillers into the multilayer structure. Until now, very few works have taken advantage from the multinanolayer process in order to disperse and/or orientate fillers (microtalc [19], phosphate glass [20], carbon black [21], montmorillonite [22], carbon nanotubes [23], graphene [24,25]) in the aim to improve mechanical and/or barrier properties. For example, an increase of the stiffness (Young's modulus from 2.5 GPa to 5 GPa) with the addition of 20% of phosphate glass fillers into a PP-g-MA multilayer film [20] or a mechanical reinforcement of the layers by 118% with the incorporation of 2% of graphene into PMMA multilayer structure [24] have been reported. Concerning gas barrier properties, a decrease of around 30% for the oxygen permeability coefficient has been measured with the multilayer film of PP-g-MA containing phosphate glass fillers [20] (around 20 times after bi-drawing treatment, which allows us to lengthen the fillers, changing from a spherical to a lamellar form). For a multilayer film composed of alternating layers of LDPE and LDPE-g-MA filled with 5 wt% of montmorillonite C20A [22], a reduction of the oxygen permeability coefficient from 0.84 to 0.14 Barrer was obtained after an annealing step leading to an orientation of clay particles in the plane of the layers.

In this work, PBSA nanocomposite films and PLA/PBSA (80/20 wt%) multinanolayer films loaded with organo-modified montmorillonite (C30B) were investigated. The intent of this work is to show how the incorporation of lamellar clays and their dispersion in PBSA nanolayers confined by PLA layers influence the mechanical and barrier properties, given that, to our knowledge, such a biodegradable nanomaterial has never been developed and studied in literature. The water transport properties of these nanomaterials were analyzed from permeation and sorption measurements, as well as their thermal and mechanical properties in relation with their microstructure.

2. Materials and Methods

2.1. Materials

Poly(lactic acid) (PLA), under the trade of 4060D corresponding to an amorphous polymer, was provided by Resinex (France). Semicrystalline poly(butylene succinate-*co*-butylene adipate) (PBSA), referenced as PBE001, was supplied by NaturePlast (France). Organo-modified montmorillonite Cloisite 30B (noted C30B in this study) was supplied from BYK Additives. This filler was selected owing to its better dispersion into the PBSA matrix [26] compared to native montmorillonite filler due to the organomodification.

2.2. Films Preparation

The incorporation of C30B clays into PBSA matrix was carried out in two steps. The first step consists in preparing a masterbatch containing 25 wt% of filler. Briefly, PBSA pellets were dried at 80 °C overnight and melt-mixed with clays by using a Coperion ZSK 26 twin-screw extruder at a temperature of 210 °C with a screw speed of 250 rpm. At the end of the die, the extrudate was cooled down in liquid water, pelletized and then dried at 80 °C to remove residual water. The second step concerns the dilution of the masterbatch with unfilled PBSA by twin-screw extrusion in order to obtain nanocomposite films with 2 wt% and 5 wt% of fillers. The same extrusion profile was used with pelletization and drying. The resulting C30B-PBSA pellets were thereafter dried before the elaboration of the nanocomposite films, as explained below.

The monolayer films of PBSA with 2% and 5 wt% of fillers were prepared using a classical single-screw extruder (20/20D Scamex extruder) with a 120/150/150/160/170 °C temperature profile and a screw speed of 29 and 33 rpm, respectively. The final thickness of films was obtained around 200 µm. Using the same parameters, the film of neat PBSA was also extruded. The films were thereafter noted as PBSA, PBSA2 and PBSA5 for the films containing 0, 2 and 5 wt% of fillers, respectively.

Multilayer films were prepared by applying the multilayer co-extrusion process which consists of melt PLA and PBSA into two separate single-screw extruders that converge in a three-layer feedblock to form an A/B/A structure, which is then multiplied by using the multiplying elements, as detailed in a previous paper [18]. The temperature profiles were 165/180/180/190/190/190 °C and 120/150/150/160/170 °C for the PLA and the PBSA, respectively. In this study, 10-layer multiplying elements (LME) at a temperature of 170 °C were used in order to obtain a film with theoretically 2049 layers. The PLA/PBSA multilayer films were composed of 80 wt% of PLA and 20 wt% of PBSA (unfilled or filled). The resulting films were noted as PLA/PBSA, PLA/PBSA2 and PLA/PBSA5 for the multilayer films containing 0, 2 and 5 wt% of C30B into the PBSA layers, which correspond at a final filler content of 0.5% and 1.25% into the two multilayer films. Considering a final thickness of 300 µm for the multilayer film, the individual thickness of the PBSA layers should be around 60 nm in theory.

In order to better evaluate the impact of the fillers within the PLA/PBSA multilayer film, a monolayer film of PBSA was investigated. The nanocomposite films of PBSA were thereafter named reference films in the second part of the characterization of the multinanolayer films.

2.3. Morphological Characterization

For Transmission Electronic Microscopy (TEM), the cross-section of the films has been prepared with an ultramicrotome (Leica UC7 cryomicrotome) at −140 °C using a diamond knife cryo immuno from Diatome (Switzerland) at a cutting speed of 1 mm/s to obtain ultrathin smooth slices of 80–90 nm thick. The films were imaged with a FEI Tecnai 12 Biotwin equipped with a thermo-ionic electron gun in LaB6 and operating at a tension of 80 KeV. TEM is equipped with CCD erlanghen ES500W camera. The different images were presented with the Digital Micrograph Software.

XRD analysis were performed on a Bruker AXS D8 Advance diffractometer with a Cobalt radiation source at a length wave of 1.789 Å. WAXS diffractograms were obtained in normal direction over the

2θ range from 5 to 40° with an angle increment fixed to 0.04° per step and a scan speed set to 1 sec/step. All scattering peaks and amorphous halos were fitted assuming Lorentzian equation.

2.4. Thermal Analyses

Thermal gravimetric analysis (TGA) was used in order to quantify the filler amount into the PBSA matrix. The analyses were performed at a heating rate of 10 °C/min from 30 to 550 °C under nitrogen atmosphere on a Q500 TGA from TA Instruments.

DSC Q2000 from TA instruments was used to perform DSC and MT-DSC analyses. DSC mode was used to determine the characteristic temperatures (T_g, T_c and T_m) and the degree of crystallinity (X_c) of polymers and MT-DSC mode to determine the mobile amorphous fraction (MAF) and rigid amorphous fraction (RAF). The DSC measurements were made with sample of ~6 mg at a heating/cooling rate of 10 °C/min from −60 to 200 °C, and the MT-DSC measurements were conducted in "Heat-only" mode (oscillation amplitude: 0.21 °C, oscillation period: 80 s, heating rate: 1 °C/min) with ~3 mg of polymer from −90 to 200 °C.

The degree of crystallinity was determined from the following equation:

$$X_c(\%) = \frac{\Delta H_m - \Delta H_c}{\Delta H_m^0 (1 - \phi)} \times 100 \quad (1)$$

where DHm is the melting enthalpy, DHc is the enthalpy of crystallization and ΔH_m^0 is the melting enthalpy of a 100% theoretical crystalline polymer. For the PBSA, ΔH_m^0 is equal to 113.4 J/g [27]. In the case of loaded PBSA, ϕ is the mass fraction of fillers.

The mobile amorphous fraction (MAF) was calculated from the glass transition by using the equation:

$$X_{MAF}(\%) = \frac{\Delta C_p}{\Delta C_p^0} \times 100 \quad (2)$$

where DCp is the specific heat capacity of PBSA and ΔC_p^0 is the specific heat capacity for the 100% amorphous PBSA polymer. The ΔC_p^0, equal to 0.614 J·g^{-1}·°C^{-1}, was previously determined by Bandyopadhyay et al. [28].

The rigid amorphous fraction (RAF) was obtained from the following equation:

$$X_{RAF}(\%) = 100 - (X_c - X_{MAF}) \quad (3)$$

2.5. Mechanical Tests

Uniaxial mechanical tests were performed on at least ten samples of 30 mm in length and 4 mm in width. The tests were conducted at room temperature (23 °C) on an Instron 5543 traction machine equipped with a 500 N load cell and with a speed of 25 mm/min. From the stress-strain curves, the mechanical parameters Young's modulus, tensile strength and the elongation at break were determined.

2.6. Barrier Properties

Water permeation experiments were performed at 25 °C on a lab-made apparatus, as detailed in a previous paper [29]. The principle is based on the monitoring of the dew point of the permeate gas using a chilled mirror hygrometer (Elcowa, France, General Eastern Instruments). The apparatus is composed of a two-compartment measurement cell and the studied film is placed between the two compartments. A preliminary drying step by nitrogen flux is conducted to obtain a low constant dew point close to −70 °C. The measurement was started when liquid water was introduced into the upstream compartment of the measurement cell and the diffusion of water through the studied film is monitoring at downstream from the variation of the dew point temperature as a function of time.

The resulting permeation flux curve allows for determining the stationary flux J_{st} (mmol·cm^{-2}·s^{-1}) at the stationary state as follows:

$$J_{st} = \frac{f \times 10^{-6} \times (x^{out} - x^{in}) \times p_t}{A \times R \times T} \quad (4)$$

with f the flow rate (560 mL·min^{-1}), A the exposed area (2.5 cm^2), R the ideal gas constant (0.082 atm·cm^3·K^{-1}·mmol^{-1}), T the temperature of the experiment (298.15 K), p_t the total pressure (1 atm), x^{in} et x^{out} the inlet and outlet water contents (ppm) present in the sweeping. These contents were calculated with the following equation:

$$x = e^{(-b/T_{dp}+c)} \quad (5)$$

where b and c are empirical constants (b: 6185.66 K and c: 31.38) valid for a dew temperature range from −75 °C to −20 °C [30] and T_{dp} the dew point temperature.

At the stationary state, where the stationary flux, J_{st}, is proportional to the permeability coefficient P, and considering that $\Delta a = 1$, the permeability coefficient can be determined from:

$$P = \frac{J_{st} \times L}{\Delta a} \quad (6)$$

with J_{st} the stationary flux, L the thickness of the film and Δa the difference in water activity between the two faces of the film (in our case $\Delta a = 1$).

In terms of diffusivity, the water diffusion coefficient is calculated from the slope of the permeation curve by plotting the normalized water flux J/J_{st} as a function of the reduced time t·L^{-2}. In the case of vapors, the variation in diffusion with water concentration during permeation course is usually fitted by an exponential law reflecting a water concentration-dependence of the diffusion coefficient. This dependence is due to the water plasticization phenomenon inducing generally an increase in free volumes within the films. This well-known phenomenon, considered as a Fickian process of type B, conforms to:

$$D = D_0 \times e^{(\gamma C)} \quad (7)$$

where D_0 is the diffusion coefficient when the water concentration is close to 0 at the starting of measurement, γ is the plasticization factor, and C is the local water concentration. The parameters were determined from the mathematical approach developed in [31]. At the stationary state, when C is equal to C_{eq}, the maximum coefficient diffusion D_m can be determined from:

$$D_m = D_0 \times e^{(\gamma C_{eq})} \quad (8)$$

where γC_{eq} is known as the plasticization coefficient.

Water vapour sorption measurements were performed at 25 °C on a gravimetric Dynamic Vapor Sorption analyser (Surface Measurement Systems Ltd., London, UK) equipped with an electronic Cahn 200 microbalance (mass resolution of 0.1 µg). The water activity ranging from 0 to 0.95 was adjusted by mixing dry and saturated nitrogen gases using electronic mass flow controllers. The film sample placed in a pan was predried at 0% RH by exposure to dry nitrogen flux until no further change in dry weight was measured and was subsequently hydrated. For each water activity tested, the film mass was monitored as a function of time until a constant value was obtained. The correspondence between water activity and the corresponding water gain mass at equilibrium state (expressed in g of water per 100 g of polymer film) led to plotting the sorption isotherm curves. The isotherms curves for the tested films were modelled using two sorption modes, the Henry-law sorption and the aggregation (clustering) sorption, as represented in Equation (9). This combination of the two sorption modes is appropriate to fit the sorption data since the mean deviation moduli MDM calculated by the root sum square method (RSS method) are lower than 10% [32], attesting to the good fitting. The curve profiles (linear then exponential) is so characteristic of a random dissolution of water molecules in the polymer amorphous phase (Henry-law sorption), followed by an exponential increase of water mass gain related to the aggregation phenomenon of water molecules. The water concentration at an equilibrium state depending on the water activity according to the two sorption modes, Henry-law and aggregation sorptions, can be expressed as follows [33]:

$$C = k_D \times a + n \times K_a \times k_D^n \times a^n \tag{9}$$

where k_D (Henry's constant) is the constant related to dissolution of permeants in the polymer matrix, K_a is the equilibrium constant associated with the water aggregation reaction and n is the mean number of water molecules constituting the aggregates.

From water vapor sorption kinetics, two diffusion coefficients were calculated to evidence the transfer of water molecules by random molecular motions through the films. Based on Fick model, the analytical expression of the sorption advancement M_t/M_{eq} as a function of dimensionless time τ ($=Dt/L^2$) is:

$$\frac{M_t}{M_{eq}} = 1 - \frac{8}{\pi^2} \sum_{n=0}^{\infty} \frac{e^{-(2n+1)^2 \times \pi^2 \times \tau}}{(2n+1)^2} \tag{10}$$

The resolution of Equation (10) led to the determination of two coefficients as a function of the extended time of the sorption process. In that case, the diffusion coefficient noted D_1 is related to the diffusion for short time (i.e., when $M_t/M_{eq} < 0.5$) (Equation (11)) and the diffusion coefficient noted D_2 is related to the diffusion for longer time (i.e., when $M_t/M_{eq} > 0.5$) (Equation (12)).

$$\frac{M_t}{M_{eq}} \approx \frac{4}{L} \sqrt{\frac{D_1}{\pi}} \sqrt{t} \tag{11}$$

$$\ln(1 - \frac{M_t}{M_{eq}}) \approx \frac{-\pi^2 \times D_2}{L^2} \times t - \ln\left(\frac{\pi^2}{8}\right) \tag{12}$$

3. Results and Discussion

3.1. Impact of Montmorillonite C30B Fillers on the PBSA Matrix

3.1.1. Microstructure Examination by Microscopy and WAXS

In order to determine the state of dispersion of montmorillonite C30B fillers into the PBSA matrix, observations by transmission electron microscopy (TEM) were done. The obtained images are gathered in Figure 1. The dispersion of 2 wt% of C30B fillers into PBSA seems to be homogeneous with only very small aggregates in some places. As expected, when the filler content increased up to 5 wt%, the C30B fillers were less individualized and the number of aggregates appeared to be more important. In both cases, the orientation of C30B fillers specifically in the extrusion flow direction is obviously confirmed. The PBSA based films filled with C30B can be considered as nanocomposites with a coexistence of intercalated and exfoliated inorganic structures.

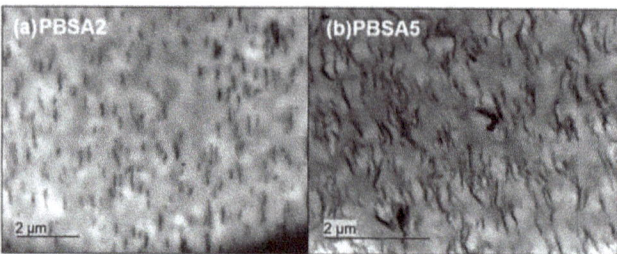

Figure 1. TEM observations for PBSA filled with (**a**) 2% and (**b**) 5 wt% of montmorillonite C30B.

To complete these observations, WAXS measurements were conducted on the neat and the nanocomposite PBSA films to evaluate the structure and microstructure of films. The DRX diffractograms for the films are presented as a function of 2θ range in Figure 2.

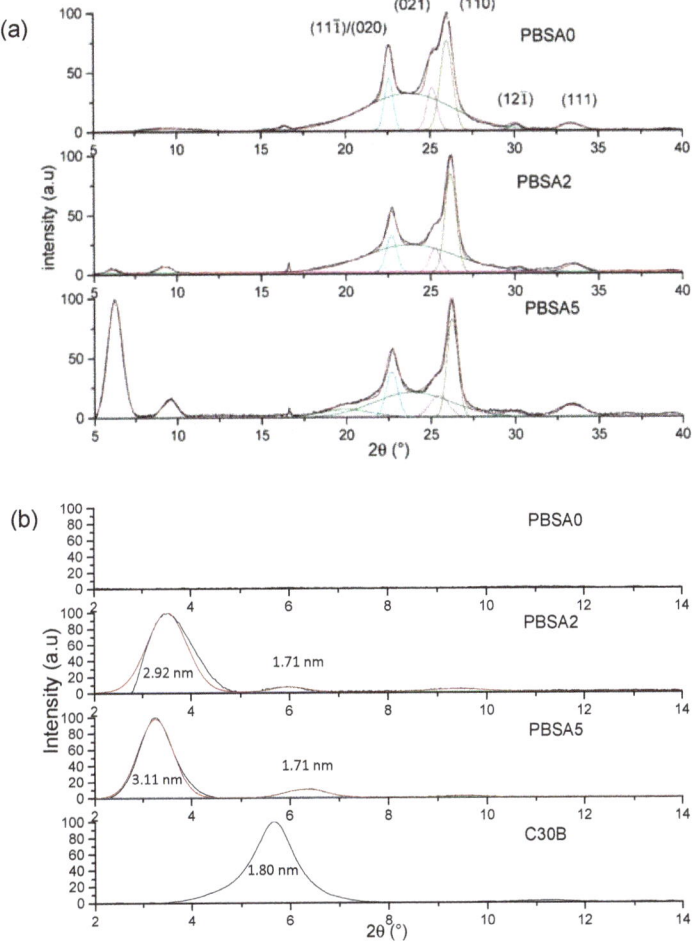

Figure 2. XRD spectra of neat PBSA film and filled PBSA films (**a**) in the 15–40° range and (**b**) in the 2–14° range.

First, PBSA being a semicrystalline polymer, the crystalline structure is evaluated specifically in the 15–40° range (Figure 2a) in agreement with the appearance of mean diffraction peaks relative to polymer crystals. Crystallizing in monoclinic crystal lattice [34], the four strong diffraction peaks are located at 2θ = 22.6°, 25.2°, 26.0°, 30.1° and 33.4° relative to (11$\bar{1}$)/(020), (021), (110), (12$\bar{1}$) and (111) diffraction planes, respectively, as previously reported [35] with the amorphous halo centered on 2θ = 24°. The diffractograms of the filled PBSA films can be superposed to the neat PBSA one, meaning that the filler incorporation has any impact on the crystalline phase of PBSA. A small difference in terms of peak intensity is only noted due likely to a certain anisotropy of crystals within the films.

In the diffraction angle 2–15° range, the dispersion and exfoliation level of fillers in the PBSA is evaluated (Figure 2b) in accordance with the diffraction patterns of nanoclays. The interlayer distances (inserted in Figure 2b) relative to diffraction peaks were calculated by using the Bragg law. As indicated in the literature and reported in this work, the interlayer distance of C30B is equal to 1.80 nm. For the filled PBSA films, the diffraction peaks detected at 3.5° and 3.3° for 2% and 5% of C30B, respectively, indicates exfoliation and intercalation of nanoclays in PBSA matrix. The interlayer distance is found to be higher than that of the C30B powder and increases with the filler content, attesting an improvement of dispersion and exfoliation levels. Indeed, the peak at $2\theta = 6.0°$ was assigned to a second intercalated structure with lower interlayer space (1.71 nm) arising from the degradation of C30B surfactant [36] and/or platelet re-aggregation by clay contact. As a result, this is larger with the highest filler content. This result agrees with the TEM observations. One can conclude that C30B nanoclays are well dispersed and exfoliated within the PBSA matrix with an increase in interlayer distances as filler content increases, even if one can observe aggregates from TEM observations.

3.1.2. Thermal and Mechanical Analyses

A thermogravimetric analysis was performed on the PBSA films filled with montmorillonite C30B fillers to quantify the exact proportion of fillers incorporated in PBSA. The value fits the residual percentage after total degradation of polymer at high temperature. The TGA curves reported in Figure 3 confirm the filler content of the PBSA2 and PBSA5 films.

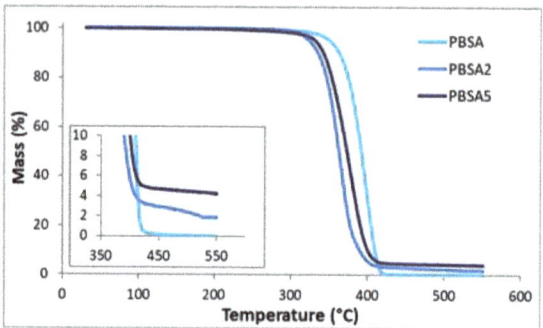

Figure 3. TGA curves obtained for the neat PBSA film and the filled PBSA films.

Then, considering the TGA profiles, it is worth noting that the degradation temperature of the PBSA (determined at 5% of mass loss) decreased with the filler content. Indeed, the degradation temperature of 350 °C for the neat PBSA film was reduced to approximately 330 °C for the nanocomposite films. To explain this phenomenon, two hypotheses can be given: the first one concerns the presence of surfactant included in the organo-modified montmorillonite that has induced a degradation temperature of 174 °C for C30B [37], lower than that of the PBSA polymer and reducing the degradation temperatures of the filled-PBSA films. The second one is linked with the fact that two steps were necessary to obtain the loaded films, inducing longer residence times in the extruder that may cause of a reduction in size of macromolecular chains, leading to a faster degradation compared to neat PBSA.

DSC and MT-DSC analyses were performed in order to determine the characteristic temperatures (T_g, T_c and T_m), the degree of crystallinity (X_c) and the specific heat capacity ΔC_p, to quantify the mobile amorphous fraction (MAF) and the rigid amorphous fraction (RAF) for the films using Equations (2) and (3). The results are gathered in Table 1. It is worth noting that similar crystallization and melting temperatures are measured before and after incorporation of the fillers and that the crystallinity degree remains unchanged (~38%). However, a slight increase of the glass transition temperature is measured, meaning the macromolecular chains involved in the amorphous phase of PBSA are less mobile, probably due to constraints exerted by the nanofillers.

Table 1. Thermal properties of the neat filled PBSA films.

	T_g PBSA (°C)	T_c PBSA (°C)	T_m PBSA (°C)	ΔC_p PBSA (J g^{-1} °C^{-1})	X_c PBSA (%)	RAF (%)
PBSA	−46	67	90	0.330	38 ± 2	9 ± 3
PBSA2	−43	66	91	0.275 *	38 ± 2	17 ± 3
PBSA5	−42	66	91	0.274 *	37 ± 2	18 ± 3

* The value of ΔC_p is calculated taking into account the proportion of PBSA in the composite.

Considering that the incorporation of fillers into the PBSA matrix did not affect the crystallization, it may be of interest to evaluate the effect of the fillers on the amorphous phase through the change in ΔC_p calculated at the glass transition. Indeed, the heat capacity step normalized to the polymer fraction is directly linked to the mobile amorphous fraction MAF according to Equation (2). A decrease of ΔC_p value would mean that there is less mobile amorphous fraction (MAF) and consequently more rigid amorphous fraction (RAF) in the filled films. The presence of the C30B fillers in PBSA led to increase the percent of RAF by a factor two (Table 1). Such an increase in RAF by incorporation of fillers has already been observed in PBSA [38], in PU [39] or in an hyperbranched polyester [40] for example. This significant decrease of ΔC_p values, evidencing an increase in local rigidity, can be due to the reduction in free volumes in polymer by addition of fillers and the resulting mobility restrictions of macromolecular chains coming from the presence of fillers. From our results, we can infer that the C30B fillers dispersed in the PBSA matrix cause geometrical constraints to amorphous polymer chains, resulting in the increase of the glass transition temperature and the increase of the RAF.

Mechanical analyses were performed and the data are gathered in Table 2. The values obtained for the neat and filled PBSA films are convenient with values presented in the literature [18,38,41,42]. The high values of elongation at break for films reveal the ductile behaviour of PBSA and are in the same order of magnitude for the neat and filled films [18]. It is worth noting an increase of the Young's modulus and a decrease of the yield stress and the elongation at break when PBSA is filled with montmorillonite. This trend is classically observed with nanocomposites [43] and reported in the case of films of PLA [44], PBS [45,46], PBSA [47] or PU [48]. A high increase in Young's modulus (in our case around 50% for 5 wt% of C30B) is usually correlated with a good exfoliation and dispersion of fillers within the polymeric matrix. The result is thus an indirect way to evidence the good dispersion and exfoliation of C30B within PBSA. This increase of stiffness could be associated with the RAF increase.

Table 2. Influence of the filler content on the mechanical properties of PBSA film.

	Young's Modulus (MPa)	Strength at Break (MPa)	Elongation at Break (%)
PBSA	241 ± 19	35 ± 3	1360 ± 148
PBSA2	276 ± 33	31 ± 2	1405 ± 67
PBSA5	362 ± 19	27 ± 2	1242 ± 97

3.1.3. Water Barrier Properties

The permeation kinetic curves of the neat and filled PBSA films are plotted in Figure 4 in the reduced scale $J*L = f(t/L^2)$ in order to overcome film thickness effects. The water permeation flux is found to be faster for the PBSA film until a higher constant value at the steady state is measured. As a result, the permeability coefficient was found to higher than the values of the filled films. Considering the steady state of water permeation, the permeation flux shifted to lower values with the increase in filler content, indicating a reduction of the water diffusivity through the nanocomposite films. These results clearly attest for an increase of water barrier properties by incorporation of impermeable fillers within the semicrystalline polyester. This barrier effect is usually attributed to the tortuosity effect induced by the presence of nanoclays, but could also be correlated, in part, to the increase of RAF for the nanocomposite films (9% vs. 18%, Table 1). The mobility of the amorphous chains at the vicinity of crystals was locally reduced by geometrical constraints which hinder passage of water molecules. A shift of the reduced permeation flux curves towards longer time is so obtained, highlighting a delay time in diffusion, which suggests an increase in the diffusion pathways with the filler incorporation. These observations are in line with the decrease of diffusion coefficients D_0 and D_M determined from Equations (7) and (8). The incorporation of C30B within PBSA has thus induced a decrease of permeability and diffusion coefficients, as shown in Table 3. As thoroughly reviewed by Tan and Thomas [49], one can explain these results by a strong improvement of water barrier effect due to tortuosity induced by the dispersion and exfoliation levels of lamellar fillers within the polymer matrix. In such a case, permeability and diffusion coefficients are accordingly reduced, as obtained in this work.

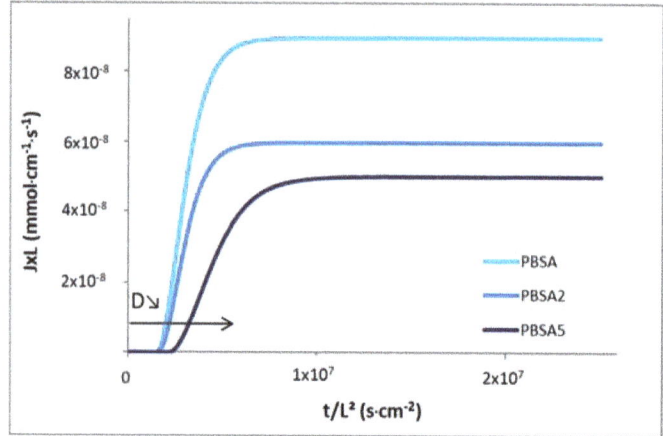

Figure 4. Reduced water permeation curves for the neat PBSA and the filled PBSA films.

As commonly observed with water as diffusing molecules for polyester films [18,35,50], a plasticization effect of the polymer by sorbed water is also obtained. The water diffusion coefficients are accordingly dependent to the sorbed water concentration and this dependence is usually described through the exponential law $D = D_0 \cdot e^{\gamma C}$ [51]. The experimental water permeation fluxes of films were successfully fitted and the calculated parameters are gathered in Table 3. The water-induced plasticization is clearly observed since D_0 (diffusion coefficient at water concentration close to 0) is found to be lower than D_m (diffusion coefficient at water concentration at equilibrium state) for the three films (Table 3). The difference between D_0 and D_m values is smaller for the nanocomposite films with the increase of filler content. This result highlights a reduction of plasticization effect with the filler incorporation, which is correlated with the reduction of the plasticization coefficient γC_{eq}, even if the water concentration at equilibrium state is rather constant (~0.7–0.8 mmol·cm^{-3}). It seems that the plasticization ability of PBSA which leads to increase the free volumes is reduced. This behaviour could be due to the reduction in macromolecular chain segment mobility, as shown from the increase of stiffness and by RAF increase, but without changing the sorbed water content.

Table 3. Water permeation parameters for the PBSA, PBSA2 and PBSA5 films.

	P (Barrer *)	D_0 (10^{-8} cm^2·s^{-1})	D_M (10^{-8} cm^2·s^{-1})	γC_{eq}	γ (cm^3·mmol^{-1})	$C_{éq}$ (mmol·cm^{-3})
PBSA	8312 ± 177	1.8 ± 0.1	36 ± 3	3.0 ± 0.1	3.9 ± 0.4	0.78 ± 0.05
PBSA2	5628 ± 127	1.7 ± 0.1	25 ± 6	2.7 ± 0.3	3.9 ± 0.9	0.69 ± 0.09
PBSA5	4821 ± 189	1.5 ± 0.1	15 ± 1	2.3 ± 0.1	2.7 ± 0.3	0.85 ± 0.03

* 1 Barrer = 10^{-10} cm^3 (STP) cm·cm^{-2}·s^{-1}·cmHg^{-1}.

In the case of the water vapor sorption phenomenon, the sorbed water concentration in films was measured as a function of time for water activity range from 0 to 0.95. The size and the thickness of the film samples are unchanged after sorption measurements. The corresponding water vapor sorption isotherms (equilibrium state of sorption kinetics) for the neat and the filled PBSA films are plotted in Figure 5. The shape of the isotherm curves is rather similar whatever the tested film: a linear increase of water mass gain as water activity a_w increases, followed by a sharp rise of water mass gain at $a_w > 0.6$. As largely reported in the literature [52–54], C30B filler sorb a higher water content than polyester or polyamide films within the whole water activity range, because of water-sensitive groups, such as silicate and hydroxyethylene groups, included in the surfactant of C30B [55]. From Figure 5, the water vapor sorption behaviour of the filled PBSA films deviates from that of the neat PBSA film at $a_w > 0.5$. The C30B incorporation in PBSA clearly contributes to this deviation in water solubility from $a_w = 0.5$, testifying to the hydrophilic nature of clays, even if there is an organomodification. This deviation in water mass gain is greater with the filler content increase, in correlation with their hydrophilic nature. It is worth noting that the increase in water solubility shows an opposite trend compared to the reduction of water permeability coefficient obtained by permeation measurement. One can specify here that the permeability coefficient results of two contributions, one is the solubility and the other the diffusivity. For this latter contribution, the reduction of water diffusivity is expected by increasing of tortuous diffusing pathways owing to the presence of impermeable fillers in the PBSA matrix. In addition, in the case of sorption process, the increase in water solubility by increase in water mass gain is mainly due to the hydrophilic nature of C30B in comparison with the sorption capacity of PBSA polymer. In terms of water transport properties, one can infer that C30B filler is responsible for an increase of tortuosity relative to the film microstructure and filler organisation, and for an increase in solubility relative to the presence of water-sensitive groups.

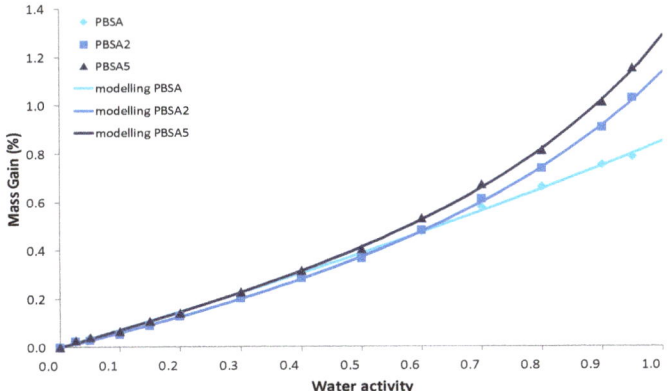

Figure 5. Water permeation curves for the neat PBSA and the filled PBSA films fitted using a model combining two sorption modes, the Henry-law sorption and the aggregation (clustering) sorption, as represented in Equation (9).

Regarding isotherm curves, such a curve profile can be appropriately fitted using Equation (9) which refers to the combination of Henry-law and aggregation sorption modes, as shown by the fitted curves reported in Figure 5 as well as by the mean deviation modulus MDM lower than 10% (Table 4), both attesting to the convenient modelling of sorption data. The linear increase of water mass gain as water activity a_w increases is consistent with the random dissolution of water molecules in the amorphous phase of the polymer. At $a_w > 0.6$, the exponential increase in water mass gain conforms to cluster formation of water molecules. The modelling parameters are summarized in Table 4. The parameters analysis points out some effect from filler incorporation. The k_D parameter is slightly higher for the highest filled PBSA film: the more the water solubility in films, the more the

k_D constant. Although a larger number of aggregates appears at this filler content, this observation reflects the effect of larger nanoclay-polymer interfaces, which can facilitate the random dissolution of water molecules in the nanocomposite films by modifying the number and the size of micro-voids inherent of PBSA polymer. The values of K_a and n indicate the formation of water aggregates at high water activities ($a_w > 0.6$). Both values increase as filler content increases, confirming an increase of interactions with sorbed water molecules to initiate a multilayer water sorption. Water aggregates are formed on polar sorption sites existing in the films. However, the increase of these two parameters is not so high (Table 4) indicating that the increase of free volumes by water plasticization is limited owing to a certain rigidity of polymer chains in the vicinity of C30B fillers, but also to good compatibility between organo-modified clays and PBSA chains.

Table 4. Modelling parameters for the neat PBSA and the filled PBSA films.

	k_D g/g	n	K_a (g/g)$^{(1-n)}$	MDM (%)
PBSA	0.583	1.4	0.407	5.10
PBSA2	0.582	2.6	0.803	5.54
PBSA5	0.702	3.1	0.811	5.56

The water diffusion coefficients in the neat and the filled PBSA films for the first-half sorption, noted D_1, and for the second-half sorption, noted D_2, are plotted in a semi-logarithmic scale (log D) as a function of water activity in Figure 6. It appears that the evolution of the diffusion coefficients was similar and in a same range of values whatever the film, except for the neat PBSA film where the evolution of D_2 coefficient is strongly decreasing with water activity. It seems that PBSA is differently impacted by the water sorption phenomenon likely due to the plasticization effect of water vapor. It is surprising that the D_2 coefficients (at longer extent of water sorption, higher water concentration) are lower than the D_1 coefficients in the activity range for all films, whereas the inverse trend is commonly reported [56] due to water plasticization effects. However, such a trend has been recently mentioned in the literature for PBSA-based nanocomposites [53]. The authors outline two antagonist effects occurring for longer time that would balance each to give similar coefficients. The former increasing water diffusion is related to plasticization effects of water and the latter decreasing the water diffusion is related to the formation of water aggregates which restricts their mobility and as a result their diffusivity within the films. The general curve profiles for the filled PBSA films show a linear part and a reduction of diffusion at higher water activities, meaning that diffusion coefficients are rather constant and are then reduced by water clustering phenomenon linked to the increase in water cluster size which makes water molecules less mobile [57]. In addition, the evolution of the diffusion coefficients is in good agreement with the BET III sorption mechanism.

For the whole water activity range, it is observed that the diffusion coefficients, D_1 and D_2, are lower for the filled PBSA films, highlighting the impact of fillers to the diffusion of water vapor in the PBSA films by tortuosity effects. However, this decrease in diffusion is inversely correlated with the filler content. In fact, the D_1 coefficient is found to be reduced as the filler content increases while the D_2 coefficient is found to be increased for the PBSA2 film and then slightly reduced for the PBSA5 film. D_2 coefficient being considered representative of the water diffusion in the core of film and D_1 coefficient representative of diffusion from the film surface, the latter result can be related to the filler presence in PBSA which prevents the water diffusion by tortuosity at the film surface, as observed with the water diffusion by permeation measurement. And, in addition, this can be the result of the RAF increase in the filled films generating geometrical constraints and restriction in mobility of macromolecular chains. When water molecules have penetrated within the films, the plasticization effect of sorbed water molecules occurs accompanied by an easier formation of water aggregates and by the creation of preferential diffusion pathways at the filler-polymer interfacial areas due to the hydrophilic character of the clays which both can help the diffusion of water in the core of the films. This effect is slowed down at the highest filler content due to larger water cluster sizes, as calculated by the modeling approach, and to larger tortuosity effects which both resulted in a reduction of the water cluster mobility. In all cases, the D_2 coefficient is found lower than the D_1 coefficient for the whole range of water activity, in a lesser extent for the filled PBSA nanocomposite films. This can be related to the filler incorporation in PBSA since acting as obstacles to diffusion which limits the water plasticization effect in the film, and hence diffusion in the film.

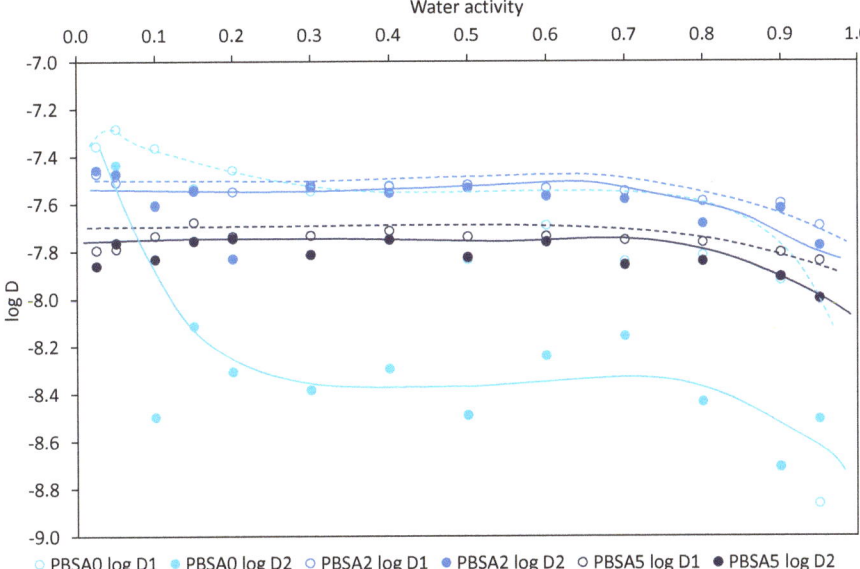

Figure 6. Water vapor diffusion coefficients D_1 and D_2 for the neat PBSA and the filled PBSA films.

For this first part of the work, one can infer that the incorporation of C30B fillers within the PBSA matrix has induced a slight increase in glass transition temperature associated with an increase in RAF for a constant degree of crystallinity, and an increase in Young's modulus because of good filler exfoliation and dispersion level, even if some filler aggregates appear in some places. The C30B fillers are also seen as oriented in the extrusion direction of the film. From these results, the increase in stiffness of macromolecular chains and the presence of larger nanoclay-polymer interfaces have caused reduction of free volumes and restrictions in mobility of macromolecular chains by geometrical constraints by close contact with fillers. The water permeability and diffusion coefficients have been accordingly reduced. The water permeation flux is shifted to longer time evidencing an increase of tortuosity due to oriented lamellar structure of C30B in PBSA matrix. The reduction of water diffusion and water sorption phenomena are in good agreement with the morphology and the thermal and mechanical properties displayed by the films. The water barrier properties were improved with an increased stiffness of the nanocomposite films.

3.2. Elaboration of PLA/PBSA Multilayer Film with Montmorillonite C30B

3.2.1. Morphological Characterization and Nanofiller Dispersion

The structure of the multilayer PLA/PBSA (80/20) film was described using AFM observation in our previous work [18] and shown in Figure 7. TEM observation was considered to evidence the clay morphology and its location into the multilayer structure, as well as the multilayer structure itself. A selection of TEM images characteristics of the loaded multilayer films is gathered in Figure 7. The thickness of the PBSA layers varies from 50 to 70 nm in the neat multilayer film, as expected by calculation resulting from the number of of layers and the final thickness of the co-extruded film. A wider thickness is found for the filled PBSA layers in loaded multilayer films compared to the neat PBSA layers in the PLA/PBSA multilayer film: around 100–200 nm and 150–250 nm for the PLA/PBSA2 and PLA/PBSA5 multilayer films, respectively. The integrity of ultrathin layers of PBSA is more difficult to preserve when they are filled with clays. Some layer breakups due to clay aggregates have been observed. Nevertheless, rather homogeneous and continuous PBSA layers (black layers in the micrographs) within the multilayer PLA/PBSA structure are overall observed. One can note the presence of some filler aggregates in the PBSA layers.

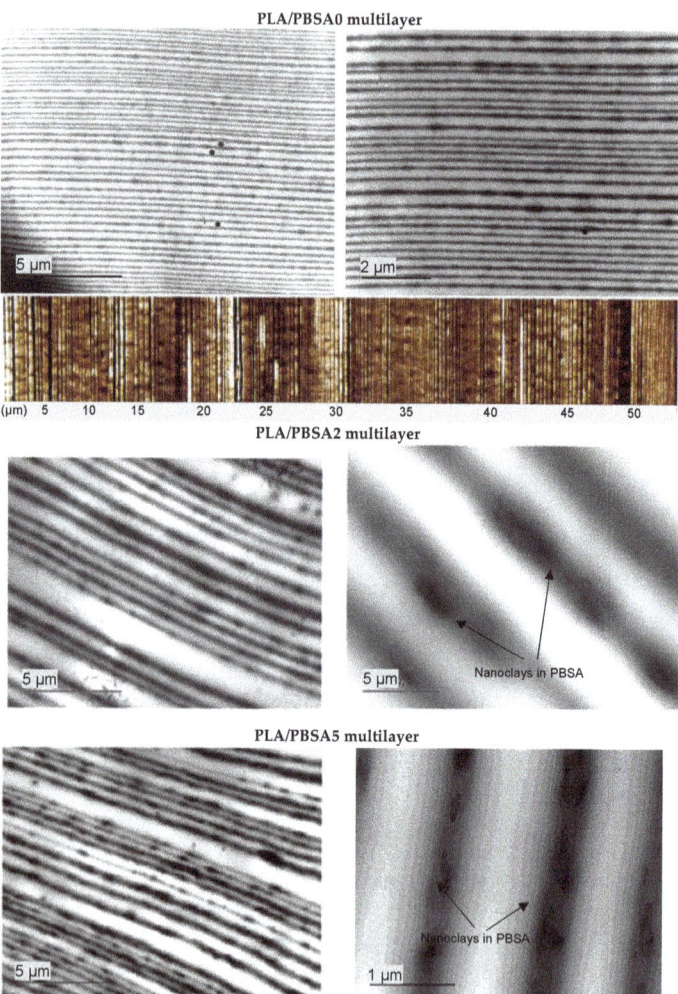

Figure 7. TEM observations of the unfilled PLA/PBSA and filled PLA/PBSA2 and PLA/PBSA5 multilayer films (PLA in white, PBSA in dark). (AFM image of unfilled PLA/PBSA multilayer films).

As a complementary experiment, XRD measurements were conducted on the different multilayer films to evaluate the microstructure of films. The XRD diffractograms for the multilayer films are presented in Figure 8. Recently, the diffractogram of the PLA/PBSA has been simulated from the diffractograms of each pure polymer considering the composition weight ratio [18]. The authors have pointed out a good agreement between the experimental and simulated XRD profiles, reflecting a similarity of structures for the two polymers in monolayer and multilayer organisation, especially the crystalline organisation. This result in line with the present work with a similarity in degree of crystallinity between PBSA in multilayer film (40%) and PBSA in monolayer film (38%). The PLA/PBSA0 film exhibits an amorphous halo centered on 20° relative to amorphous phase of PLA and PBSA layers and crystalline diffraction peaks relative to PBSA layers (Figure 8a). The characteristic strong peaks of PBSA at 18.2°, 22.6°, 26.2°, 31.5° and 33.5° attributed to the $(11\bar{1})/(020)$, (021), (110), $(12\bar{1})$ and (111) planes, respectively, are found by extraction of the crystalline contribution from the XRD profile. These two contributions are observed for the filled-PBSA layers multilayer films.

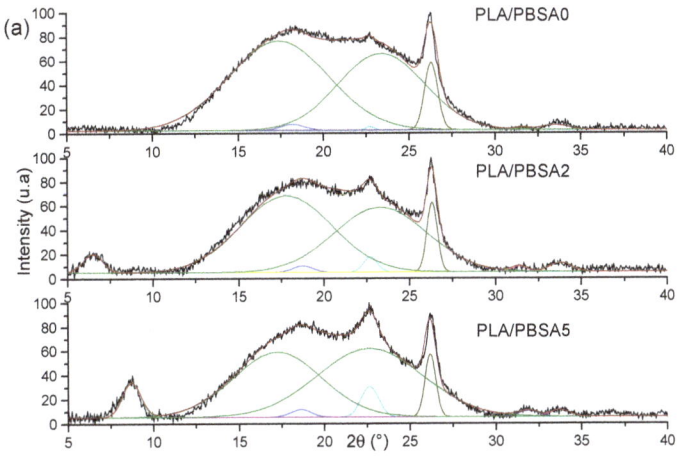

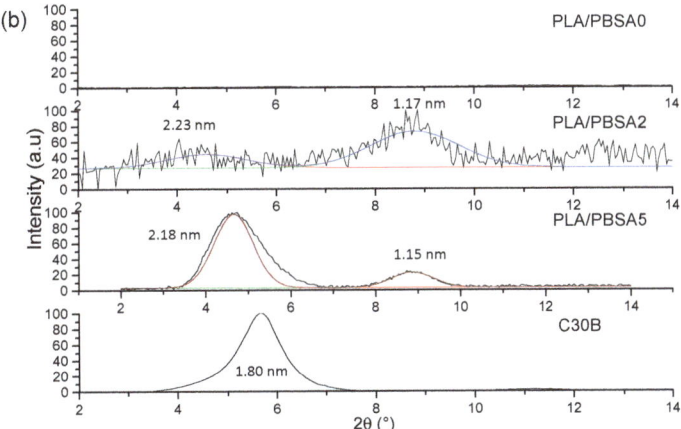

Figure 8. XRD spectra for the PLA/PBSA, PLA/PBSA2 and PLA/PBSA5 multilayer films in the 5–40° range (**a**) and for C30B powder in 2–14° range (**b**).

In the 2–14° range (Figure 8b), the characteristic peaks for the filler were detected at $2\theta = 4.7°$ and $8.9°$. One can note an increase of the interlayer distance corresponding to an intercalated structure within the PBSA layers of the PLA/PBSA multilayer structure compared to the value of C30B powder. Indeed, the peak at $2\theta = 8.9°$ assigned to a second intercalated structure with lower interlayer space is also visible. The relative lower amplitude of the two diffractions peaks for the PLA/PBSA2 film compared to the PLA/PBSA5 film is an indication of a better exfoliation of C30B within the PBSA layers. This observation can be explained by the difference in thickness of PBSA layers and by some filler aggregates which seem to be more present in PLA/PBSA5 multilayer film.

3.2.2. Thermal and Mechanical Analyses

The characteristic temperatures for the multilayer films were measured by using MT-DSC measurements to well separate the PBSA crystallization observed in the non-reversible heat flow signal from the PLA glass transition phenomenon observed in the reversible heat flow signal because both events are present in the same temperature range. The values are gathered in Table 5. The thermal events are calculated from the first heat cycle to ensure that the films are in the same thermal and structural states as for permeation and sorption analyses. Those MT-DSC thermograms are a practical way to compare the thermal properties of the PLA/PBSA multilayer films as filler content increases compared to the neat PLA/PBSA multilayer film.

Table 5. Thermal properties of the neat and filled PLA/PBSA multilayer films.

	T_g PBSA (°C)	T_g PLA (°C)	T_c PBSA (°C)	T_m PBSA (°C)	ΔC_p PBSA (J g^{-1} °C^{-1})	X_c PBSA (%)	RAF (%)
PLA/PBSA (PBSA thickness ~60 nm)	−46	53	71	91	0.250 *	40	21
PLA/PBSA2 (PBSA thickness ~100 nm)	−42	54	71	90	0.204 *	34	33
PLA/PBSA5 (PBSA thickness ~200 nm)	−36	54	71	91	0.237 *	38	23

* The value of ΔC_p is calculated given the proportion of PBSA in the multilayer film.

It appears a significant increase of the glass transition temperature of the PBSA up to 10 °C deviation as filler content increases (Table 5). In comparison with the reference filled PBSA films (Table 1), this deviation is greater for the multilayer films (10 °C vs. 4 °C), highlighting a greater reduction of the macromolecular chain mobility. This result suggests that the constraining effect due to the clays is more pronounced in a thinner layer (100 nm on average vs. 300 μm for the reference PBSA film). As a consequence, the macromolecular chains of PBSA exhibit less mobility, requiring more energy to switch from glass state to rubbery state, which causes a higher glass transition temperature. The other characteristic temperatures are quite similar to those of the reference films. The glass transition temperature of PLA is unchanged, meaning that the clays are well contained in the PBSA layers or located at the polymer interfaces without effect on PLA chain mobility.

For the neat and loaded multilayers films, the degree of crystallinity of PBSA in the multilayer films was evaluated (Table 5). It is worth noting a decrease of the crystallinity of PBSA with clays when the PBSA layers contains 2% of fillers. This trend is opposed to that displayed by the reference filled PBSA films (i.e., in monolayer structure) where the clays incorporation did not impact the degree of crystallinity. Such an opposite trend could be the result of confinement effect of the PBSA layers by reduction of the layer thickness generated during the forced assembly coextrusion process. The multiplication of layers by using the multiplying element device during coextrusion has drastically reduced the layer thickness (100 nm on average vs. 300 μm), restricting as a result the space available to macromolecular chains of PBSA to be properly realigned to form crystals, and also significantly increasing the local rigidity of macromolecular chains by close contact with clays in a thinner space. This is particularly noticeable with the multilayer PLA/PBSA2 film (layer thickness of 100 nm on average), for which the crystallinity is equal to 34%. In that case, the RAF is found to be of 33%, higher than the RAF of the neat PLA/PBSA, confirming the local rigidification of macromolecular chains by close clay contact in a thinner thickness. For the film with higher clay content (PLA/PBSA5), one can note an increase of the degree of crystallinity and a decrease of RAF that are similar to the unloaded PLA/PBSA multilayer. This could be attributed to a greater ease to form crystalline phase owing to a larger mobility of macromolecular chains in a thicker PBSA layer thickness (increase of 100 nm).

Mechanical analyses (Table 6) were also performed to evaluate the impact of the clays in the PBSA layers within the multilayer structure on the mechanical performances. As the multilayer film is mainly composed of PLA (around 80 wt%), the final filler content is extremely low (0.5% and 1.25% for the two films). Therefore, no significant variations were observed. Despite a rather high standard deviation, one can only note for PLA/PBSA2 a slight decrease in the Young's modulus and strength at break as well as an increase of the elongation at break, which could be correlated with the slight decrease of crystallinity measured by DSC.

Table 6. Mechanical parameters for the multilayer film PLA/PBSA films.

	Young's Modulus (MPa)	Strength at Break (MPa)	Elongation at Break (%)
PLA/PBSA	1723 ± 79	38 ± 3	37 ± 10
PLA/PBSA2	1541 ± 61	34 ± 2	55 ± 12
PLA/PBSA5	1686 ± 66	38 ± 3	39 ± 7

3.2.3. Water Barrier Properties

Water permeation measurements were performed on the multilayer films and the resulting water flux curves are represented in Figure 9 in a reduced scale $J \times L = f(t/L^2)$ to overcome the film thickness effect. Regarding the steady state of the permeation, the water flux is shifted to lower values with filler content increases, indicating a reduction in permeability coefficients, as reported in Table 7. This improved water barrier effect is surprisingly not so high as it was expected from the oriented nanoclays more or less dispersed in nanolayers of PBSA. It must however remind that only 1.25% of C30B clays were incorporated in the multilayer film. If the water flux is reduced with the presence of fillers, a shift of the reduced water flux curves to lower time is obtained, reflecting a faster diffusion of water molecules through the filled multilayer films, as shown by the increase of diffusion coefficients D_0 (Table 7). With the tortuosity effect induced by lamellar structures and the stiffness and RAF increase, it could be expected that the diffusion coefficients were accordingly reduced. This is not the case. It seems that the presence of layer breaks and of thicker PBSA layers in the filled multilayer structure has impacted the water diffusivity, regardless of the fact that C30B fillers are assumed to be impermeable entities. One can infer that the presence of very few defects in the microstructure are at the origin of a loss of macroscopic properties. Nevertheless, this increase in diffusion can be considered as low if we note the small deviation between values in comparison with the diffusion coefficient of the neat PLA/PBSA multilayer film. The low filler fraction of C30B (0.25 and 1.25 wt%) in multilayer films are probably not sufficient to highly increase the tortuous diffusion pathway of water molecules.

Again, it is worth noting that diffusion coefficient D is not constant with permeation time since $D_0 < D_M$, confirming the water plasticization effect. Even in multilayer form, the water diffusion coefficients are dependent to the sorbed water concentration. The experimental water fluxes for the multilayer films were fitted by applying the exponential law (Equation (7)) and the calculated parameters are summarized in Table 7. The deviation of D_0 and D_M between the values of the unfilled multilayer film with those of the filled multilayer films again confirms the faster diffusion of water in the filled multilayer films. Also, this is correlated with the filler content increase. Regarding the plasticization coefficient and the water concentration at equilibrium state, the values are quite similar, that clearly shows no real influence of the low clay fraction on the water plasticization effect. Nevertheless, in comparison with the parameters of the reference films (Table 3), it can be evidenced a large decrease in permeability and diffusion coefficients and a reduction of plasticization coefficient for the multilayer films. This effect is maintained even in presence of layer breaks in the multilayer structure. One can infer that the multilayer structure by multiplying the number of polymer layers, and hence interfaces between a semicrystalline polymer and an amorphous polymer, is at the origin of the large reduction in water transfer and its diffusion. The increase in RAF for the multilayer PLA/PBSA2 film does not seem to have impacted the barrier properties more than that.

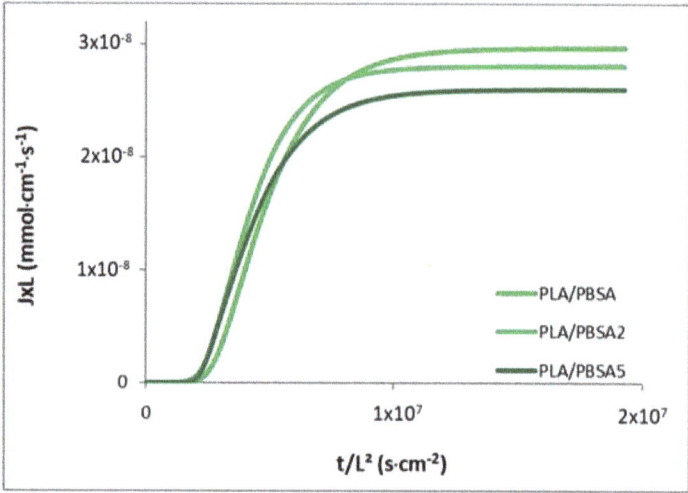

Figure 9. Reduced water permeation curves for the PLA/PBSA multilayer film and the filled PLA/PBSA multilayer films.

Considering the opposite variation of permeability and diffusion coefficients, it seems that the solubility is a key parameter in the improvement of barrier properties to water for the multilayer films where the semicrystalline polymer layers are filled with C30B. The contribution of solubility has been investigated through water vapor sorption measurements.

Table 7. Water permeation parameters for the PLA/PBSA, PLA/PBSA2 and PLA/PBSA5 multilayer films. Experimental and predicted values.

	Water Permeability (Barrer)	Calculated Permeability (Series Model) (Barrer)	Calculated P of PBSA Layers (Barrer)	D_0 (10^{-8} cm²·s⁻¹)	D_M (10^{-8} cm²·s⁻¹)	γC_{eq}	γ (cm³·mmol⁻¹)	C_{eq} (mmol·cm⁻³)
PLA/PBSA	2765 ± 123	2917	4658	1.48 ± 0.04	12.7 ± 0.4	2.1 ± 0.1	3.8 ± 0.1	0.56 ± 0.01
PLA/PBSA2	2717 ± 136	2917/2823 *	4055	1.61 ± 0.06	16.0 ± 1.0	2.3 ± 0.1	5.1 ± 0.6	0.47 ± 0.05
PLA/PBSA5	2659 ± 81	2917/2776 *	3487	1.71 ± 0.1	13.0 ± 1.1	2.0 ± 0.2	4.1 ± 0.6	0.50 ± 0.05
PLA monolayer	2510 ± 177			0.87 ± 0.07	11.6 ± 0.7	2.6 ± 0.2	4.0 ± 0.1	0.67 ± 0.03
PBSA monolayer	8312 ± 177			1.8 ± 0.1	36 ± 3	3.0 ± 0.1	3.9 ± 0.4	0.78 ± 0.05

* Calculated from loaded PBSA monolayer (Table 3).

By applying the series model equation as follows:

$$\frac{L}{P} = \frac{L_{PLA}}{P_{PLA}} + \frac{L_{PBSA}}{P_{PBSA}} \quad \text{or} \quad \frac{1}{P} = \frac{\varphi_{PLA}}{P_{PLA}} + \frac{\varphi_{PBSA}}{P_{PBSA}} \qquad (13)$$

where L, P, φ are the thickness, the permeability and the volume fraction of PLA or PBSA, respectively.

The permeability coefficient of the neat or filled PLA/PBSA multilayer films was calculated. In case of filled PBSA, the experimental permeability coefficients of PBSA2 and PBSA5 were taken from Table 3. Doing so, we can see that the experimental values of water permeability of the neat and filled multilayers are slightly lower than the calculated ones (Table 7). This result seems to show a slight barrier effect induced by the multinanolayer structure. Assuming that the water permeability of PLA is unchanged before and after loading due to its amorphous state and containing no fillers, it was then interesting to deduce the predicted water permeability coefficients of the PBSA layers in multilayer structures from Equation (13) and to see how both the nanostratification and the presence of fillers improve the water resistance of PBSA nanolayers. Results are gathered in Table 7 and are represented in Figure 10 with the experimental permeability coefficients of reference films for the sake of comparison.

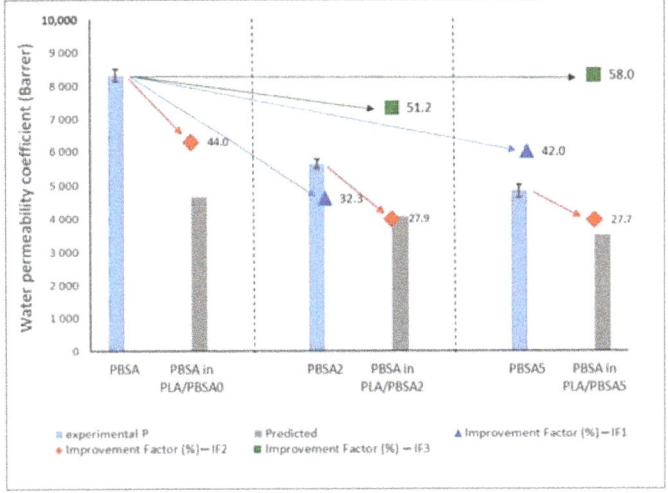

Figure 10. Comparison of the experimental and predicted permeabilities of PBSA under monolayer and multilayer films. ▲ IF1: calculated from neat PBSA—effect of loading ♦ IF2: calculated from PBSA (neat of filled)—effect of multilayer ■ IF3: calculated from neat PBSA—effect of loading and multilayer.

First, one can point out that the predicted permeability coefficients of PBSA layers in neat and filled multilayer films deduced from Equation (13) are lower than those of neat and loaded PBSA monolayer films. As can be speculated from the improvement factor IF (%) calculated from the difference between experimental and predicted values (($P_{exp} - P_{predicted})/P_{ex}$), it was possible to highlight the effect of the presence of clays and their confinement in nanolayers on the barrier effect (Figure 10). Indeed, first by comparing IF1 and IF2, which correspond to improvement factors for the loading effect and for the multinanolayer effect, respectively, we can see that the highest values for the nanocomposite and the multilayer are quite similar, 42 and 44%, respectively. Also, when the PBSA is filled, the gain in barrier effect due to the nanostratification is ~30% and comparable to the gain when 2% of fillers are incorporated in the PBSA monolayer. In other words, it seems that the improvement of water barrier by incorporation of fillers in PBSA is similar to the PBSA layers confined by PLA layers. Consequently, it can be expected a high increase in the water barrier effect in PBSA with the presence of fillers and when confined in very thin layers. Indeed, as revealed by IF3, improvement factor that corresponds to cumulative effect of both loading and nanolayering effects, the highest improvement factor (58%) is obtained for PBSA in PLA/PBSA5.

Complementary to water permeation, after measuring water vapor sorption kinetics of the multilayer film samples, the corresponding water vapor sorption isotherms were plotted and are presented in Figure 11. For comparison, water sorption isotherm of PLA and PBSA were also plotted in Figure 11. First, it can be seen that PLA and PBSA water sorption behaviors are similar and characteristic of Henry mode sorption (linear curve) with a low water sorption capacity not exceeding 1% and slightly higher for PBSA. In general, the PLA/PBSA multilayer have a linear increase of mass gain with the water activity comparable to PLA, except for PLA/PBSA film from $a_w > 0.5$, showing a minor water clustering phenomenon. If we can observe that the water sorption capacity of multilayer films is governed by the major phase of PLA, it is however surprising to see that higher the presence of clays, lower the mass gain (in opposite to PBSA composites, Figure 5) and the presence of water aggregation in PLA/PBSA0 multilayer which is not so pronounced for PLA and PBSA. This result is clearly opposite to the sorption behavior of the filled PBSA films where the hydrophilic contribution of clays in PBSA matrix has induced a water uptake. In the case of multilayer films, the sorption curve profile indicates a reduction in water clustering. We can infer that the C30B acts as obstacles to water diffusion by tortuosity effects. The contribution of C30B to re-up the water mass gain at higher water activities due to water-sensitive groups is stopped by the layered microstructure. This effect is also a function of filler content. The confinement to filled PBSA layers by amorphous PLA layers prevents an increase in water solubility. The layer breaks in multilayer films as observed in TEM micrographs have not altered the barrier contribution of clays for water vapor diffusing molecules.

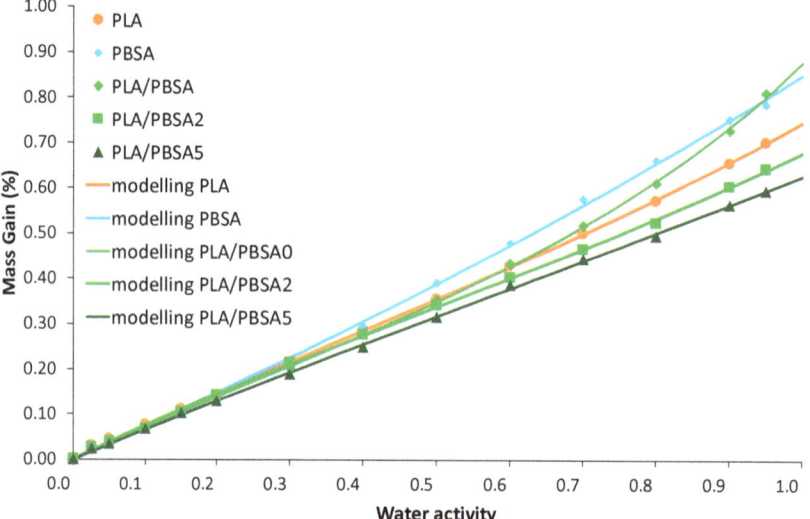

Figure 11. Water vapor sorption isotherms for the PLA/PBSA, PLA/PBSA2 and PLA/PBSA5 multilayer films modelled by two consecutive sorption modes, the Henry-law sorption and the aggregation (clustering) sorption, as represented in Equation (9).

The combination of two sorption modes—Henry-law and aggregation modes—was applied to simulate the isotherm curves and the modelling parameters are summarized in Table 8. The MDM values lower than 10% testifies to the appropriate fitting of sorption data. The k_D parameter can be indicated as a constant value for the three multilayer films. So, the water solubility is not impacted by the C30B incorporation within the layered structure. The reduction in water clustering formation is clearly evidenced from the variation of K_a and n values. With the C30B content increase, the aggregation reaction is reduced and the water cluster size is largely impacted. In the case of vapor sorption, the preferential diffusion pathways occurring at the clay-polymer interfacial areas, as discussed from permeation data, seem to be blocked by the polymer alternating layers structure of films. The modelling approach is correlated with the sorption isotherms. The water plasticization effect is, as a consequence, limited due to the reduced water concentration. As the degree of crystallinity is not really changed, one can report this result to increase stiffness and to the RAF increase, which has impacted the local macromolecular chains rigidity at the clay-polymer interfacial areas. Also, we can suppose that a specific orientation of PBSA crystals induced by the confinement effect in multilayer could hinder the passage of water molecules. From a previous study on PLA/PBSA multilayer films without fillers [18], it was shown from WAXS patterns recorded on the face and on the cross section a slight crystal orientation on both transverse and extrusion view patterns. Thus, it is possible that in filled PLA/PBSA multilayers, this slight orientation is still present, as the filler content is relatively low.

Table 8. Modelling parameters for the PLA/PBSA, PLA/PBSA2 and PLA/PBSA5 multilayer films.

	k_D g/g	n	K_a (g/g)$^{(1-n)}$	MDM (%)
PLA/PBSA	0.677	3.83	0.209	3.25
PLA/PBSA2	0.614	0.64	0.109	3.65
PLA/PBSA5	0.621	0.06	0.012	4.87

The diffusivity behavior in sorption measurement for the PLA/PBSA multilayer films was also evaluated in a semi-logarithmic scale (log D) as a function of water activity, as presented in Figure 12. The diffusion curve profiles are rather similar for all the films, irrespective of the D coefficient. As for PBSA nanocomposites, in unfilled and filled PLA/PBSA multilayers, a decrease in diffusion is observed at high water activities due to the water clustering phenomenon, even though this phenomenon is not so significant. Again, like for the PBSA-based films, the D_2 coefficients are found to be lower than the D_1 coefficients in the whole range of water activity. The balance between increase and reduction of water diffusion is again considered to explain this curve profile. Regarding the variation of D_1 and D_2 coefficients as C30B content increases, one can note that both D_1 and D_2 coefficients are reduced, highlighting tortuosity effects induced by C30B on water diffusion at the film surface and in the core of the film, respectively. In addition to the tortuosity effect, geometrical constraints and reduced chain segment mobility contribute to water mobility restriction.

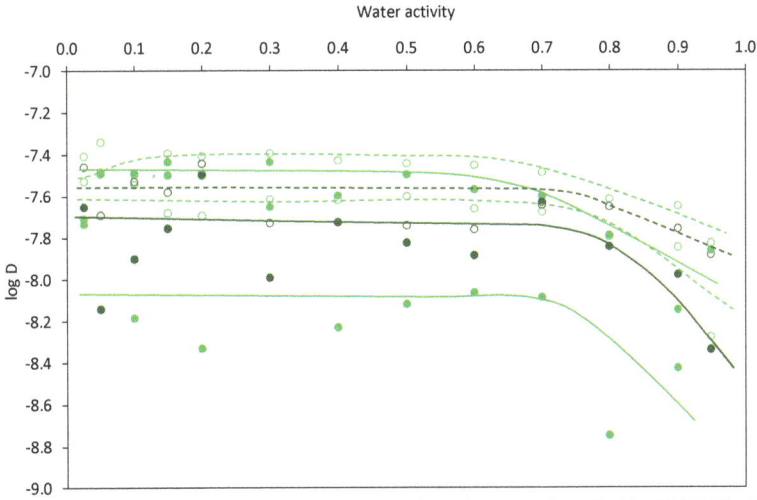

Figure 12. Water vapor diffusion coefficients D_1 and D_2 for the PLA/PBSA, PLA/PBSA2 and PLA/PBSA5 multilayer films.

For this second part of the work, one can infer that the incorporation of C30B fillers within PBSA layers of PLA/PBSA multilayer film has induced an increase of the glass transition temperature, while the degree of crystallinity and RAF is unchanged or slightly for PLA/PBSA2, showing less chain mobility by confinement effect and a local rigidification of chains by close clay contact in a thin layer thickness (150 nm on average). However, wider thicknesses for the filled PBSA layers are obtained compared to the neat PBSA layers in the layered structure due to the clay presence. In addition, some defects as layer breaks are observed without polymer interdiffusion at the polymer-polymer interfaces. From TEM micrographs; clays are located in the PBSA layers or at the polymer-polymer interfaces without an effect on PLA mobility since glass transition is unchanged.

One can point out an additional improvement of water barrier performances by tortuosity effects compared to the behaviour of the neat PLA/PBSA multilayer film. By permeation, a shift of the water flux curves to lower time is measured indicating a faster water diffusion through preferential diffusing pathways generated at the polymer-C30B interfacial areas while by sorption a decrease in D coefficients is measured evidencing greater difficulty in forming water clusters and their diffusion in the multilayer films due to geometrical constraints and chain rigidification.

It should be remembered that the boundary conditions are different between water permeation and water vapor sorption measurements. For water permeation, only one side of the tested film is in contact with liquid water and the water diffusion is deduced from the quantity of water molecules passing through the film thickness, while for water vapor sorption both sides of the film are in contact with water vapor and the water diffusion results from water molecules penetrating the film in opposite directions inside the tested film. Therefore, the duration of the measurement is longer for sorption process carried out at various water activities compared with the permeation process. It seems that the formation of water clusters is considered a slow phenomenon and that leads to reduce water mobility (decrease of D) has not sufficient time to occur in permeation as it can be clearly observed with water sorption.

4. Conclusions

In this work, we successfully prepared biodegradable PLA/PBSA multinanolayer films loaded with lamellar nanoclays in confined PBSA layers. It was observed that the incorporation of the organo-modified montmorillonite led to some layer breakups, especially in the presence of aggregates. As a consequence, irregularities in the thickness of layers (which doubled or tripled compared to the unfilled PLA/PBSA multilayer film) may appear. Despite this, the integrity and cohesion of the loaded multilayered structure was overall maintained, leading to the same mechanical and water barrier properties as the unfilled multilayer. Besides, we have shown that inclusion of clays in PBSA (monolayer comparing or nanolayer confined by PLA layers in a multilayer structure) leads to an improvement of the water barrier properties. The highest improvement factor (IF ~ 58%) was obtained when fillers are oriented and confined in very thin layers and for the highest filler content. However, this improvement is not as high as expected, probably because of the presence of some defects and aggregates and the absence of impact on the crystallinity. Consequently, we consider that these biodegradable PLA/PBSA multilayer films loaded with lamellar clays are promising materials. Our next goal will be to investigate the impact on barrier properties of a subsequent stretching step, and of a higher filler content.

Author Contributions: T.M.: Ph D student, conception of materials, analysis and interpretation of data for the work, drafting of the work; N.F.: co-supervisor, analysis, interpretation, drafting; Q.L.: structural analysis, interpretation of data; A.G.: conception of materials, analysis of data; N.D.: structural analysis, interpretation of data; J.S.: conception of materials; C.S.: analysis, interpretation of data, revising; S.M.: Supervisor, analysis, interpretation, drafting. All authors have read and agreed to the published version of the manuscript.

Funding: This research was funded by Upper Normandy Region (France)—Grant EOTP No. 427341.

Acknowledgments: The authors thank the GRR Crunch supported by Upper Normandy region (France) for the financial support of the of Tiphaine Messin.

Conflicts of Interest: The authors declare no conflict of interest.

References

1. Inagaki, N.; Tasaka, S.; Nakajima, T. Preparation of oxygen gas barrier polypropylene films by deposition of SiOx films plasma-polymerized from mixture of tetramethoxysilane and oxygen. *J. Appl. Polym. Sci.* **2000**, *78*, 2389–2397. [CrossRef]
2. Marais, S.; Hirata, Y.; Cabot, C.; Morin-Grognet, S.; Garda, M.R.; Atmani, H.; Poncin-Epaillard, F. Effect of a low-pressure plasma treatment on water vapor diffusivity and permeability of poly(ethylene-co-vinyl alcohol) and polyethylene films. *Surf. Coat. Technol.* **2006**, *201*, 868–879. [CrossRef]

3. Rahnama, M.; Oromiehie, A.; Ahmadi, S.; Ghasemi, I. Oxygen barrier films based on low-density polyethylene/ethylene vinyl alcohol/polyethylene-grafted-maleic anhydride compatibilizer. *Polyolefins J.* **2017**, *4*, 137–147.
4. Kit, K.M.; Schultz, J.M. Morphology and barrier properties of oriented blends of poly(ethylene terephthalate) and poly(ethylene 2, 6-naphthalate) with poly(ethylene-co-vinyl alcohol). *Polym. Eng. Sci.* **1995**, *35*, 680–692. [CrossRef]
5. Aït-Kadi, A.; Bousmina, M.; Yousefi, A.A.; Mighri, F. High performance structured polymer barrier films obtained from compatibilized polypropylene/ethylene vinyl alcohol blends. *Polym. Eng. Sci.* **2007**, *47*, 1114–1121. [CrossRef]
6. Mittal, V. Gas permeation and mechanical properties of polypropylene nanocomposites with thermally-stable imidazolium modified clay. *Eur. Polym. J.* **2007**, *43*, 3727–3736. [CrossRef]
7. Picard, E.; Espuche, E.; Fulchiron, R. Effect of an organo-modified montmorillonite on PLA crystallization and gas barrier properties. *Appl. Clay Sci.* **2011**, *53*, 56–65. [CrossRef]
8. Huang, H.D.; Ren, P.G.; Chen, J.; Zhang, W.Q.; Ji, X.; Li, Z.M. High barrier graphene oxide nanosheet/poly(vinyl alcohol) nanocomposite films. *J. Memb. Sci.* **2012**, *409*, 156–163. [CrossRef]
9. Jin, J.; Rafiq, R.; Gill, Y.Q.; Song, M. Preparation and characterization of high performance of graphene/nylon nanocomposites. *Eur. Polym. J.* **2013**, *49*, 2617–2626. [CrossRef]
10. Ponting, M.; Hiltner, A.; Baer, E. Polymer nanostructures by forced assembly: Process, structure, and properties. *Macromol. Symp.* **2010**, *294*, 19–32. [CrossRef]
11. Wang, H.; Keum, J.K.; Hiltner, A.; Baer, E.; Freeman, B.; Rozanski, A.; Galeski, A. Confined crystallization of polyethylene oxide in nanolayers assemblies. *Science* **2009**, *323*, 757–760. [CrossRef] [PubMed]
12. Wang, H.; Keum, J.K.; Hiltner, A.; Baer, E. Confined crystallization of PEO in nanolayered films impacting structure and oxygen permeability. *Macromolecules* **2009**, *42*, 7055–7066. [CrossRef]
13. Carr, J.M.; Langhe, D.S.; Ponting, M.T.; Hiltner, A.; Baer, E. Confined crystallization in polymer nanolayered films: A review. *J. Mater. Res.* **2012**, *27*, 1326–1350. [CrossRef]
14. Ponting, M.; Lin, Y.; Keum, J.K.; Hiltner, A.; Baer, E. Effect of substrate on the isothermal crystallization kinetics of confined poly(ε-caprolactone) nanolayers. *Macromolecules* **2010**, *43*, 8619–8627. [CrossRef]
15. Fernandes Nassar, S.; Delpouve, N.; Sollogoub, C.; Guinault, A.; Stoclet, G.; Regnier, G.; Domenek, S. Impact of nanoconfinement on polylactide crystallization and gas barrier properties. *ACS Appl. Mater. Interfaces* **2020**, *12*, 9953–9965. [CrossRef]
16. Carr, J.M.; Mackey, M.; Flandin, L.; Hiltner, A.; Baer, E. Structure and transport properties of polyethylene terephthalate and poly(vinylidene fluoride-co-tetrafluoroethylene) multilayer films. *Polymer* **2013**, *54*, 1679–1690. [CrossRef]
17. Messin, T.; Follain, N.; Guinault, A.; Miquelard-Garnier, G.; Sollogoub, C.; Delpouve, N.; Gaucher, V.; Marais, S. Confinement effect in PC/MXD6 multilayer films: Impact of the microlayered structure on water and gas barrier properties. *J. Memb. Sci.* **2017**, *525*, 135–145. [CrossRef]
18. Messin, T.; Follain, N.; Guinault, A.; Sollogoub, C.; Gaucher, V.; Delpouve, N.; Marais, S. Structure and barrier properties of multinanolayered biodegradable PLA/PBSA: Confinement effect via forced assembly coextrusion. *ACS Appl. Mater. Interfaces* **2017**, *9*, 29101–29112. [CrossRef]
19. Sekelik, D.; Stepanov, E.; Nazarenko, S.; Schiraldi, D.; Hiltner, A.; Baer, E. Oxygen barrier properties of crystallized and talc-filled poly(ethylene terephthalate). *J. Polym. Sci. Part B Polym. Phys.* **1999**, *37*, 847–857. [CrossRef]
20. Gupta, M.; Lin, Y.; Deans, T.; Baer, E.; Hiltner, A.; Schiraldi, D.A. Structure and gas barrier properties of poly(propylene-graft-maleic anhydride)/phosphate glass composites prepared by microlayer coextrusion. *Macromolecules* **2010**, *43*, 4230–4239. [CrossRef]
21. Wen, M.; Sun, X.; Su, L.; Shen, J.; Li, J.; Guo, S. The electrical conductivity of carbon nanotube/carbon black/polypropylene composites prepared through multistage stretching extrusion. *Polymer* **2012**, *53*, 1602–1610. [CrossRef]
22. Decker, J.J.; Meyers, K.P.; Paul, D.R.; Schiraldi, D.A.; Hiltner, A.; Nazarenko, S. Polyethylene-based nanocomposites containing organoclay: A new approach to enhance gas barrier via multilayer coextrusion and interdiffusion. *Polymer* **2015**, *61*, 42–54. [CrossRef]

23. Miquelard-Garnier, G.; Guinault, A.; Fromonteil, D.; Delalande, S.; Sollogoub, C. Dispersion of carbon nanotubes in polypropylene via multilayer coextrusion: Influence on the mechanical properties. *Polymer* **2013**, *54*, 4290–4297. [CrossRef]
24. Li, X.; McKenna, G.B.; Miquelard-Garnier, G.; Guinault, A.; Sollogoub, C.; Renier, G.; Rozanski, A. Forced assembly by multilayer coextrusion to create oriented graphène reinforced polymer nanocomposites. *Polymer* **2014**, *55*, 248–257. [CrossRef]
25. Gao, Y.; Picot, O.T.; Tu, W.; Bilotti, E.; Peijs, T. Multilayer coextrusion of graphene polymer nanocomposites with enehanced structural organization and properties. *J. Appl. Polym. Sci.* **2018**, *135*, 46041. [CrossRef]
26. Ray, S.S.; Bandyopadhyaya, J.; Bousmina, M. Influence of degree of intercalation on the crystal growth kinetics of poly[(butylene succinate)-co-adipate] nanocomposites. *Eur. Polym. J.* **2008**, *44*, 3133–3145.
27. Ahn, B.D.; Kim, S.H.; Kim, Y.H.; Yang, J.S. Synthesis and characterization of biodegradable copolymers from succinic acid and adipic acid with 1, 4-butanediol. *J. Appl. Polym. Sci.* **2001**, *82*, 2808–2826. [CrossRef]
28. Bandyopadhyay, J.; Al-Thabaiti, S.A.; Ray, S.S.; Basahel, S.N.; Mokhtar, M. Unique cold-crystallization behavior and kinetics of biodegradable poly[(butylene succinate)-co adipate] nanocomposites: A high speed differential scanning calorimetry study. *Macromol. Mater. Eng.* **2014**, *299*, 939–952. [CrossRef]
29. Métayer, M.; Labbé, M.; Marais, S.; Langevin, D.; Chappey, C.; Dreux, F.; Brainville, M.; Belliard, P. Diffusion of water through various polymer films: A new high performance method of characterization. *Polym. Test.* **1999**, *18*, 533–549. [CrossRef]
30. Gray, D.E. (Ed.) *American Institute of Physics Handbook*; Mc Graw-Hill: New York, NY, USA, 1957.
31. Marais, S.; Métayer, M.; Nguyen, Q.T.; Labbé, M.; Langevin, D. New methods for the determination of the parameters of a concentration-dependent diffusion law for molecular penetrants from transient permeation or sorption data. *Macromol. Theor. Simul.* **2000**, *9*, 207–214. [CrossRef]
32. Lomauro, C.J.; Bakshi, A.S.; Labuza, T.P. Evaluation of Food Moisture Sorption Isotherm Equations—Part, I.; Fruit, Vegetable and Meat-Products. *Lebensm. Wiss Technol.* **1985**, *18*, 111–117.
33. Bungay, P.M.; Lonsdale, H.K.; de Pinho, M.N. Synthetic membranes: Science, Engineering and Applications. *ASI Ser. Ser. C Math. Phys. Sci.* **1986**, *181*, 93–96.
34. Ren, M.; Song, J.; Song, C.; Zhang, H.; Sun, X.; Chen, Q.; Zhang, H.; Mo, Z. Crystallization kinetics and morphology of poly(butylene succinate-co-adipate). *J. Polym. Sci. Pol. Phys.* **2005**, *43*, 3231–3241. [CrossRef]
35. Charlon, S.; Follain, N.; Chappey, C.; Dargent, E.; Soulestin, J.; Sclavons, M.; Marais, S. Improvement of barrier properties of bio-based polyester nanocomposite membranes by water-assisted extrusion. *J. Memb. Sci.* **2015**, *496*, 185–198. [CrossRef]
36. Yoon, P.J.; Hunter, D.L.; Paul, D.R. Polycarbonate nanocomposites. Part 1. Effect of Organoclays Structure on the Morphology and Properties. *Polymer* **2003**, *44*, 5323–5339. [CrossRef]
37. Cervantes-Uc, J.M.; Cauich-Rodríguez, J.V.; Vásquez-Torres, H.; Garfias-Mesías, L.F.; Paul, D.R. Thermal degradation of commercially available organoclays studied by TGA-FTIR. *Thermochimica Acta* **2007**, *457*, 92–102. [CrossRef]
38. Charlon, S.; Follain, N.; Dargent, E.; Soulestin, J.; Sclavons, M.; Marais, S. Poly[(butylene succinate)-co-(butylene adipate)]-montmorillonite nanocomposite prepared by water-assisted extrusion: Role of the dispersion level and of the structure-microstructure on the enhanced barrier properties. *J. Phys. Chem.* **2019**, *120*, 13234–13248. [CrossRef]
39. Esposito Corcione, C.; Maffezzoli, A.; Cannoletta, D. Effect of a nanodispersed clay fillers on glass transition of thermosetting polyurethane. *Macromol. Symp.* **2009**, *286*, 180–186. [CrossRef]
40. Meyers, K.P.; Decker, J.J.; Olson, B.G.; Lin, J.; Jamieson, A.M.; Nazarenko, S. Probing the confining effect on clay particles on an amorphous intercalated dendritic polyester. *Polymer* **2017**, *112*, 76–86. [CrossRef]
41. Ray, S.S.; Bousmina, M.; Okamoto, K. Structure and properties of nanocomposites based on poly(butylene succinate-co-adipate) and organically modified montmorillonite. *Macromol. Mater. Eng.* **2005**, *290*, 759–768. [CrossRef]
42. Zhang, X.; Zhang, Y. Poly(butylene succinate-*co*-butylene adipate)/Cellulose nanocrystal composites modifies with phthalic anhydride. *Carbohydr. Polym.* **2015**, *134*, 52–59. [CrossRef] [PubMed]
43. Jordan, J.; Jacob, K.I.; Tannenbaum, R.; Sharaf, M.A.; Jasiuk, I. Experimental trends in polymer nanocomposites—A review. *Mat. Sci. Eng. A* **2005**, *393*, 1–11. [CrossRef]

44. Zaidi, L.; Bruzaud, S.; Bourmaud, A.; Mérédic, P.; Kaci, M.; Grohens, Y. Relationship between structure and rheological, mechanical and thermal properties of polylactide/cloisite 30B nanocomposites. *J. Appl. Polym. Sci.* **2010**, *116*, 1357–1365. [CrossRef]
45. Someya, Y.; Nakazato, T.; Teramoto, N.; Shibata, M. Thermal and mechanical properties of poly(butylene succinate) nanocomposites with various organo-modified montmorillonites. *J. Appl. Polym. Sci.* **2004**, *91*, 1463–1475. [CrossRef]
46. Hwang, S.Y.; Yoo, E.S.; Im, S.S. Effect of the urethane group on treated clay surfaces for high-performance poly(butylene succinate)/montmorillonite nanocomposites. *Polym. Degrad. Stab.* **2009**, *94*, 2163–2169. [CrossRef]
47. Chen, G.; Yoon, J.S. Nanocomposites of poly[(butylene succinate)-co-(butylene adipate)] (PBSA) and twice-functionalized organoclay. *Polym. Int.* **2005**, *54*, 939–945. [CrossRef]
48. Finnigan, B.; Martin, D.; Halley, P.; Truss, R.; Campbell, K. Morphology and properties of thermoplastic polyurethane nanocomposites incorporating hydrophilic layered silicates. *Polymer* **2004**, *45*, 2249–2260. [CrossRef]
49. Tan, B.; Thomas, N.L. A review of the water barrier properties of polymer/clay and polymer/graphene nanocomposites. *J. Memb. Sci.* **2016**, *514*, 595–612. [CrossRef]
50. Tenn, N.; Follain, F.; Fatyeyeva, K.; Poncin-Epaillard, F.; Labrugère, C.; Marais, S. Impact of hydrophobic plasma treatments on the barrier properties of poly(lactic acid) films. *RSC Adv.* **2014**, *4*, 5626. [CrossRef]
51. Follain, N.; Valleton, J.M.; Lebrun, L.; Alexandre, B.; Schaetzel, P.; Metayer, M.; Marais, S. Simulation of kinetic curves in mass transfer phenomena for a concentration-dependent diffusion coefficient in polymer membranes. *J. Memb. Sci.* **2010**, *349*, 195–207. [CrossRef]
52. Follain, N.; Alexandre, B.; Chappey, C.; Colasse, L.; Médéric, P.; Marais, S. Barrier properties of polyamide 12/montmorillonite nanocomposites: Effect of clay structure and mixing conditions. *Compos. Sci. Technol.* **2016**, *136*, 18–28. [CrossRef]
53. Charlon, S.; Follain, N.; Soulestin, J.; Sclavons, M.; Marais, S. Water transport properties of poly(butylene succinate) and poly[(butylene succinate)-co-(butylene adipate)] nanocomposite films: Influence of the water-assisted extrusion process. *J. Phys. Chem. C* **2017**, *121*, 918–930. [CrossRef]
54. Tenn, N.; Follain, N.; Soulestin, J.; Crétois, R.; Bourbigot, S.; Marais, S. Effect of nanoclay hydration on barrier properties of PLA/Montmorillonite based nanocomposites. *J. Phys. Chem. C* **2013**, *117*, 12117–12135. [CrossRef]
55. Alexandre, M.; Dubois, P. Polymer-silicate nanocomposites: Preparation, properties and use of a new class of materials. *Mater. Sci. Eng.* **2000**, *28*, 1–63. [CrossRef]
56. Follain, N.; Belbekhouche, S.; Bras, J.; Siqueira, G.; Marais, S.; Dufresne, A. Water transport properties of bio-nanocomposites reinforced by luffa cylindrica cellulose nanocrystals. *J. Membr. Sci.* **2013**, *427*, 218–229. [CrossRef]
57. Rouse, P.E. Diffusion of vapors in films. *J. Am. Chem. Soc.* **1947**, *69*, 1068–1073. [CrossRef]

Publisher's Note: MDPI stays neutral with regard to jurisdictional claims in published maps and institutional affiliations.

© 2020 by the authors. Licensee MDPI, Basel, Switzerland. This article is an open access article distributed under the terms and conditions of the Creative Commons Attribution (CC BY) license (http://creativecommons.org/licenses/by/4.0/).

Article

Multifunctional PLA Blends Containing Chitosan Mediated Silver Nanoparticles: Thermal, Mechanical, Antibacterial, and Degradation Properties

Agueda Sonseca [1,2,3,*], Salim Madani [4], Gema Rodríguez [1,2], Víctor Hevilla [1,2], Coro Echeverría [1,2], Marta Fernández-García [1,2], Alexandra Muñoz-Bonilla [1,2], Noureddine Charef [4] and Daniel López [1,2,*]

1. MacroEng Group, Instituto de Ciencia y Tecnología de Polímeros, ICTP-CSIC, C/ Juan de la Cierva 3, 28006 Madrid, Spain; gema@ictp.csic.es (G.R.); v.hevilla@ictp.csic.es (V.H.); cecheverria@ictp.csic.es (C.E.); martafg@ictp.csic.es (M.F.-G.); sbonilla@ictp.csic.es (A.M.-B.)
2. Interdisciplinary Plataform for "Sustainable Plastics towards a Circular Economy" (SUSPLAST-CSIC), 28006 Madrid, Spain
3. Instituto de Tecnología de Materiales, Universitat Politècnica de València (UPV), Camino de Vera s/n, 46022 Valencia, Spain
4. Laboratory of Applied Biochemistry, University Ferhat Abbas, Setif 19000, Algeria; madanisalim79@gmail.com (S.M.); charefnr@hotmail.com (N.C.)
* Correspondence: agsonol@posgrado.upv.es (A.S.); daniel.l.g@csic.es (D.L.); Tel.: +34-915622900 (D.L.)

Received: 20 November 2019; Accepted: 16 December 2019; Published: 20 December 2019

Abstract: Poly(lactic acid) (PLA) is one of the most commonly employed synthetic biopolymers for facing plastic waste problems. Despite its numerous strengths, its inherent brittleness, low toughness, and thermal stability, as well as a relatively slow crystallization rate represent some limiting properties when packaging is its final intended application. In the present work, silver nanoparticles obtained from a facile and green synthesis method, mediated with chitosan as a reducing and stabilizing agent, have been introduced in the oligomeric lactic acid (OLA) plasticized PLA in order to obtain nanocomposites with enhanced properties to find potential application as antibacterial food packaging materials. In this way, the green character of the matrix and plasticizer was preserved by using an eco-friendly synthesis protocol of the nanofiller. The X-ray diffraction (XRD) and differential scanning calorimetry (DSC) results proved the modification of the crystalline structure as well as the crystallinity of the pristine matrix when chitosan mediated silver nanoparticles (AgCH-NPs) were present. The final effect over the thermal stability, mechanical properties, degradation under composting conditions, and antimicrobial behavior when AgCH-NPs were added to the neat plasticized PLA matrix was also investigated. The obtained results revealed interesting properties of the final nanocomposites to be applied as materials for the targeted application.

Keywords: poly(lactic acid); oligomeric lactic acid; eco-friendly silver nanoparticles; biopolymer properties; antimicrobial activity; packaging

1. Introduction

Although biodegradable polymers are not meant to fully substitute the non-biodegradable synthetic polymers, when applying them in packaging and medical industries it is still one of the most hopeful approaches to face polymer waste's environmentally harmful problems [1]. Poly(lactic acid) (PLA) is a well-known biodegradable synthetic polymer that is considered one of the most attractive materials for packaging applications [2] thanks to its biocompatibility, biodegradability, renewable character (can be obtained from the fermentation of 100% renewable and biodegradable plant sources such as corn, rice, and sugar feedstocks), as well as commercial availability [3,4]. PLA has also been approved by the American Food and Drug Administration (FDA) for direct contact with biological fluids, and it is also highly transparent and possesses good water vapor barrier properties [5,6]. In general, PLA possesses higher mechanical strength and is easier to process in comparison with other biopolymers and, additionally, is naturally degraded in soil or compost [7,8] in products completely assimilated by microorganisms [2]. Despite such strengths, PLA toughness, thermal stability, and mechanical properties are still limited when compared with their petroleum-based counterparts [9]. Despite its strong resistance, it is very brittle (elongation at break lower than 10%) [10] and another important shortcoming limiting its practical application is the relatively slow crystallization rate [4]. In this regard, to overcome these drawbacks, the addition of plasticizers, blending [11], copolymerization [12], as well as the development of micro- or nano-composites have been often employed as promising strategies to enhance the mechanical performance of PLA while preserving its transparency, which are both useful properties in the packaging sector [13,14].

Nowadays, food packaging is in constant development in order to meet the consumer and industry demands that trend towards minimally processed food products with prolonged shelf-life and controlled quality [15]. The packaging is known as the cornerstone of the food processing industry as almost 100% of the food and drinks that we buy and consume are packaged at a higher or lower level [16,17]. Packaging functions are mainly related to preservation and protection of the food, maintaining its quality and safety as well as the reduction of food waste [16]. Therefore, the only way to respond to this demand is by developing improved packaging concepts to guarantee food safety, quality, and traceability. In this sense, the development of active food-packaging can represent a big step forward to provide additional functions that prevent food waste in comparison with traditional non-active (passive) materials.

Food-borne pathogens are of utmost concern in food safety [18]. Each year, millions of people get sick or even die due to the ingestion of unsafe food and water, mainly caused by bacteria, viruses, and parasites [19], and as a direct consequence, antimicrobial active-packaging materials are attracting the attention from food and packaging industries [20]. Nowadays, the introduction of silver particles (Ag NPs) into polymers able to imprint antimicrobial properties, is widely accepted as silver is considered a potent broad-spectrum antimicrobial agent non-toxic to human cells and with proven long-lasting biocidal activity, high temperature stability, and low volatility [21–24]. Therefore, Ag NPs have been approved for use in food-contact polymers in the USA and the European Union, after previous consideration of the use conditions [25]. Recently, synthesis under eco-friendly conditions of silver (Ag) and gold (Au) nanoparticles have been gaining interest for researchers [25,26], as most common reducing agents employed for the preparation of nanoparticles are $NaBH_4$ (sodium borohydride), citrate, or ascorbate, which are associated with environmental toxicity or biological hazards. In this context, chitosan is one of the most used biopolymers, thanks to their large amount of free amino and hydroxyl groups, and under the proper thermal conditions can be employed as reducing and stabilizing agents for the synthesis of Ag and Au nanoparticles preventing their aggregation [27–33]. In proper thermal conditions, the hydroxyl groups are converted to carboxyl groups by air oxidation, which reduces the silver ions [34].

With this background, in this study eco-friendly silver nanoparticles synthesized with a green protocol, using non-toxic biodegradable chitosan as the reducing agent, have been employed for obtaining plasticized PLA nanocomposites suitable for packaging applications thanks to its biocide properties. The aim of the present work is to study in detail the morphological, mechanical, thermal, and antibacterial properties as well as the crystallinity and the degradation profile in composting conditions of degradable plasticized PLA nanocomposites containing chitosan-based silver nanoparticles, in order to assess the prospective approach offered by these new systems. Nanocomposites have been prepared by melt compounding in a twin-screw extruder and their mechanical and thermal responses have been related to the nanoparticle content. The oligomeric lactic acid employed as a plasticizer agent together with the chitosan-based silver nanoparticles imposes enhanced ductility, toughness, and antibacterial activity to the developed nanocomposites.

2. Materials and Methods

2.1. Materials

Polylactic acid (PLA3051D, 3% of D-lactic acid monomer, molecular weight 142×10^4 g/mol, density 1.24 g/cm^3) and Lactic acid oligomer (Glyplas OLA8, ester content >99%, density 1.11 g/cm^3, viscosity 22.5 mPa.s, molecular weight 1100 g/mol) were supplied from NatureWorks LLC (Minnetonka, MN, USA) and Condensia Química SA (Barcelona, Spain), respectively. Silver nitrate (AgNO$_3$) and Chitosan from shrimp shells with a deacetylation degree >75% were purchased from Sigma-Aldrich (St. Quentin Fallavier, France). Acetic acid and sodium hydroxide were purchased from Fluka (Seelze, Germany).

2.2. Synthesis of Based Chitosan Silver Nanoparticles

Chitosan-based silver nanoparticles (AgCH-NPs) were synthesized by a method described elsewhere [33,35]. Typically, 4.5 mL of a solution of 52.0 mM of AgNO$_3$ and 10 mL of a solution of chitosan of concentration 6.92 mg/mL in 1% acetic acid were mixed and heated at 95 °C under stirring for 12 h. The color of the mixture changed from colorless to yellow and, finally, to brownish which indicates the formation of the nanoparticles. The dispersions obtained were dialyzed for three days against distilled water using dialysis membranes with a MWCO of 12,000–14,000 Da. AgCH-NPs were then isolated by lyophilization.

2.3. Preparation of Oligomeric Lactic Acid Plasticized Poly(Lactic Acid) Nanocomposites Containing AgCH-NPs (PLA/OLA AgCH-NPs Nanocomposites)

Appropriate amounts of PLA, AgCH-NPs, and OLA were mixed in a microextruder equipped with two twin conical co-rotating screws (Thermo Scientific MiniLab Haake Rheomex CTW5). PLA pellets and OLA were previously dried overnight at 80 °C in order to avoid the presence of moisture. AgCH-NPs were dried at 40 °C for 4 h. PLA was first processed in the MiniLab mixer at 180 °C and a rotation speed of 100 rpm; after 2 min OLA and AgCH-NPs were added once PLA had reached the melt state to avoid oligomer and nanoparticle thermal degradation. The total time of mixing was 3 min. Films were then obtained by processing blends by compression molding at 180 °C in a hot press (Collin P-200-P, Collin Lab & Pilot Solutions GmbH, Maitenbeth, Germany). Blends were kept between the hot plates at atmospheric pressure for one minute at 180 °C until reaching the melt state. Then, they were submitted to the following pressure cycles, 5 MPa for one minute at 180 °C and then cooling down to room temperature at 5 MPa for one minute. The different obtained formulations are gathered in Table 1.

Table 1. Poly(lactic) acid/oligomeric lactic acid (PLA/OLA) AgCH-NPs formulations.

Sample	PLA (wt%)	OLA (wt%)	AgCH-NPs (wt%)
PLA/OLA	80	20	0
PLA/OLA-AgCH-0.5%	79.6	20	0.4
PLA/OLA-AgCH-1%	79.2	20	0.8
PLA/OLA-AgCH-3%	77.6	20	2.4
PLA/OLA-AgCH-5%	76	20	4

2.4. Characterization Techniques

A Philips XL30 scanning electron microscope (SEM, Philips, Mahwah, NJ, USA), with an accelerating voltage of 10 kV, a work distance of 10–15 mm was used to record SEM micrographs of samples cross-section and observe changes produced by the different amount of filler (AgCH-NPs). Nanocomposites were cryo-fractured with the aid of liquid N$_2$ and each specimen was gold-coated (~5 nm thickness) in a Polaron SC7640 Auto/Manual Sputter (Quorum Technologies LTD, Kent, UK).

Fourier transmission spectra (FTIR) were recorded for all the samples using a Spectrum-Two (Perkin Elmer) spectrometer between 650 and 4500 cm^{-1} spectral range with a 4 cm^{-1} resolution and an attenuated total reflectance (ATR) cell. A background spectrum was acquired before every sample and all samples were vacuum-dried prior to measurement.

Thermal transitions and stability of neat PLA/OLA, as well as obtained nanocomposites with different amounts of AgCH-NPs, were studied by differential scanning calorimetry (DSC) and thermogravimetric analysis (TGA) in a DSC Q2000 and TA-Q500 apparatus both from TA Instruments (New Castle, DE, USA), respectively. In the DSC, samples were heated from −60 °C up to 180 °C at 10 °C/min under an N_2 atmosphere (flow rate of 50 mL/min). Glass transition temperatures (T_g), calculated as the midpoint of the transition, melting temperatures (T_m), cold crystallization (ΔH_{cc}), and melting enthalpies (ΔH_m) were calculated by analyzing the thermograms in the TA Universal Analysis software. The degree of crystallinity (X_c%) was therefore obtained from Equation (1) using 93.6 J/g as crystallization enthalpy value for pure crystalline PLA ($\Delta H^0{}_m$) [36] W_f represents the weight fraction of pure PLA present in the sample ($W_f = 1 - 0.2 - m_f$, where m_f is the weight fraction of the nanoparticles in the nanocomposite).

$$X_c\% = 100 \times ((\Delta H_m - \Delta H_{cc})/\Delta H^0{}_m) \times (1/W_f) \quad (1)$$

Tensile tests were carried out on an Instron instrument (Instron, Norwood, MA, USA). equipped with a 50 N load cell, operated at room temperature and at a crosshead speed of 10 mm/min (strain rate). The initial length between clamps was set at 10 mm, samples of 6 mm width and ~100 μm of average thickness were measured and results from five to ten specimens were averaged. Young's modulus (slope of the curve from 0–2% deformation), maximum stress, ultimate tensile strength, and elongation at break were calculated.

To study the crystalline structure of the plasticized PLA nanocomposite films, a Bruker D8 Advance X-ray diffractometer (Bruker Scientific LLC, Billerica, MA, USA) equipped with a CuKα (λ = 0.154 nm) source was employed. Samples were mounted on an appropriate holder and scanned between 2° and 60° (2θ) with a scanning step of 0.02°, a collection time of 10 s per step, and 40 kV of operating voltage to obtain X-ray diffraction patterns (XRD). Percentages of crystallinity were calculated with the aid of PeakFit software (Systat Software Inc. V4.12, San Jose, CA, USA) by a fitting process using a deconvolution method.

Antimicrobial activity of the prepared nanocomposites was determined following the E2149-13a standard method of the American Society for Testing and Materials (ASTM) (West Conshohocken, PA, USA) [37] against *Staphylococcus aureus* (*S. aureus*, ATCC 29213) and *Escherichia coli* (*E. coli*, ATCC 25922) bacteria. Each nanocomposite was placed in a sterile falcon tube and then 10 mL of the bacterial suspension (ca. 10^6 colony forming units (CFU)/mL) were added. Falcon tubes with only the inoculum and neat plasticizer PLA were also prepared as control experiments. The samples were shaken at room temperature at 150 rpm for 24 h. Bacterial concentrations at time 0 and after 24 h were calculated by the plate count method. The contact killing experiment was done per triplicate on different days and the plate counting by duplicate.

The biodegradation test for all the samples was conducted under aerobic composting conditions in a laboratory-scale reactor following the ISO-20200 standard. [38] Briefly, samples were cut into square geometries of 15 × 15 mm and buried 4–6 cm in depth inside reactors containing solid biodegradation media −10 wt% of compost (Compo GmbH, Münster, Germany), 30 wt% of rabbit food, 10 wt% of starch, 5 wt% of sugar, 1 m wt% of urea, 4 wt% of corn oil, 40 wt% of sawdust and water in a 45:55 wt% ratio, and incubated at 58 °C for 40 days. Samples were kept into textile meshes to allow easy removal from the composting medium, while when buried access of microorganisms and moisture was ensured. Water was periodically added to the reaction containers to maintain the relative humidity in the medium, and the aerobic conditions were guaranteed by the regular mixing of the compost medium. Samples were recovered from the disintegration medium at different time intervals (7, 17, 21, 28, 36, and 44 days), washed with distilled water, and dried in an oven at 37 °C until constant weight.

The mass loss weight % was calculated by normalizing the sample weight at different incubation times to the initial weight value. Photographs were taken from samples once extracted from the composting medium. The test was done at least with two samples of each composition.

3. Results

3.1. Microstructure and Morphology of the PLA/OLA AgCH-NPs Nanocomposites:

The morphology of AgCH-nanoparticles and PLA/OLA AgCH-NPs nanocomposites was studied by scanning electron microscopy (SEM). Figure 1 shows representative SEM micrographs of the synthesized AgCH-NPs, revealing a spherical morphology with diameters of ~8 ± 1 nm in agreement with previously reported values for similar procedures [35].

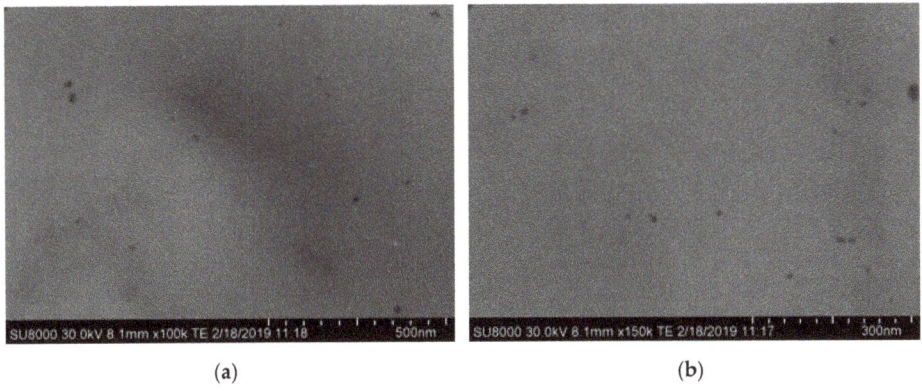

Figure 1. Scanning electron microscopy (SEM) images of AgCH-NPs at different magnifications, (**a**) ×100,000 and (**b**) ×150,000.

Figure 2 depicts scanning electron microscopy (SEM) images of the cross-sectional areas of neat PLA/OLA matrix and nanocomposites with 0.5 wt%, 1 wt%, 3 wt%, and 5 wt% of AgCH-NPs. Clearly, the incorporation of AgCH-NPs into the polymer matrix induces a change in the morphology of the cross-section surface. The unfilled material shows more uniform and plane fracture while for nanocomposites a surface with shear-yield and plastic deformation in different directions can be observed. The rougher surface morphology into the nanocomposites is homogeneously distributed along the cross-section and no large agglomerations of nanoparticles seem to be present. Additionally, voids and more stratified structures appeared and increased with the AgCH-NPs content probably due to more restricted mobility (higher rigidity) of the polymer chains at the interface with the nanoparticles.

FTIR spectra of PLA/OLA, AgCH-NPs and PLA/OLA AgCH-NPs reinforced nanocomposites are shown in Figure 3, where characteristic bands of PLA/OLA and AgCH-NP can be identified. Neat PLA/OLA and nanocomposites show a weak band at 3518 cm^{-1} that can be attributed to the vibration of the hydroxyl groups in the terminal chain of PLA/OLA. At 1748 cm^{-1} there is a carbonyl group stretching vibration band, and at 1454 cm^{-1}, 1383 cm^{-1}, and 1366 cm^{-1} appears –CH$_3$ groups, -CH deformation, and asymmetric bands. The –C-O- stretching bands appear at 1180 cm^{-1}, 1130 cm^{-1}, and 1085 cm^{-1}. AgCH-NPs spectrum exhibits the characteristics absorption bands of chitosan at 1639 cm^{-1} (Amide I), 1551 cm^{-1} (Amide II), and 1324 cm^{-1} (Amide III). These bands are not detectable in the nanocomposite's spectra due to the low percentage of nanoparticles in the nanocomposites and the weakness of these bands.

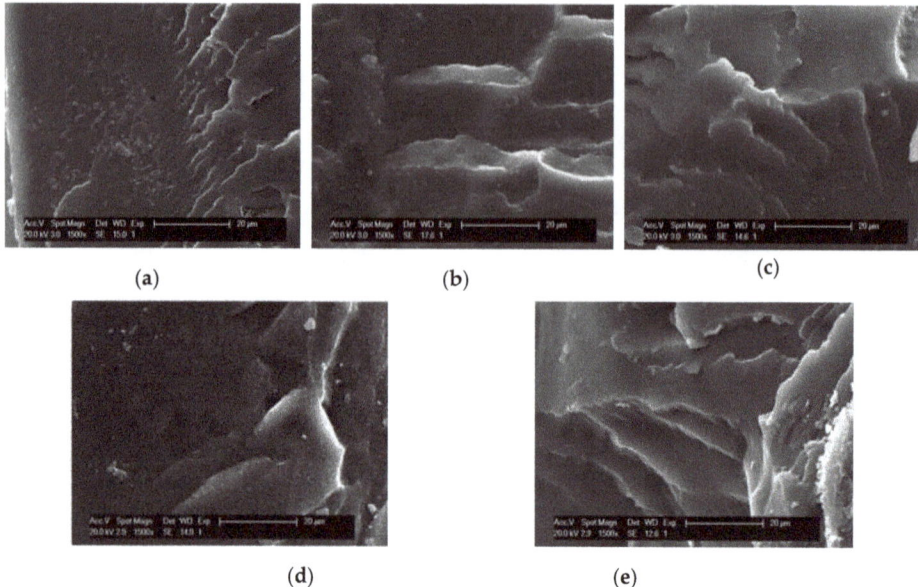

Figure 2. SEM images of the cross-section of PLA/OLA and PLA/OLA AgCH-NPs reinforced nanocomposites at ×1500 magnification, with different nanoparticles concentration. (**a**) PLA/OLA, (**b**) 0.5 wt% AgCH-NPs, (**c**) 1 wt% AgCH-NPs, (**d**) 3 wt% AgCH-NPs, (**e**) 5 wt% AgCH-NPs.

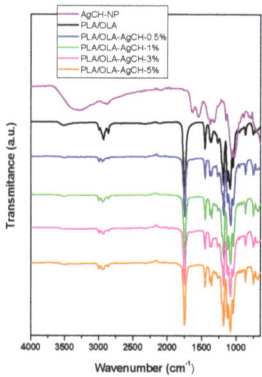

Figure 3. Fourier transmission spectra (FTIR) of PLA/OLA, AgCH-NPs and PLA/OLA AgCH-NPs reinforced nanocomposites.

XRD patterns were recorded for all the samples and are shown in Figure 4 (see supporting information Figure S1 for AgCH-NPs XRD spectrum). Neat PLA/OLA exhibits a broad reflection indicative of its amorphous nature. The addition of 0.5 wt% of AgCH-NPs did not produce a considerable difference in the PLA/OLA matrix crystalline structure; however, the addition of 1 wt% of AgCH-NPs starts to induce crystallinity to the PLA/OLA matrix as evidenced by the appearance of a diffraction peak at $2\theta = 16.7°$ corresponding to (110/200) planes of PLA. With increasing the AgCH-NPs content, this peak characteristic of the α and α' type crystals become more evident. Additionally, in samples with 3 wt% and 5 wt% of AgCH-NPs loads, new peaks at $2\theta = 15°$ and $2\theta = 19.2°$ corresponding to (010) and (203) plane reflections of PLA chains belonging to α or α' type crystals were visible, what is indicative of the nanoparticles nucleating effect [39,40].

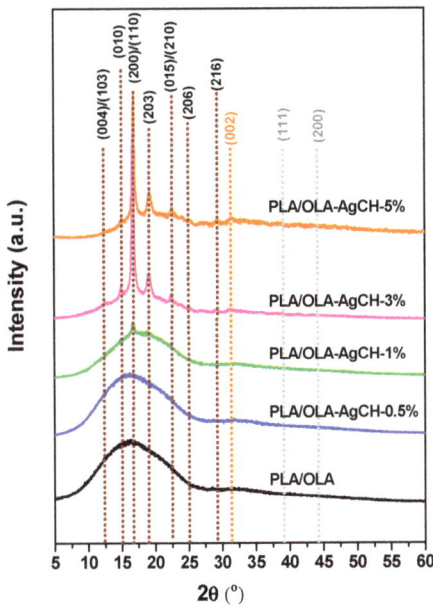

Figure 4. XRD patterns of PLA/OLA and PLA/OLA AgCH-NPs reinforced nanocomposites.

Besides these diffraction peaks, samples with contents of 3 and 5 wt% of AgCH-NPs, start to exhibit reflections at around 32° (002), 38° (111), and 44° (200) corresponding to the presence of AgCH-NPs (see Supporting Information Figure S2) being a clear evidence of their presence into the nanocomposites. Interestingly, with increasing the AgCH-NPs content to 5 wt%, reflections from (010), (200/110), and (203) planes very slightly shifted to higher 2θ, and intensity of peaks at 15° and 22.4° corresponding to (010) and (015) crystalline planes of PLA, increased (Table 2). This fact can be indicative of a higher degree of order in the structure with increasing the AgCH-NPs content to 5 wt% from more distorted orthorhombic crystals (α' form) to more perfect crystals (α form). Calculated interplanar distance from Braggs' law evidenced lower distance between planes in the sample with the highest AgCH-NPs content (Table 2) [41]. It is also possible to note that small diffraction peaks at 12.6° and 25.3° appeared in 1, 3, and 5 wt% of AgCH-NPs containing samples that are assigned to the reflections of (004/103) and (016) planes of crystalline PLA respectively [42,43].

Table 2. Distance between planes of nanocomposites (1–5 wt% of AgCH-NPs) calculated from the most intensive diffraction peaks.

Sample	2θ (Angle)			Distance between Planes		
	(010)	(200)/(110)	(203)	d(A)	d(A)	d(A)
PLA/OLA AgCH1%	–	16.7	–	–	5.30	–
PLA/OLA AgCH3%	14.8	16.7	19.1	5.98	5.30	4.64
PLA/OLA AgCH5%	14.9	16.8	19.2	5.94	5.27	4.62

3.2. Thermal Analysis

In order to study the effect of the addition of AgCH-NPs over the PLA/OLA matrix thermal stability as well as over the thermal degradation mechanism, thermogravimetric analysis was conducted under nitrogen atmosphere. Figure 5 shows weight loss vs. temperature thermogravimetric curves (Figure 5a) and their corresponding derivative curves (Figure 5b), for neat PLA/OLA and AgCH-NPs loaded

PLA/OLA nanocomposites. Table 3 collects temperatures at different percentages of weight loss for each sample.

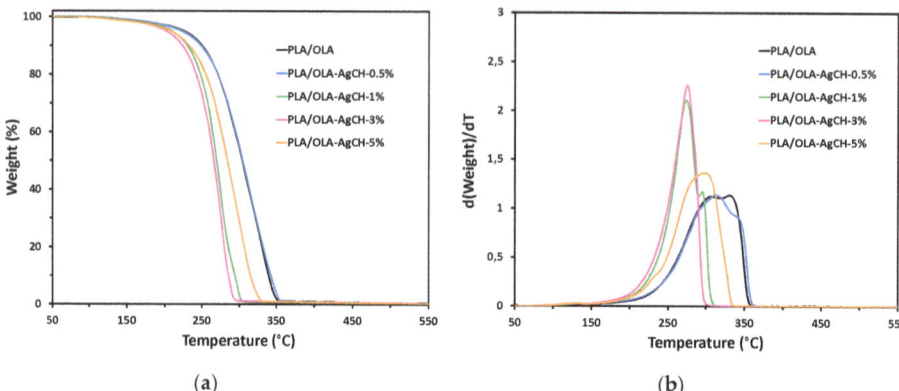

Figure 5. Thermogravimetric curves of neat PLA/OLA and its nanocomposites with different AgCH-NPs content: (**a**) Weight loss vs. temperature curves; (**b**) derivative curves.

It is worth noticing that derivative thermogravimetric curves (DTG) of nanocomposites containing 1 and 3 wt% of AgCH-NPs are sharper than the rest (neat matrix and the other nanocomposites), which the main degradation peak presents a shoulder, evidencing a two-step degradation process. In general, the thermal stability of the samples seems to be dependent on the amount of AgCH-NPs. Maximum degradation temperature decreases between 20–56 °C with increasing AgCH-NPs content from 1 to 3 wt%. Interestingly, with the addition of 5 wt% of AgCH-NPs, the shape of the degradation curve changes significantly, with a temperature at maximum weight loss rate closer to the nanocomposites with 0.5 wt% of AgCH-NPs instead of the one with 3 wt% of AgCH-NPs. This phenomenon could be related with the high amount of AgCH-NPs that can affect the mass transport barrier, resulting in more intricate paths for the volatile products (i.e., low molecular weight OLA plasticizer) to scape during thermal decomposition and therefore, promoting slightly better thermal resistance for sample with AgCH-NPs loading of 5 wt% [44,45]. Thus, in general, TGA results show that the addition of low amounts of AgCH-NPs (0.5 wt%) did not significantly affect the thermal degradation of the PLA/OLA matrix, while amounts closer to 5 wt% of AgCH-NPs could even better prevent the thermal degradation of the matrix than the addition of 1–3 wt% of nanoparticles. Additionally, it is important to remark that no degradation occurred from room temperature to 180 °C, the range where nanocomposites were processed.

Table 3. Temperatures at different weight losses of neat PLA/OLA and the different formulations containing AgCH-NPs.

Sample	Temperature at Maximum Weight Loss Rates (°C)		Temperature at Different Weight Losses (°C)			
	T_{max1}	T_{max2}	T_5	T_{30}	T_{50}	T_{70}
PLA/OLA	306	331	227	284	304	321
PLA/OLA AgCH0.5%	312	341	224	285	304	322
PLA/OLA AgCH1%	273	296	211	258	270	279
PLA/OLA AgCH3%	275	–	203	253	266	276
PLA/OLA AgCH5%	303	–	210	268	284	299

Differential scanning calorimetry (DSC) was employed to study the glass transition, melting, and crystallization of plasticized PLA and its silver-containing nanocomposites. Figure 6 shows the thermograms of neat PLA/OLA and its AgCH-NPs nanocomposites. Glass transition (T_g), cold crystallization (T_{cc}, ΔH_{cc}), and melting temperatures and enthalpies (T_m, ΔH_m), as well as crystallinity values (X_c%) are summarized in Table 4. All the samples showed a single glass transition temperature that is indicative of good miscibility between polymer matrix and plasticizer as no evidence of T_g from free OLA was observed. The addition of AgCH-NPs into PLA/OLA resulted in significant differences over T_g values with respect to the neat matrix. The addition of 0.5–1 wt% of AgCH-NPs produced a significant decrease in T_g values from 32 °C to 24–25 °C. This reduction due to the addition of nanoparticles can be related to the higher mobility of the polymer macromolecules due to an increase in the free volume of the matrix probably due to the plasticizer. On the contrary, the addition of 3–5 wt% of AgCH-NPs increased the T_g towards typical values of non-plasticized PLA probably due to the presence of high content of nanoparticles within the PLA plasticized matrix, that hinders the movements of the macromolecular chains in agreement with TGA results [46].

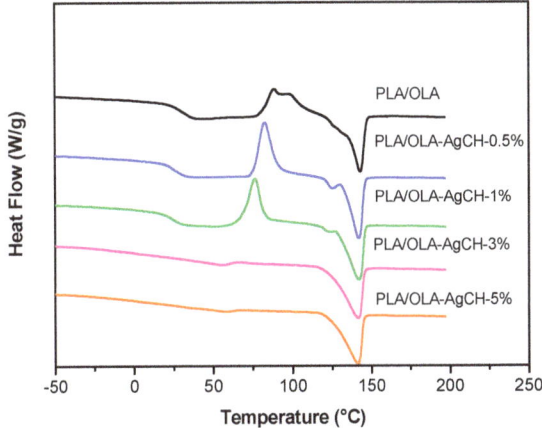

Figure 6. Differential scanning calorimetry (DSC) curves of neat PLA/OLA and its nanocomposites with different AgCH-NPs content.

Accordingly, and as can be seen in Figure 7, T_{cc} values decreased as well as the area under the peak (ΔH_{cc}) with the addition of AgCH-NPs that acts as nucleus promoting/favoring crystallization of PLA/OLA at lower temperatures. Moreover, the cold crystallization in samples with 3 and 5 wt% of AgCH-NPs is barely visible compared to the samples with lower AgCH-NPs content, which agrees with their higher crystallinity just after processing as confirmed by XRD. T_m values did not show notable differences when nanoparticles are present, and all the samples showed a broad transition with multiple shoulders which appear well-defined as two individual peaks in samples with 0.5 and 1 wt% of AgCH-NPs contents. As previously reported for some research, this phenomenon can be attributed to the melting of different crystals (α and α') mainly formed during heating in DSC analysis as the total enthalpy (ΔH_{Total}) of the continuous transitions for these samples in the range between crystallization and melting is near zero. The fact that ΔH_{Total} value in 1 wt% of AgCH-NPs sample is slightly higher is in agreement with its higher crystallinity compared to neat PLA/OLA and nanocomposite containing 0.5 wt% of AgCH-NPs. In correlation with XRD results, the clearly visible first melting peak can be due to the presence of α' form just after processing, or crystallites with different lamellar thickness. In contrast, it is clear that higher AgCH-NPs content (3–5 wt%) induces crystallinity during processing as evidenced by the low ΔH_{cc} values and subsequent melting peak.

Table 4. Thermal properties and crystallinity calculated from DSC scan for neat PLA/OLA and formulations containing AgCH-NPs.

Sample	T_g	T_{cc}	ΔH_{cc}	T_m	ΔH_m	ΔH_{Total}	$X_{c\text{-DSC}}$ (%)	$X_{c\text{-XRD}}$ (%)
PLA/OLA	32	88	25	143	27	2	2.8	–
PLA/OLA AgCH0.5%	25	83	27	142	27	0	0.0	–
PLA/OLA AgCH1%	24	76	23	142	29	6	9.2	3.3
PLA/OLA AgCH3%	50	66	2	142	30	28	38.0	26.2
PLA/OLA AgCH5%	53	68	1	141	27	26	37.5	21.9

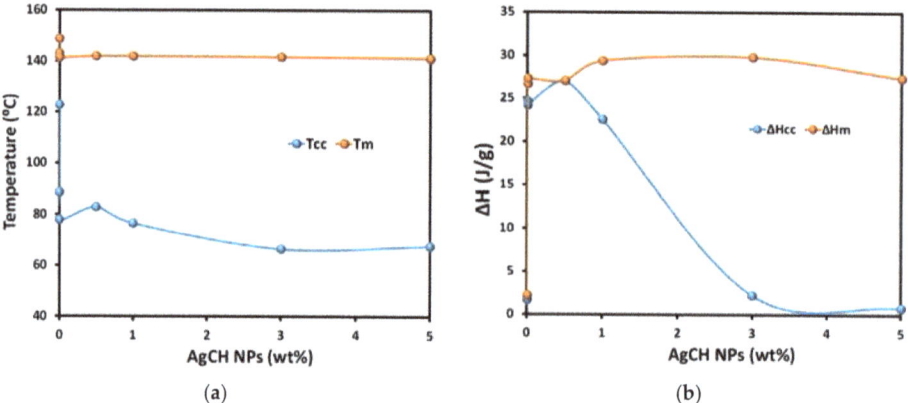

Figure 7. (a) T_m and T_{cc}, and (b) ΔH_{cc} and ΔH_m evolution at different AgCH-NPs contents in the PLA/OLA nanocomposites.

3.3. Mechanical Properties

The characteristic mechanical behavior of all processed samples is shown in Figure 8 and values are summarized in Table 5. Neat PLA/OLA showed low elongation at break, (ε), (neat PLA/OLA ε ~108%; AgCH-NPs nanocomposites ε ~338–371%) and high Youngs' or elastic modulus, (E), (neat PLA/OLA E = 783 MPa; AgCH-NPs nanocomposites E ~88–256 MPa) compared to nanocomposites. All the nanocomposites possessed higher elongation at break while a reduction in the elastic modulus and maximum tensile strength (σ_{max}) was observed similarly as previously reported in other studies [41]. Interestingly, the best balance of mechanical behavior into the nanocomposites was obtained for the least amount of AgCH-NPs tested (0.5 wt%) having the highest Youngs' modulus and maximum tensile strength of the whole nanocomposites while retaining similar elongation at break. This sample (0.5 wt% of AgCH-NPs) is totally amorphous, therefore, the low amount of filler helps to retain mechanical properties closer to the neat PLA/OLA matrix that also possesses the lowest crystallinity. Thus, it seems that the nanoparticles did not disturb the chain mobility allowing for an increased elongation at break also enhanced with the loss of crystallinity of this system. In the rest of the nanocomposites, AgCH-NPs started to induce crystallinity producing an increase of the Youngs' modulus with the % of crystallinity and, therefore, with the number of nanoparticles. This fact can also be related to the observation from XRD results that evidenced the presence of crystals with a higher perfection degree into the 5 wt% of the AgCH-NPs sample, in comparison to the crystalline samples with a lower number of nanoparticles. In this regard, it is well known that the toughness of semicrystalline polymers is strongly influenced by the variation of crystalline structure and therefore by the solid density, affecting not only the elastic modulus but also the elongation at break. In the nanocomposites, toughness decreased with crystallinity, being lower for nanocomposites with 1–5 wt% of AgCH-NPs that possess some crystallinity in comparison with 0.5 wt% AgCH-NPs nanocomposites.

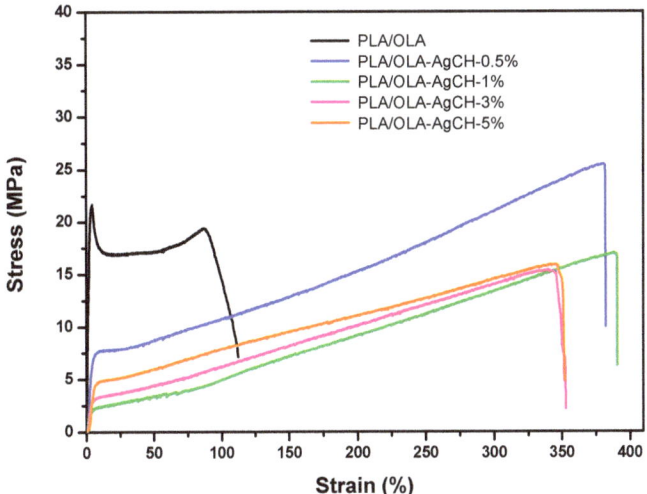

Figure 8. Representative tensile stress vs. strain curves obtained for neat PLA/OLA and PLA/OLA AgCH-NPs nanocomposites.

Table 5. Mechanical properties for neat PLA/OLA and formulations containing AgCH-NPs.

Sample	E (MPa)	ε (%)	σ_{max} (MPa)	Toughness (MJ/m^3)
PLA/OLA	783 ± 102	108 ± 6	23 ± 2	1.8 ± 0.1
PLA/OLA AgCH0.5%	256 ± 29	372 ± 26	23 ± 2	5.2 ± 0.7
PLA/OLA AgCH1%	88 ± 13	368 ± 32	16 ± 1	3.1 ± 0.6
PLA/OLA AgCH3%	123 ± 36	369 ± 50	16 ± 2	3.3 ± 0.5
PLA/OLA AgCH5%	132 ± 29	338 ± 51	14 ± 3	3.1 ± 0.9

3.4. Antibacterial Activity

The effectiveness of these nanocomposites against Gram-positive *S. aureus* and Gram-negative *E. coli* bacteria was evaluated by the ASTM standard method of [37]. The chitosan is well-known to have antimicrobial activity depending on its structure and molecular weight. For comparison purposes, one nanocomposite using chitosan at in-between compositions (2%) was obtained by the same procedure. Figure 9 displays the antibacterial activity of all nanocomposites represented as the percentage of bacteria kill, which was calculated by the difference between the CFU after contact with control substrates (PLA/OLA and none) and CFUs after contact with the polymeric nanocomposites.

When chitosan is introduced it confers activity to the matrix, but this effect is more evident when AgCH-NPs are introduced. The behavior against Gram-positive is more powerful than against Gram-negative since the latter presents the outer lipid cell wall, which is more difficult to destroy. In the case of AgCH-NPs low content, the nanocomposites are not effective against *E. coli* but are against *S. aureus* bacteria.

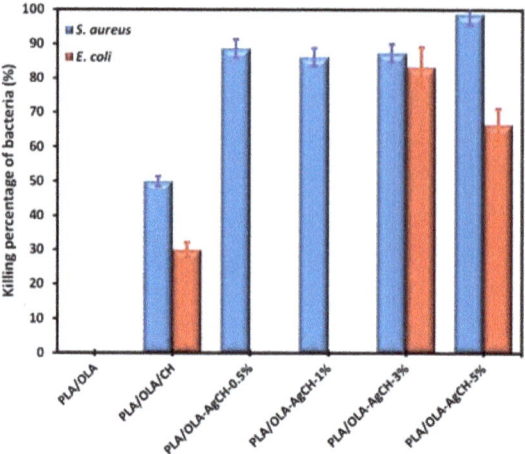

Figure 9. Percentage of killing bacteria for the different nanocomposites.

3.5. Disintegration under Composting Conditions of PLA/OLA AgCH-NPs Nanocomposites:

In order to evaluate the ability of PLA/OLA and PLA/OLA AgCH-NPs active nanocomposites to undergo disintegration, firstly, a visual examination of samples at different times when subjected to composting conditions was performed and results are collected in Figure 10. After seven days of incubation, nanocomposites containing 1–5 wt% of AgCH-NPs already exhibited fractures, while at 17 days of incubation all samples broke down and changed color, becoming totally opaque. In general, a change in the color is due to a variation in the refraction index due to water absorption and/or presence of products formed by the hydrolytic process being a clear indicator of degradation processes occurring [47,48].

All the samples were visibly disintegrated after 28 days. Thus, the nanocomposites containing 1–5 wt% of AgCH-NPs started to disintegrate faster than neat PLA/OLA and PLA/OLA containing a low number of nanoparticles (0.5 wt%). These results were confirmed by the weight loss values (Figure 11a) that remained almost constant for all the samples until seven days of incubation while, at 36 days, values near 50% and 80% were reached for nanocomposites containing 0.5 wt%, 3 wt%, and 1 wt%, 5 wt% of AgCH-NPs, respectively. Interestingly, all the nanocomposites degraded faster than the neat PLA/OLA matrix; this fact might be attributed to the release of Ag ions that can catalyze the disintegration of the samples accelerating the degradation process. Figure 11b shows, as an example, the FTIR spectra in the 650–4000 cm^{-1} range of sample PLA/OLA AgCH-3% at different composting times. The carbonyl group (–C=O) and –C–O group stretching from PLA and OLA are clearly visible at 1748 cm^{-1} and 1180 cm^{-1}, respectively. After seven days of composting, a band, whose intensity increases over composting time, appeared at 1590 cm^{-1} due to carboxylate ions indicating the consumption of lactic acid by the microorganisms [49].

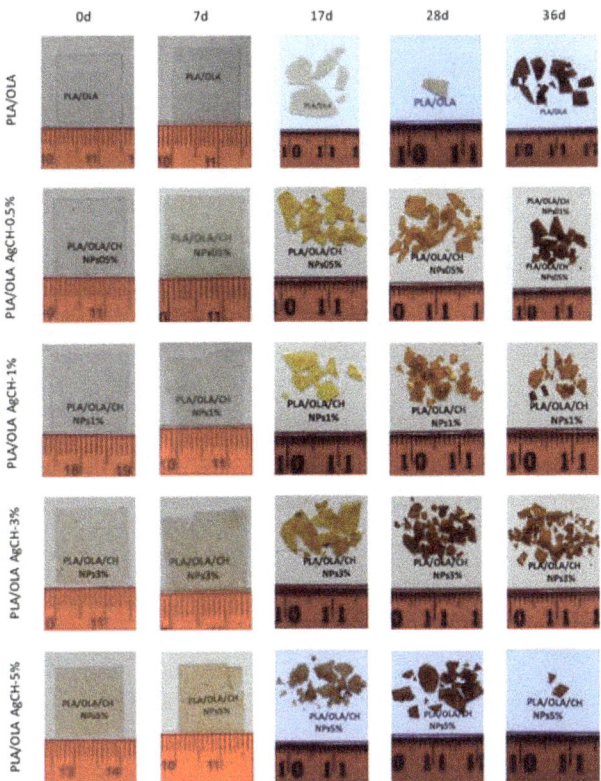

Figure 10. Disintegrated samples under composting conditions.

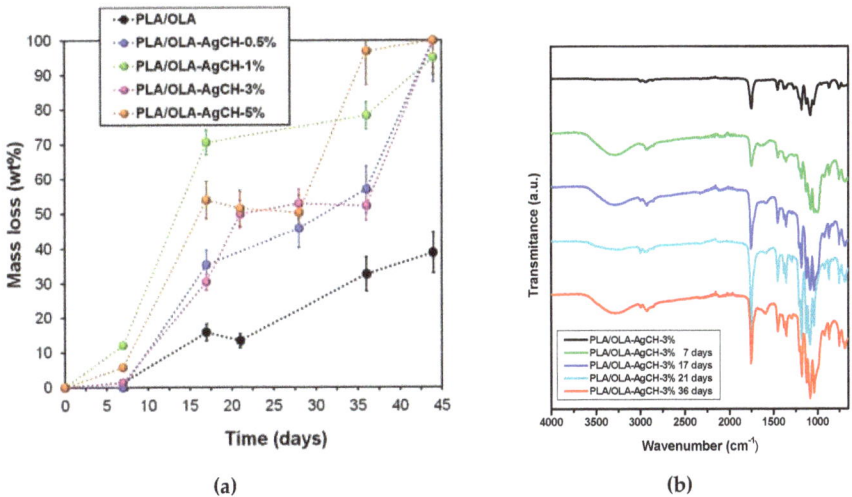

Figure 11. (a) Weight loss percentage values of PLA and PLA/OLA AgCH-NPs nanocomposites at different incubation times under composting conditions. (b) Infrared spectra of PLA/OLA nanocomposites containing 3 wt% of AgCH-NPs, before composting and after different disintegration times.

4. Conclusions

In this work, novel plasticized PLA nanocomposites containing 0.5 wt%, 1 wt%, 3 wt%, and 5 wt% of synthesized chitosan mediated silver nanoparticles (AgCH-NPs) were prepared by melt compounding in a twin-screw extruder and profusely characterized. The presented research depicts a new utilization of antimicrobial AgCH-NPs synthesized using nontoxic biodegradable chitosan as a reducing agent of silver nitrate. The AgCH-NPs obtained from a facile green synthesis method, clearly affected the crystal structure as well as the ability of neat PLA/OLA matrix chains to organize enhancing its crystallinity and lowering the cold crystallization temperature. In fact, as proven by XRD, nanoparticles induced a change in the lattice spacing (lowering the distance between crystalline planes) of the neat matrix, indicating the formation of crystals with higher perfection for the nanocomposite containing 5 wt% of nanoparticles. In accordance with XRD, a detailed analysis of DSC thermograms showed that the crystallinity degree of PLA/OLA was significantly affected when increasing the AgCH-NPs reaching values of ~37% for 3 wt% and 5 wt% of nanoparticle loads. A great enhancement in toughness and elongation at break was noticed for all the nanocomposites as compared with the neat PLA/OLA, while Youngs' modulus did not change significantly upon the increase of AgCH-NPs wt%. In general, the properties of PLA were improved with the AgCH-NPs, mechanical and thermal properties showed balanced results despite the concentration of nanoparticles, and interestingly, the incorporation of even the smallest amount of AgCH-NPs increased significantly the disintegration rate under composting conditions. PLA/OLA AgCH-NPs nanocomposites exhibited antimicrobial activities against Gram-positive *S. aureus* and Gram-negative bacteria as compared to the pristine matrix. A dual mechanism of action combining the bactericidal effect of silver with the cationic effect of chitosan is probably the underlying and responsible mechanism for the enhanced antibacterial properties of the nanocomposites. The bactericide effect together with the enhanced degradation rate, ductility, toughness, and crystallinity of the developed nanocomposites as well as the possibility to be processed under the same conditions than neat PLA without thermal degradation, are good indications for their potential utilization in antimicrobial active food packaging applications.

Supplementary Materials: The following are available online at http://www.mdpi.com/2079-4991/10/1/22/s1, Figure S1: UV–vis absorption spectra of AgCH-NPs, Figure S2: XRD spectrum of AgCH-NPs.

Author Contributions: Conceptualization, D.L., A.S. and S.M.; methodology, D.L., A.S., G.R., M.F.-G., C.E., S.M., A.M.-B., V.H., N.C.; validation, D.L., A.S. and M.F.-G.; investigation, D.L., A.S., G.R., M.F.-G., C.E., S.M., A.M.-B., V.H., N.C.; writing—original draft preparation, A.S.; writing—review and editing, A.S., D.L., G.R., M.F.-G.; supervision, D.L.; project administration, D.L. and M.F.-G.; funding acquisition, D.L. and M.F.-G. All authors have read and agreed to the published version of the manuscript.

Funding: This research was funded by Spanish Ministry of Science, Innovation and Universities (AEI-MICINN/FEDER); Projects MAT2016-78437-R and MAT2017-88123-P.

Acknowledgments: A.S. acknowledges her "APOSTD/2018/228" postdoctoral contract from the Education, Research, Culture and Sport Council from the Government of Valencia. C.E. acknowledges IJCI-2015-26432 contract from MICINN.

Conflicts of Interest: The authors declare no conflict of interest.

References

1. Yang, H.S.; Yoon, J.S.; Kim, M.N. Dependence of biodegradability of plastics in compost on the shape of specimens. *Polym. Degrad. Stab.* **2005**, *87*, 131–135. [CrossRef]
2. Liu, L.; Li, S.; Garreau, H.; Vert, M. Selective enzymatic degradations of poly(L-lactide) and poly(∈-caprolactone) blend films. *Biomacromolecules* **2000**, *1*, 350–359. [CrossRef] [PubMed]
3. Garlotta, D. A Literature Review of Poly(Lactic Acid). *J. Polym. Environ.* **2001**, *9*, 63–84. [CrossRef]
4. Rasal, R.M.; Janorkar, A.V.; Hirt, D.E. Poly(lactic acid) modifications. *Prog. Polym. Sci.* **2010**, *35*, 338–356. [CrossRef]

5. Mattioli, S.; Peltzer, M.; Fortunati, E.; Armentano, I.; Jiménez, A.; Kenny, J.M. Structure, gas-barrier properties and overall migration of poly(lactic acid) films coated with hydrogenated amorphous carbon layers. *Carbon N. Y.* **2013**, *63*, 274–282. [CrossRef]
6. Rhim, J.W.; Hong, S.I.; Ha, C.S. Tensile, water vapor barrier and antimicrobial properties of PLA/nanoclay composite films. *LWT—Food Sci. Technol.* **2009**, *42*, 612–617. [CrossRef]
7. Tsuji, H.; Ikada, Y. Blends of aliphatic polyesters. II. Hydrolysis of solution-cast blends from poly(L-lactide) and poly(E-caprolactone) in phosphate-buffered solution. *J. Appl. Polym. Sci.* **1998**, *67*, 405–415. [CrossRef]
8. Fukushima, K.; Abbate, C.; Tabuani, D.; Gennari, M.; Camino, G. Biodegradation of poly(lactic acid) and its nanocomposites. *Polym. Degrad. Stab.* **2009**, *94*, 1646–1655. [CrossRef]
9. Petersen, K.; Nielsen, P.V.; Olsen, M.B. Physical and mechanical properties of biobased materials—Starch, polylactate and polyhydroxybutyrate. *Starch/Staerke* **2001**, *53*, 356–361. [CrossRef]
10. Hiljanen-Vainio, M.; Varpomaa, P.; Seppälä, J.; Törmälä, P. Modification of poly(L-lactides) by blending: Mechanical and hydrolytic behavior. *Macromol. Chem. Phys.* **1996**, *197*, 1503–1523. [CrossRef]
11. Davoodi, S.; Oliaei, E.; Davachi, S.M.; Hejazi, I.; Seyfi, J.; Heidari, B.S.; Ebrahimi, H. Preparation and characterization of interface-modified PLA/starch/PCL ternary blends using PLLA/triclosan antibacterial nanoparticles for medical applications. *RSC Adv.* **2016**, *6*, 39870–39882. [CrossRef]
12. Khakbaz, M.; Hejazi, I.; Seyfi, J.; Davachi, S.M.; Jafari, S.H.; Khonakdar, H.A. Study on the effects of non-solvent and nanoparticle concentrations on surface properties of water-repellent biocompatible l-lactide/glycolide/trimethylene carbonate terpolymers. *Colloids Surfaces A Physicochem. Eng. Asp.* **2016**, *502*, 168–175. [CrossRef]
13. Mathew, A.P.; Oksman, K.; Sain, M. Mechanical properties of biodegradable composites from poly lactic acid (PLA) and microcrystalline cellulose (MCC). *J. Appl. Polym. Sci.* **2005**, *97*, 2014–2025. [CrossRef]
14. Davachi, S.M.; Shiroud Heidari, B.; Hejazi, I.; Seyfi, J.; Oliaei, E.; Farzaneh, A.; Rashedi, H. Interface modified polylactic acid/starch/poly ε-caprolactone antibacterial nanocomposite blends for medical applications. *Carbohydr. Polym.* **2017**, *155*, 336–344. [CrossRef]
15. Cierpiszewski, R.; Korzeniowski, A.; Dobrucka, R. Intelligent food packaging—Research and development. *LogForum* **2015**, *11*, 7–14.
16. Chaudhry, Q.; Scotter, M.; Blackburn, J.; Ross, B.; Boxall, A.; Castle, L.; Aitken, R.; Watkins, R. Applications and implications of nanotechnologies for the food sector. *Food Addit. Contam. Part A Chem. Anal. Control. Expo. Risk Assess.* **2008**, *25*, 241–258. [CrossRef] [PubMed]
17. Mahalik, N. Advances in Packaging Methods, Processes and Systems. *Challenges* **2014**, *5*, 374–389. [CrossRef]
18. Koedrith, P.; Thasiphu, T.; Tuitemwong, K.; Boonprasert, R.; Tuitemwong, P. Recent advances in potential nanoparticles and nanotechnology for sensing food-borne pathogens and their toxins in foods and crops: Current technologies and limitations. *Sensors Mater.* **2014**, *26*, 711–736.
19. Amaya-González, S.; de-los-Santos-Álvarez, N.; Miranda-Ordieres, A.J.; Lobo-Castañón, M.J. Aptamer-based analysis: A promising alternative for food safety control. *Sensors* **2013**, *13*, 16292–16311. [CrossRef]
20. Vermeiren, L.; Devlieghere, F.; Van Beest, M.; De Kruijf, N.; Debevere, J. Developments in the active packaging of foods. *Trends Food Sci. Technol.* **1999**, *10*, 77–86. [CrossRef]
21. Oloffs, A.; Grosse-Siestrup, C.; Bisson, S.; Rinck, M.; Rudolph, R.; Gross, U. Biocompatibility of silver-coated polyurethane catheters and silvercoated Dacron® material. *Biomaterials* **1994**, *15*, 753–758. [CrossRef]
22. Muñoz-Bonilla, A.; Fernández-García, M. The roadmap of antimicrobial polymeric materials in macromolecular nanotechnology. *Eur. Polym. J.* **2015**, *65*, 46–62. [CrossRef]
23. Álvarez-Paino, M.; Muñoz-Bonilla, A.; Fernández-García, M. Antimicrobial polymers in the nano-world. *Nanomaterials* **2017**, *7*, 48. [CrossRef] [PubMed]
24. Muñoz-Bonilla, A.; Echeverria, C.; Sonseca, Á.; Arrieta, M.P.; Fernández-García, M. Bio-based polymers with antimicrobial properties towards sustainable development. *Materials* **2019**, *12*, 641. [CrossRef]
25. Quintavalla, S.; Vicini, L. Antimicrobial food packaging in meat industry. *Meat Sci.* **2002**, *62*, 373–380. [CrossRef]
26. Chowdhury, N.R.; Cowin, A.J.; Zilm, P.; Vasilev, K. "Chocolate" gold nanoparticles—one pot synthesis and biocompatibility. *Nanomaterials* **2018**, *8*, 496. [CrossRef]
27. Slepička, P.; Slepičková Kasálková, N.; Pinkner, A.; Sajdl, P.; Kolská, Z.; Švorčík, V. Plasma induced cytocompatibility of stabilized poly-L-lactic acid doped with graphene nanoplatelets. *React. Funct. Polym.* **2018**, *131*, 266–275. [CrossRef]

28. Murugadoss, A.; Chattopadhyay, A. A "green" chitosan-silver nanoparticle composite as a heterogeneous as well as micro-heterogeneous catalyst. *Nanotechnology* **2008**, *19*. [CrossRef]
29. Sanpui, P.; Murugadoss, A.; Prasad, P.V.D.; Ghosh, S.S.; Chattopadhyay, A. The antibacterial properties of a novel chitosan-Ag-nanoparticle composite. *Int. J. Food Microbiol.* **2008**, *124*, 142–146. [CrossRef]
30. Abdel-Mohsen, A.M.; Abdel-Rahman, R.M.; Fouda, M.M.G.; Vojtova, L.; Uhrova, L.; Hassan, A.F.; Al-Deyab, S.S.; El-Shamy, I.E.; Jancar, J. Preparation, characterization and cytotoxicity of schizophyllan/silver nanoparticle composite. *Carbohydr. Polym.* **2014**, *102*, 238–245. [CrossRef]
31. Abdel-Mohsen, A.M.; Hrdina, R.; Burgert, L.; Krylová, G.; Abdel-Rahman, R.M.; Krejčová, A.; Steinhart, M.; Beneš, L. Green synthesis of hyaluronan fibers with silver nanoparticles. *Carbohydr. Polym.* **2012**, *89*, 411–422. [CrossRef] [PubMed]
32. Abdel-Mohsen, A.M.; Aly, A.S.; Hrdina, R.; El-Aref, A.T. A novel method for the preparation of silver/chitosan-O-methoxy polyethylene glycol core shell nanoparticles. *J. Polym. Environ.* **2012**, *20*, 459–468. [CrossRef]
33. Kalaivani, R.; Maruthupandy, M.; Muneeswaran, T.; Hameedha Beevi, A.; Anand, M.; Ramakritinan, C.M.; Kumaraguru, A.K. Synthesis of chitosan mediated silver nanoparticles (Ag NPs) for potential antimicrobial applications. *Front. Lab. Med.* **2018**, *2*, 30–35. [CrossRef]
34. Venkatesham, M.; Ayodhya, D.; Madhusudhan, A.; Veera Babu, N.; Veerabhadram, G. A novel green one-step synthesis of silver nanoparticles using chitosan: Catalytic activity and antimicrobial studies. *Appl. Nanosci.* **2014**, *4*, 113–119. [CrossRef]
35. Wei, D.; Sun, W.; Qian, W.; Ye, Y.; Ma, X. The synthesis of chitosan-based silver nanoparticles and their antibacterial activity. *Carbohydr. Res.* **2009**, *344*, 2375–2382. [CrossRef]
36. Fischer, E.W.; Sterzel, H.J.; Wegner, G. Investigation of the structure of solution grown crystals of lactide copolymers by means of chemical reactions. *Kolloid-Zeitschrift Zeitschrift Für Polym.* **1973**, *251*, 980–990. [CrossRef]
37. ASTM International, ASTM E2149: Standard Test Method for Determining the Antimicrobial Activity of Antimicrobial Agents Under Dynamic Contact Conditions. Available online: https://www.astm.org/Standards/E2149.html (accessed on 18 December 2019).
38. International Standard, ISO20200:2015: Determination of the degree of disintegration of plastic materials under simulated composting conditions in a laboratory-scale test: International Standard. Available online: https://www.iso.org/standard/63367.html (accessed on 18 December 2019).
39. Chu, Z.; Zhao, T.; Li, L.; Fan, J.; Qin, Y. Characterization of antimicrobial poly (lactic acid)/nano-composite films with silver and zinc oxide nanoparticles. *Materials* **2017**, *10*, 659. [CrossRef]
40. Sasaki, S.; Asakura, T. Helix distortion and crystal structure of the α-form of poly(L-lactide). *Macromolecules* **2003**, *36*, 8385–8390. [CrossRef]
41. Bautista-Del-Ángel, J.E.; Morales-Cepeda, A.B.; Lozano-Ramírez, T.; Sanchez, S.; Karami, S.; Lafleur, P. Enhancement of crystallinity and toughness of poly (L-lactic acid) influenced by Ag nanoparticles processed by twin screw extruder. *Polym. Compos.* **2018**, *39*, 2368–2376. [CrossRef]
42. Pan, P.; Kai, W.; Zhu, B.; Dong, T.; Inoue, Y. Polymorphous crystallization and multiple melting behavior of poly(L-lactide): Molecular weight dependence. *Macromolecules* **2007**, *40*, 6898–6905. [CrossRef]
43. Miyata, T.; Masuko, T. Morphology of poly(L-lactide) solution-grown crystals. *Polymer (Guildf)* **1997**, *38*, 4003–4009. [CrossRef]
44. Liu, H.; Zhang, L.; Li, J.; Zou, Q.; Zuo, Y.; Tian, W.; Li, Y. Physicochemical and biological properties of nano-hydroxyapatite-reinforced aliphatic polyurethanes membranes. *J. Biomater. Sci. Polym. Ed.* **2010**, *21*, 1619–1636. [CrossRef] [PubMed]
45. Navarro-Baena, I.; Arrieta, M.P.M.P.; Sonseca, A.; Torre, L.; López, D.; Giménez, E.; Kenny, J.M.J.M.; Peponi, L. Biodegradable nanocomposites based on poly(ester-urethane) and nanosized hydroxyapatite: Plastificant and reinforcement effects. *Polym. Degrad. Stab.* **2015**, *121*, 171–179. [CrossRef]
46. Liu, X.; Wang, T.; Chow, L.C.; Yang, M.; Mitchell, J.W. Effects of Inorganic Fillers on the Thermal and Mechanical Properties of Poly(lactic acid). *Int. J. Polym. Sci.* **2014**, *2014*, 827028. [CrossRef] [PubMed]
47. Li, S.; Girard, A.; Garreau, H.; Vert, M. Enzymatic degradation of polylactide stereocopolymers with predominant D-lactyl contents. *Polym. Degrad. Stab.* **2000**, *71*, 61–67. [CrossRef]

48. Li, S.; McCarthy, S. Further investigations on the hydrolytic degradation of poly (DL-lactide). *Biomaterials* **1999**, *20*, 35–44. [CrossRef]
49. Khabbaz, F.; Karlsson, S.; Albertsson, A.C. Py-GC/MS an effective technique to characterizing of degradation mechanism of poly (L-lactide) in the different environment. *J. Appl. Polym. Sci.* **2000**, *78*, 2369–2378. [CrossRef]

© 2019 by the authors. Licensee MDPI, Basel, Switzerland. This article is an open access article distributed under the terms and conditions of the Creative Commons Attribution (CC BY) license (http://creativecommons.org/licenses/by/4.0/).

Article

Biodegradable and Antimicrobial PLA–OLA Blends Containing Chitosan-Mediated Silver Nanoparticles with Shape Memory Properties for Potential Medical Applications

Agueda Sonseca [1,2,*,†], Salim Madani [3], Alexandra Muñoz-Bonilla [1,2], Marta Fernández-García [1,2], Laura Peponi [1,2], Adrián Leonés [1,2], Gema Rodríguez [1,2], Coro Echeverría [1,2] and Daniel López [1,2,*]

1. MacroEng Group, Instituto de Ciencia y Tecnología de Polímeros, ICTP-CSIC, C/Juan de la Cierva 3, 28006 Madrid, Spain; sbonilla@ictp.csic.es (A.M.-B.); martafg@ictp.csic.es (M.F.-G.); lpeponi@ictp.csic.es (L.P.); aleones@ictp.csic.es (A.L.); gema@ictp.csic.es (G.R.); cecheverria@ictp.csic.es (C.E.)
2. Interdisciplinary Plataform for "Sustainable Plastics towards a Circular Economy" (SUSPLAST-CSIC), 28006 Madrid, Spain
3. Laboratory of Applied Biochemistry, Department of Biology, Faculty of Sciences, University Ferhat Abbas, Setif 19000, Algeria; madanisalim79@gmail.com
* Correspondence: agsonol@posgrado.upv.es (A.S.); daniel.l.g@csic.es (D.L.); Tel.: +34-915622900 (D.L.)
† Current address: Instituto de Tecnología de Materiales, Universitat Politècnica de València (UPV), Camino de Vera s/n, 46022 Valencia, Spain.

Received: 13 May 2020; Accepted: 27 May 2020; Published: 30 May 2020

Abstract: To use shape memory materials based on poly (lactic acid) (PLA) for medical applications is essential to tune their transition temperature (T_{trans}) near to the human body temperature. In this study, the combination of lactic acid oligomer (OLA), acting as a plasticizer, together with chitosan-mediated silver nanoparticles (AgCH-NPs) to create PLA matrices is studied to obtain functional shape memory polymers for potential medical applications. PLA/OLA nanocomposites containing different amounts of AgCH-NPs were obtained and profusely characterized relating their structure with their antimicrobial and shape memory performances. Nanocomposites exhibited shape memory responses at the temperature of interest (near physiological one), as well as excellent shape memory responses, shorter recovery times and higher recovery ratios (over 100%) when compared to neat materials. Moreover, antibacterial activity tests confirmed biocidal activity; therefore, these functional polymer nanocomposites with shape memory, degradability and biocidal activity show great potential for soft actuation applications in the medical field.

Keywords: poly (lactic acid); oligomeric lactic acid; eco-friendly silver nanoparticles; shape memory properties; antimicrobial activity; biomedical

1. Introduction

The use of shape memory polymers (SMP) for the development of active medical devices is perhaps the most promising and attractive area of application for these materials. They are potential candidates for stimuli-sensitive appliances, such as self-tightening and anchoring sutures/staples, surgical fasteners, orthopedic fixations and self-expandable vascular stents and implants, among others [1–3], with applicability in minimally invasive surgery. In this sense, these materials can be located in the body in a temporary and compressed/pre-deformed geometry through a small incision, to achieve their final/original and desired shape when heated to above the melting or glass transition temperature, causing less surgical stress. Additionally, biodegradable polymers with shape memory activation temperatures around body temperature are preferred for these types of implants, as they will cause less surgical stress and can degrade after a specific period of time, avoiding the need for their surgical removal. The activation temperature, commonly named T_{trans}, can be the melting temperature, T_m, of the crystalline phase of the polymer or the glass transition temperature, T_g, depending on the nature of the polymer. Thus, in order to implement a shape memory polymeric system for applications in medicine, it is necessary to tailor its mechanical and thermal properties to fall into the desired range, which is not easily achievable.

Poly (lactic acid) (PLA) is a thermoplastic polymer that has found wide applications in the area of biomedicine in the past three decades, thanks to its good mechanical strength (pure PLLA, elastic modulus, E = 3–4 GPa; tensile strength = 50–70 MPa), degradability, biocompatibility and non-toxicity, as well as a good thermal shape memory performance around 60 °C and its T_g [4]. In spite of its strengths, PLA possesses well-known limitations such as its relatively poor processability, high stiffness and crystallinity, as well as its low elongation at break (ε = 2–10%) compared with other commercially available polymers of medical grade [4,5]. Moreover, from a shape memory point of view, its T_g is well above the body temperature, which limits its direct application and the exploitation of such an interesting ability when creating a medical device (PDLA, T_g = 50–54 °C; PLLA, T_g = 57–60 °C) [5]. In this sense, it is important to highlight that the SMP research of today should overcome the limitations of SMPs for tailored practical applications such as biomedical uses, especially when involving potential materials as PLA. Therefore, numerous strategies have been explored in order to tune the properties of PLA for specific applications, including copolymerization [6] and blending [7]. The first strategy is usually used to enhance the glassy state elasticity of PLA through the incorporation of an amorphous elastic phase with a low T_g to form a copolymer [8,9], which usually results in complex structures and a successful balance of shape memory performance and mechanical/thermal properties is not always reached [10,11]. Therefore, physical blending, either with miscible or immiscible components, is considered as a much more practical and economic way to modulate the properties of PLA [12–14]. Considering that, and taking advantage of the intrinsic compatibility among PLA and oligomers such as lactic acid oligomers (OLA), due to their similar chemical structure, we used OLA as a plasticizer for triggering the activation temperature (T_{trans}) of the shape memory effect, as well as the thermal and mechanical properties of PLA. We have previously studied different PLA/OLA formulations and processed them by electrospinning, being able to control the T_g and the moduli (loss and storage) in the ranges of 21–60 °C and 30–91 MPa, respectively [15]. Therefore, we were able to tailor the T_g of the system to fall into the range of operating temperatures useful for medical applications (T_g = 40–45 °C) and, thus, allow the shape memory effect of the PLA to be activated by body heat or a temperature slightly above body heat for samples containing 20–30 wt.% of OLA.

When using a polymeric material in a medical field, bacterial infection usually needs to be considered as, statistically, it causes half of all hospital infections. More importantly, the biocompatibility of SMPs can be limited due to residues of microorganisms even prior to sterilization. Although shape memory polymers have been largely studied for creating parts of biomedical devices, the introduction of antibacterial activity into these polymeric systems remains an important task and has been mainly limited to non-degradable shape memory polyurethane materials. Furthermore, the limitation of microorganism growth on SMP systems must be addressed. Bionanocomposites represent an inspiring route for creating new and innovative medical materials, and silver nanoparticles (AgNPs) have been reported to bring improvements in the mechanical properties of dental materials [16] and, more interestingly, in their antibacterial activity [17], as well as their sustained release [18] and osteo-integration [19]. Therefore, in the present work, we have synthesized chitosan-mediated silver nanoparticles (AgCH-NPs) with the aim of incorporating them into a PLA–OLA SMP matrix at different weight ratios, to obtain appropriate formulations with antimicrobial activity for their use as thermoplastic medical materials. The microstructural, morphological and antibacterial properties of synthesized AgCH-NPs as well as obtained nanocomposites were studied in a previous work [20].

The obtained balanced mechanical properties (ductility and toughness), together with the bactericidal effect and the non-toxicity of the AgCH-NP synthesis method, led us to examine their potential application as functional smart materials of interest in biomedicine. Consequently, to that end, the present work explores, in detail, the thermal, dynamo-mechanical and shape memory properties, as well as the antifungal activity, of the developed nanocomposites. The resulting shape memory polymeric systems are expected to provide shape recovery of the permanent shape and, at the same time, wider biocidal activity, broadening the application of PLA in the field of medicine.

2. Materials and Methods

2.1. Materials

Chitosan from shrimp shells (deacetylation degree > 75%) and silver nitrate ($AgNO_3$) were purchased from Sigma-Aldrich (St. Quentin Fallavier, France). Sodium hydroxide and acetic acid were purchased from Fluka (Seelze, Germany). Lactic acid oligomer (OLA) (Glyplas OLA8, ester content > 99%, density 1.11 g/cm^3, viscosity 22.5 mPa·s, molecular weight 1100 g/mol) was a gift from Condensia Quimica SA (Barcelona, Spain) and polylactic acid (PLA3051D, 3% of D-lactic acid monomer, molecular weight 142×10^4 g/mol, density 1.24 g/cm^3) was provided by NatureWorks® (Naarden, The Netherlands).

2.2. Processing of Shape Memory Plasticized PLA/OLA Nanocomposites

Chitosan-mediated silver nanoparticles (AgCH-NPs) were synthesized following a previously reported protocol [21,22], obtaining spherical nanoparticles with diameters about 8 nm [20]. Nanocomposites of PLA–OLA containing the synthesized AgCH-NPs were obtained at 180 °C with a rotation speed of 100 rpm, in a microextruder equipped with two twin conical co-rotating screws (Thermo Scientific, MiniLab Haake Rheomex CTW5, Karlsruhe, Germany, 7 cm^3 capacity). Firstly, PLA was added to the MiniLab and after 2 min, when it had reached the melt state, OLA and AgCH-NPs were loaded and mixed for 1 more minute (total mixing/residence PLA time of 3 min). Obtained blends were compression molded at 180 °C in a hot press (Collin P-200-P, Maitenbeth, Germany); after 1 min, to ensure melting, materials were subjected to 5 MPa for 1 min and subsequently cooled down to room temperature while retaining the pressure for a further 1 min. All the materials were dried previous to their processing, and the amounts of polymer (PLA), plasticizer (OLA) and nanoparticles (AgCH-NPs) were calculated in order to obtain nanocomposites with approximately 0.5, 1 and 3 wt.% of AgCH-NPs with respect to the PLA amount.

2.3. Characterization Techniques

Viscoelastic measurements of the pure PLA–OLA matrix and nanocomposites were carried out in a DMA/SDTA861e Dynamic Mechanical analyzer (Mettler-Toledo, Greifensee, Switzerland). Dynamo-mechanical thermal analyses (DMTA) were carried out from −60 to 120 °C with isothermal steps of 5 °C, in the range of 0.1–1 Hz at 3 steps per decade (0.1, 0.5 and 1 Hz). Samples with 4 mm width and ~100 μm of average thickness were measured in tensile mode with 10 mm of effective length between clamps. Strain amplitude was kept constant at 15 μm. Analyses were performed at least thrice per sample and the average was taken as representative values. Storage and Loss moduli, and Tangent Delta were recorded as a function of temperature and time, and glass transition temperatures (T_{gDMTA}) were calculated as the maximum in Tangent Delta peak. Shape memory properties were quantified in cyclic-thermomechanical tensile tests consisting on a heating-stretching-cooling protocol implemented in a Q800 dynamo mechanical analyzer (TA Instruments, New Castle, DE, USA). Each single cycle included programming the temporary shape and recovering the permanent shape, as follows: (1) programming step—a temporary fixed shape was created under strain-controlled conditions. The sample was first heated up and equilibrated at 45 °C (a useful temperature for biomedical applications) and stretched to a maximum strain of 50% by applying a force ramp of

0.2 MPa/min. Then, the sample was cooled down and equilibrated at 10 °C and maintained while the stress was released to zero at 0.5 MPa/min, to fix the temporary shape; (2) recovery step—once the stress was released, the sample was heated again up to 45 °C at 3 °C/min to recover the permanent shape. Then, a subsequent cycle was started and the protocol was repeated four times. Strain fixity ratio (R_f) represents the ability to fix the mechanical deformation (ε_m) applied in the programming step; strain recovery ratio (R_r) represents the ability of the material to memorize/recover the permanent shape after the programming step. Both were quantified with the following equations [23–25]:

$$R_f (N)\% = (\varepsilon_u (N)/(\varepsilon_m (N)) \times 100 \quad (1)$$

$$R_r (N)\% = ((\varepsilon_m (N) - \varepsilon_f (N))/(\varepsilon_m (N) - \varepsilon_f (N - 1))) \times 100 \quad (2)$$

where ε_m is the maximum strain after cooling and before unloading the sample and ε_u is the fixed strain after unloading at 10 °C and in the Nth cycle during the programming step. ε_f is the residual strain of the sample after the recovery step in the Nth cycle. The shape memory temperature profiles were selected based on the results obtained from the DMTA analysis at 1 Hz, and taking into account the fact that potential active medical devices are recommended to be activated in the range of 40–55 °C, a close range to physiological temperatures, which is not harmful for body tissues and avoids premature activation at room temperature [1].

The antimicrobial properties of the nanocomposites were determined against *Candida parapsilosis* (*C. parapsilosis*, ATCC 22019) fungi, following the E2149-13a standard method of the American Society for Testing and Materials (ASTM) [26]. Briefly, each nanocomposite was placed into a sterile falcon tube, filled with 10 mL of the fungi suspension (ca. 10^5 colony forming units (CFU)/mL) and shaken at 150 rpm at room temperature for 24 h. Fungi concentrations at time 0 and after 24 h were calculated by the plate count method after a 48-h incubation period. Falcon tubes containing only the inoculum and neat, plasticized PLA were also used as control experiments.

3. Results

3.1. Glass Transition Temperature, Activation Energy and Crystallinity of the Systems

Regarding the shape memory effect in PLA-based materials, glass transition temperature can be used as the transition temperature of SMPs. Therefore, in order to better understand the shape memory behavior of the developed materials prior to studying their shape memory response, an insight into the effect of the addition of both OLA and AgCH-NPs in terms of the glass transition temperature is essential. In order to characterize the motions of pure PLA/OLA matrix on both crystalline and amorphous regions of the nanocomposites, dynamic mechanical measurements were carried out at multi-frequency temperature sweep mode at three different frequencies. The neat PLA matrix was also characterized for comparison purposes in order to better understand the transitions obtained with the incorporation of OLA. Figure 1 shows the temperature dependence of the storage modulus, E′, the loss tangent, tan δ, and the loss modulus, E″, for neat PLA and PLA/OLA at three different frequencies.

Figure 1d (PLA/OLA loss modulus) displays the existence of several transitions that fall into three main absorption regions from −30 to 10, 20 to 60 and 60 to 110 °C that have been labeled as β relaxation, α relaxation and cold crystallization region (cc), respectively. The sharp α peak, centered at around 40 °C in tan δ (Figure 1b), is accompanied by a sharp decrease in the storage modulus (E′) and corresponds to the segmental relaxation associated to the glass transition (micro-Brownian motions of long chain segments in the amorphous phase of the matrix). At temperatures below T_g, secondary relaxation processes (β transitions) are appreciated, as broad shallow peaks result from the movements of localized groups of backbone atoms in the amorphous phase due to an increase in free volume with temperature. After T_g relaxation, a slight increase in E′ and E″ is accompanied with a transition peak in tan δ, which is attributed to cold crystallization processes. As occurs with PLA/OLA [27], in PLA three distinct regions can be differentiated. Apparently, the β transitions due to secondary relaxation processes seem not to be highly affected, while glass transition relaxation is clearly diminished, probably in association with the increase in free volume between chains due to the plasticizing effect of OLA. As expected, an increase in the frequency produces a shift to higher temperatures of the transitions, which is in line with results from the literature [28].

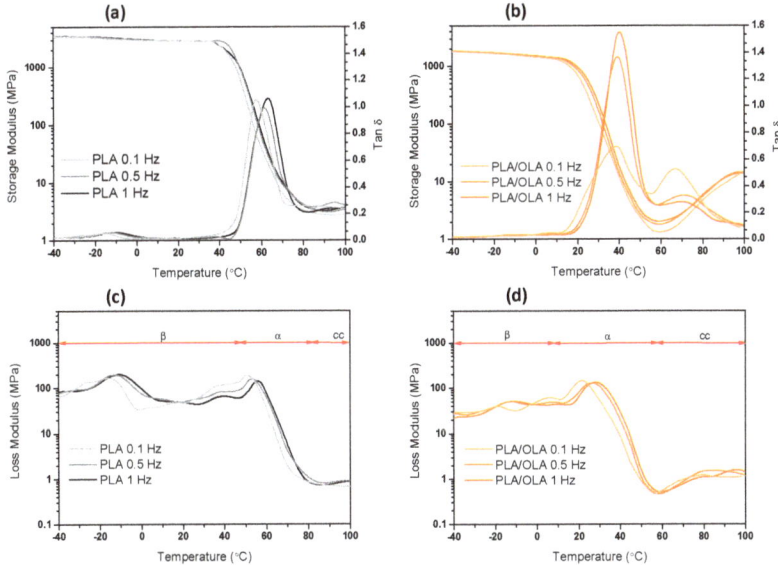

Figure 1. Temperature dependence of (**a**) poly (lactic acid) (PLA) and (**c**) PLA/lactic acid oligomer (OLA) storage modulus (E′) and loss tangent (tanδ) and (**b**) PLA and (**d**) PLA/OLA loss modulus (E″) at different frequencies (0.1, 1 and 3 Hz).

Figure 2 shows the temperature dependence of tan δ for neat and plasticized PLA and for all the nanocomposites at a frequency of 1 Hz.

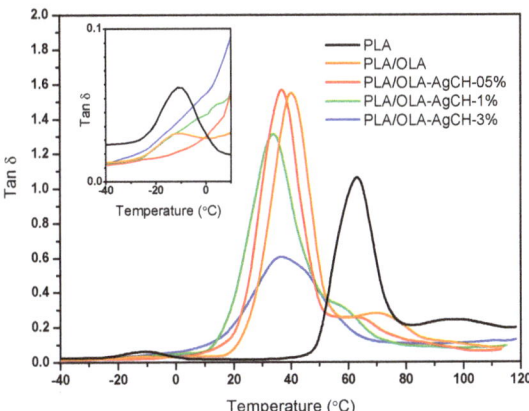

Figure 2. Temperature dependence of PLA, PLA/OLA and PLA/OLA-AgCH nanocomposite delta tangents (tan δ) at a frequency of 1 Hz.

As can be seen in Figure 2, the α relaxation corresponding to the glass–rubber transition of samples shifts to a lower temperature with the incorporation of OLA into the PLA matrix. Additionally, the β-relaxation shoulder decreases with both the incorporation of OLA and increases in AgCH-NPs. We observe the same number of absorption peaks for all samples; therefore, it seems that, qualitatively, there are no dramatic changes in the relaxation processes between PLA and PLA/OLA and the nanocomposites at a frequency of 1 Hz. To explore this effect in more detail, we have calculated the characteristic relaxation time of the glass transition relaxation process using the maximum of the delta tangent peak (T_g) for the three frequencies investigated by means of the following expression:

$$\tau(T_{\alpha,\beta,\gamma}) \propto (1/2\pi f) \tag{3}$$

where τ is the characteristic relaxation time, T is the temperature at the maximum of the relaxation peak, and f is the experimental frequency. The dependence of the characteristic relaxation time on temperature, $\tau(T)$, is demonstrated in Figure 3. This dependence can be evaluated using the Arrhenius law, that is:

$$\tau(T) \propto \exp(E_a/(k_B T)) \tag{4}$$

where E_a is the activation energy of the corresponding relaxation process. The values of E_a for α relaxation process for pure PLA, plasticized PLA/OLA matrix and the nanocomposites together with the values of T_g at studied frequencies (0.1, 0.5 and 1 Hz) are depicted in Table 1. As previously observed in neat PLA and PLA/OLA, by increasing the test frequency, an increase in T_g occurs in nanocomposite samples.

Table 1. Dynamo-mechanical thermal analyses (DMTA) results of T_g.

Sample	Frequency [Hz]	[1] T_g [°C]	[1] E_a [kJ]
PLA	1	63	
	0.5	60	435
	0.1	58	
PLA/OLA	1	41	
	0.5	38	365
	0.1	36	
PLA/OLA-AgCH-0.5%	1	37	
	0.5	34	270
	0.1	31	
PLA/OLA-AgCH-1%	1	35	
	0.5	34	215
	0.1	27	
PLA/OLA-AgCH-3%	1	37	
	0.5	37	410
	0.1	33	

[1] Standard errors (±): 1 °C for temperatures; 5 kJ for E_a.

Figure 3 shows the values of $\ln f$ and $(1/T_g)$ for T_g, determined as the maximum in tan δ. The activation energy related to the glass transition is obtained by multiplying the slopes of the 1/T vs. $\ln f$ plot with the gas constant, R = 8.314 × 10^{-3} kJ mol^{-1} K^{-1}. The E_a values for the glass transition of the neat PLA and PLA/OLA systems are estimated as 435 and 365 kJ, respectively.

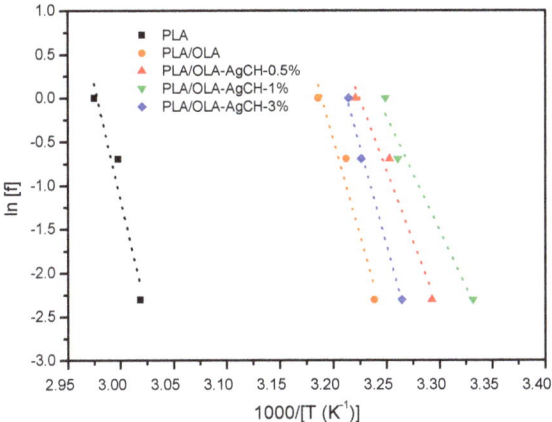

Figure 3. Plot of $(1/T_g)$ vs. $\ln f$ based on tan δ peaks for the different systems.

DMTA results evidenced lower glass transition values and E_a for PLA/OLA and its nanocomposites in comparison with neat PLA. The addition of OLA shifts the T_g towards lower values, evidencing a plasticizing effect. Furthermore, it is noticed that the glass transition temperature decreases with the addition of AgCH-NP content up to 1 wt.% and then increases with a further increase in nanoparticles (3 wt.%) being closer to the neat PLA/OLA. This increase, with a high nanoparticle load, can be attributable to the confinement effects of polymer chains between nanoparticles, restricting chain mobility near the surface of AgCH-NPs [29,30]. The activation energy trend agrees with the tendency of T_g; thus, the activation energy decreases for contents up to 1 wt.% of AgCH-NPs with respect to neat PLA/OLA, while reaching the highest value for 3 wt.% of AgCH-NP load. This fact reflects the fact that high AgCH-NP loads (3 wt.%) started to hinder the molecular motions of polymeric chains and the fact that higher E_a is needed than for neat PLA/OLA.

Underlying thermal mechanisms in triggered semicrystalline SMPs involve the glass transition temperature (T_g) and crystallinity degree. Therefore, as both properties can significantly affect the shape memory performance, it is also worth studying the extent of the effect that the addition of OLA and AgCH-NPs produces over the T_g and degree of crystallinity of the samples in detail. Table 2 summarizes the T_g values obtained from differential scanning calorimetry (DSC), as previously reported [20], and DMTA, as well as the degree of crystallinity. Both T_g values (from DMTA and DSC), along with numerical differences, follow the same trend. Similar observations have been reported widely in the literature and are usually related to the differences found in the fundamental working principle of each technique as well as in the sample scale size [31,32]. Regardless of those differences, as commented before for tan δ and taking neat PLA as reference, the addition of either OLA or AgCH-NP significantly lowers both T_{gDSC} and T_{gDMTA}. As can be deduced from Table 2, the decrease in PLA glass transition is accompanied by an increase in melting enthalpy, suggesting that OLA promotes PLA crystal nucleation, as evidenced through the calculated degree of crystallinity. This is in line with observations done in other PLA plasticized systems and probably occurs due to the lowering of the interfacial surface energy of the plasticized molecules facilitating PLA crystal nucleation [33,34].

Regarding the nanocomposites, an important observation is that amounts of AgCH-NPs from 1 to 3 wt.% induced crystallinity into the PLA/OLA matrix while retaining lower T_{gDMTA} and E_a. This, in principle, can be related to reduced entanglements/interactions occurring among nanoparticles and polymer chains into the amorphous domains, favoring their mobility and therefore reducing the T_g and E_a, as previously reported in other nanocomposite systems [35].

Table 2. Thermal properties and crystallinity calculated from differential scanning calorimetry (DSC) scan for neat PLA and PLA/OLA and nanocomposites formulations containing AgCH-NPs.

Sample	[2] T_{gDMTA} [°C]	[1,2] T_{gDSC} [°C]	[1,2] T_{cc} [°C]	ΔH_{cc} [J/g]	[1,2] T_m [°C]	ΔH_m [J/g]	ΔH_{Total} [J/g]	[1] $X_{c\text{-}DSC}$ [%]
PLA	63	62	123	2	149	2	0	–
PLA/OLA	41	32	88	25	143	27	2	2.8
PLA/OLA AgCH0.5%	37	25	83	27	142	27	0	0.0
PLA/OLA AgCH1%	35	24	76	23	142	29	6	9.2
PLA/OLA AgCH3%	37	50	66	2	142	30	28	38.0

[1] From reference [20]. [2] Standard errors (±): 1 °C for temperatures, 1 J/g in ΔH_m and 5% for X_c.

Apart from tan δ, the storage modulus and loss modulus can give a good insight into the understanding of the shape memory performance. Figure 4 shows the evolution of the storage modulus (E′) and loss modulus (E″) as a function of temperature. As expected, the addition of OLA decreases the E′ and E″ values of the system at low temperatures. The loss modulus width broadens with the increase in AgCH-NP content at the same time that the maximum moves to lower temperatures in comparison to PLA/OLA samples. These results illustrate that molecular mobility is more easily activated for nanocomposites containing 0.5–1 wt.% of AgCH-NPs in agreement with calculated E_a.

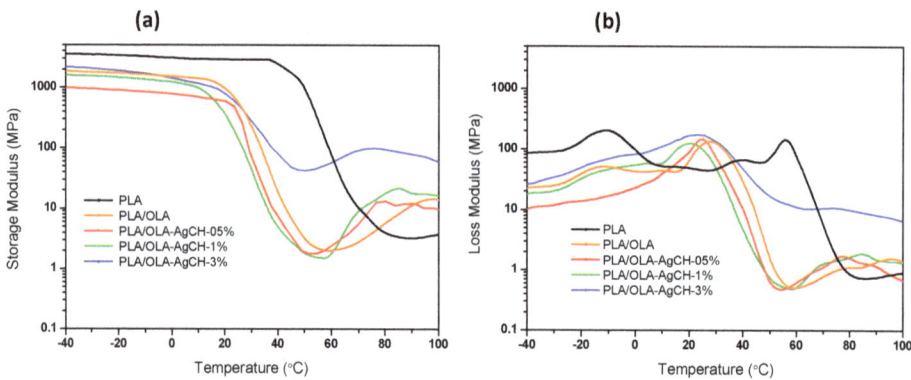

Figure 4. Evolution of (a) storage modulus (E′) and (b) loss modulus (E″) as a function of temperature.

As can be observed in Figure 5, in the glassy state (at −40 °C), PLA/OLA and its nanocomposites containing 1–3 wt.% of AgCH-NPs show similar E′ values, 1811 MPa, 1578 MPa and 2134 MPa for neat PLA/OLA, 1 and 3 wt.% for AgCH-NP nanocomposites, respectively, while, for a lower amount of filler (0.5 wt.%), a slight decrease in the modulus occurs (966 MPa), which can be related to the absence of crystallinity in this sample. At ambient temperature (25 °C), the stiffness starts to display a reduction as all the samples are close to glass transition temperature and, upon increasing the temperature to 45 °C, the stiffness displays a drastic reduction, reaching values of ~4 MPa for nanocomposites containing 0.5–1 wt.% of AgCH-NPs and ~7 MPa for neat PLA/OLA. Interestingly, the addition of 3 wt.% of AgCH-NPs leads to a substantially higher reinforcement of the E′ moduli compared to neat PLA/OLA. Moreover, this reinforcement effect is retained despite the increase in the temperature and the cold crystallization.

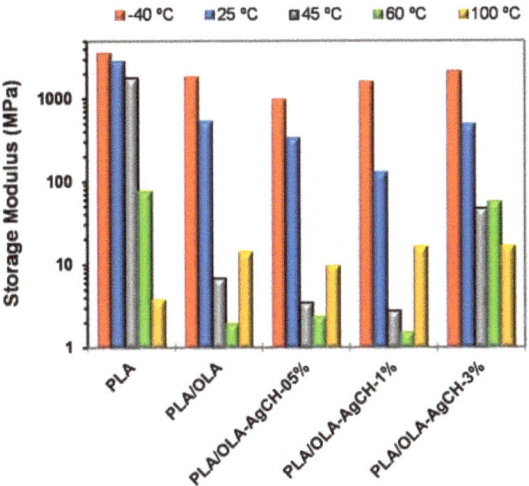

Figure 5. Storage Modulus, E′, at different temperatures.

3.2. Shape Memory Properties; Thermal Activation

As mentioned before, nowadays, PLA represents one of the most exceptional polymers in the biomedicine field for controlled drug delivery and tissue engineering. Nevertheless, in spite of its benefits, its inherent brittleness has been limiting its use. In this sense, the shape memory effect of neat PLA has been theorized by many research groups [36–38]; however, it was not observed due to its brittleness at room temperature. Therefore, in order to avoid the brittle fracture of the material, mainly polyurethanes (TPU) and poly (ε-caprolactone) polyester, have become common choices for blending or copolymerizing with PLA to create useful and flexible SMP [39,40]. Our research approach using lactic acid oligomer (OLA) as plasticizer ensures the compatibility of the blend and contributes to improving the ductility of the PLA, decreasing the glass transition to temperatures useful for biomedical applications. Thus, considering the results obtained in the DMTA analysis for the addition of AgCH-NPs, and to fulfill the requirements for further applications in biomedicine—in terms of facilitating the manipulation, storage and implantation of a possible device—a temperature of 45 °C has been selected for the activation of the shape recovery [1]. As previously explained in the experimental section, the shape memory properties were quantified through shape recovery (R_r) and shape fixity (R_f) ratios in three consecutive thermomechanical cycles in a DMTA. Temporary shape was programmed at 45 °C (T_{high} > T_{gDMTA} determined from the maximum in the tanδ curves) by elongating the sample until it reached 50% strain, then it was fixed at 10 °C (T_{low} < T_{gDMTA}) under constant stress. After removing the load, recovery was performed at 45 °C. Figure 6 shows the evolution of stress and strain as a function of time and temperature for neat PLA/OLA and its AgCH-NP nanocomposites. Table 3 collects the calculated numerical values for R_r and R_f as well as stress at maximum deformation.

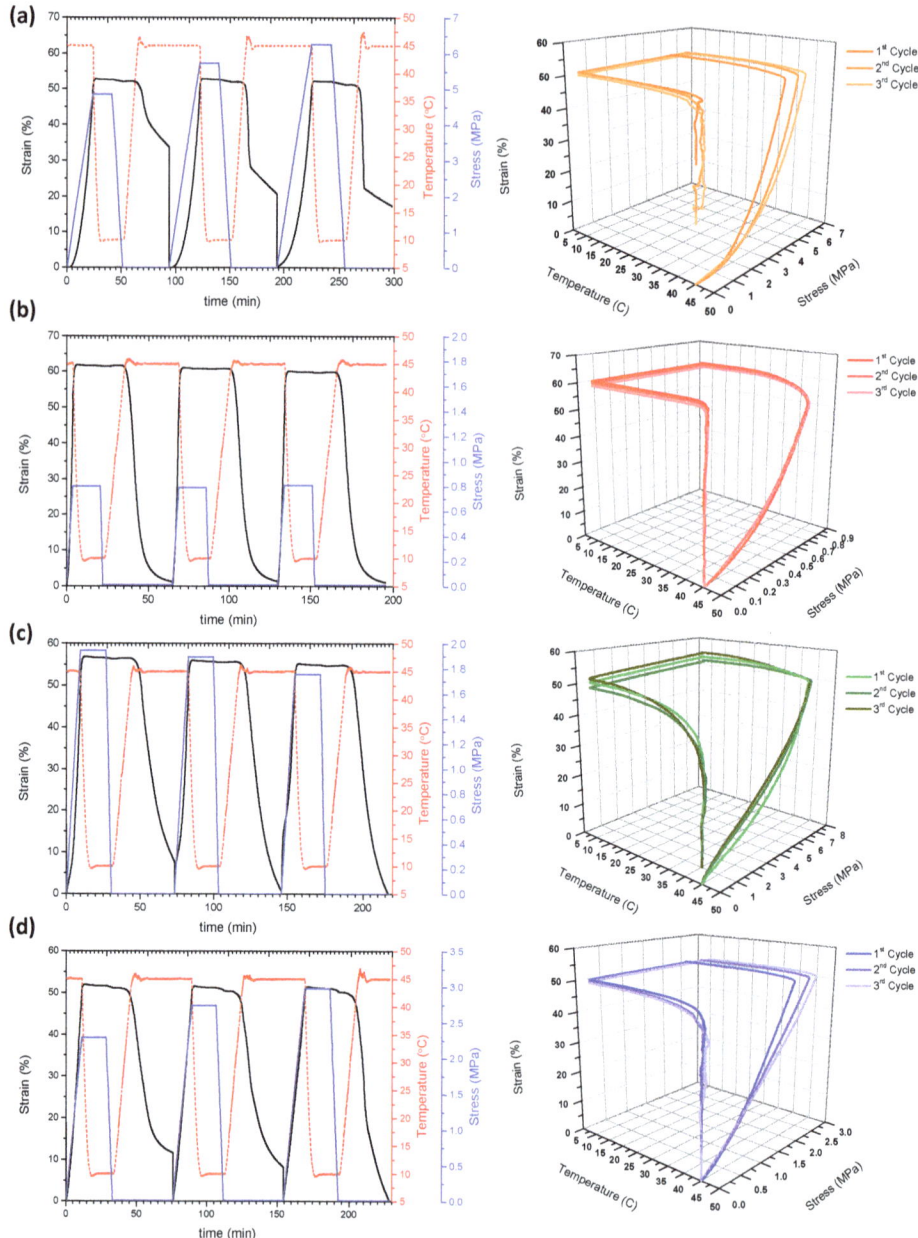

Figure 6. 2D and 3D thermo-mechanical cycles performed at 45 °C for (**a**) neat PLA/OLA and its nanocomposites containing (**b**) 0.5 wt.% of AgCH-NPs, (**c**) 1 wt.% of AgCH-NPs and (**d**) 3 wt.% of AgCH-NPs.

Table 3. Strain recovery, strain fixity and stress at maximum strain values for each shape memory cycle.

Sample	Cycle [N°]	R_r (N) [%]	R_f (N) [%]	Stress at Max. Strain [MPa]
PLA/OLA	1	36	99	4.9
	2	61	99	5.6
	3	67	99	6.0
PLA/OLA-AgCH-0.5%	1	98	100	0.8
	2	98	100	0.8
	3	98	100	0.8
PLA/OLA-AgCH-1%	1	86	100	1.9
	2	99	100	1.9
	3	100	100	1.7
PLA/OLA-AgCH-3%	1	77	99	2.3
	2	84	99	2.7
	3	100	99	3.0

Under the conditions used, shape fixity ratios remained constant for all the samples and close to unity (R_f = 99–100%), without the appreciable influence of the AgCH-NP content or the number of cycles. Conversely, shape recovery ratios (R_r) were higher for all the nanocomposites compared to neat PLA/OLA, which showed an incomplete shape recovery (lower than 50% for the first cycle). Figure 7 details the graphical evolution of shape memory parameters (R_f and R_r) as a function of AgCH-NP content, complementing the information in Table 3. By incorporating AgCH-NPs into PLA/OLA, the recovery ratio in the first cycle improves, reaching the optimum at 0.5 wt.% of AgCH-NPs with an R_r of 98% and then, despite the better performance of the nanocomposites, a reduction in R_r to 86% and 77% was observed for samples with 1 and 3 wt.% of AgCH-NP content, respectively. According to DMTA results, these lower R_r values for the first cycle, with increasing AgCH-NP content, might be attributed to the enhancement of crystallinity and the more limited movement of amorphous polymer chains. Shape recovery is driven by the entropic stresses in the amorphous phase and, as shown in Figure 2, the magnitude of the tan δ transition decreases with increasing nanoparticle content, indicating that less mobile units are involved in the relaxation process and thus higher hysteresis occurs. With the increasing cycle number in the shape memory test, R_r shows an increasing trend, stabilizing near 100% after the second or third cycle for 1 and 3 wt.% of AgCH-NPs containing samples. Besides this training effect of R_r with cycles, strain recovery showed a strong dependence on nanoparticle content, whereas this dependence seems negligible for strain fixity performance.

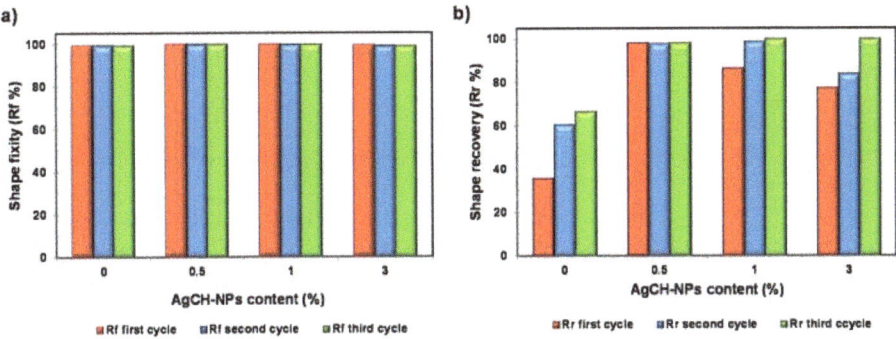

Figure 7. Shape fixity ratio (a) and shape recovery ratio (b) characteristic values for neat PLA/OLA and its nanocomposites containing 0.5, 1 and 3 wt.% of AgCH-NPs.

The stress at maximum deformation (Table 3) is lower for all the nanocomposites compared to pure PLA/OLA. This fact is in accordance with the effect of the AgCH-NP addition over the mechanical properties of PLA/OLA matrices reported in our previous work [20].

Figure 8 shows the evolution over time of the normalized strain during the recovery stage of the first cycle of the shape memory, for neat PLA/OLA and its nanocomposites at 45 °C. It could be seen that, for the same recovered strain (i.e., 80% of strain, thus 20% of strain recovered), all the nanocomposites presented faster recovery than the neat PLA/OLA, with the fastest being the nanocomposite containing 1 wt.% of AgCH-NPs. All the materials presented their T_g at lower temperatures than the set for recovery (45 °C) and, in spite of similar glass transitions to neat PLA/OLA and 3 wt.% of AgCH-NP samples, the recovery of the latter is faster. Thus, it seems that the presence of AgCH-NPs provides an extra force to the PLA/OLA matrix, helping the molecular chains to return to their original configuration.

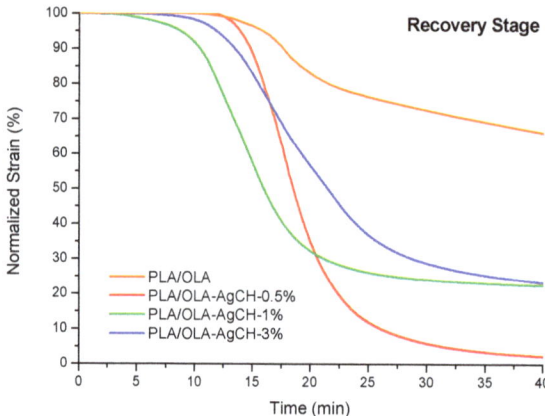

Figure 8. Normalized recovered strain with time for neat PLA/OLA and its nanocomposites containing 0.5, 1 and 3 wt.% of AgCH-NPs.

These results demonstrate that the better overall performance of the nanocomposites might also be due to improved local heating and more efficient heat transfer during heating, which leads to less dissipation of energy, a faster response and higher shape recovery.

3.3. Antifungal Activity

In a previous work, these nanocomposites were tested against Gram-positive *Staphylococcus aureus* and Gram-negative *Escherichia coli* bacteria, showing that they exhibited antibacterial activities due to the dual combination action of silver ions and the cationic chitosan [20]. In this work, we extended our analysis of their antimicrobial behavior, testing them against *C. parasilopsis* fungi. Fungi present different membrane structures than bacteria; fungal walls are composed of chitin and polysaccharides, while bacteria are mainly made of peptidoglycans, lipoproteins and others. Therefore, the antibacterial activity of nanocomposites can be different than the antifungal effectivity.

The antifungal activity of nanocomposites was tested by the ASTM standard method [26]. Table 4 summarizes the antifungal activity represented as the percentage of fungi killed, which was calculated by the difference between the CFU after contact with control substrates (PLA/OLA and none), A, and CFU after contact with the polymeric nanocomposites, B, following the equation:

$$\text{Percentage of fungi killed} = (A - B)/A \times 100 \qquad (5)$$

Table 4. Killing percentage for *C. parapsilosis* fungi by nanocomposite samples.

Sample	Killing Percentage [%]
PLA	-
PLA/OLA	-
PLA/OLA CH	90
PLA/OLA-AgCH-0.5%	99
PLA/OLA-AgCH-1%	99
PLA/OLA-AgCH-3%	99

As can be seen, all nanocomposites exhibited significant antifungal activity. It is worth noting that the incorporation of only 0.5 wt.% of chitosan-mediated silver nanoparticles (AgCH-NPs) was able to kill 99% of fungi present in the tested suspension in contact with the film.

4. Conclusions

In this study, plasticized PLA-based SMP nanocomposites with tailored functionality were prepared with different amounts of chitosan-mediated silver nanoparticles, AgCH-NPs (0.5, 1 and 3 wt.%). Nanocomposites were obtained by melt-compounding in a twin-screw extruder, and were subsequently hot pressed. A comprehensive evaluation of the performance of the neat plasticized matrix as well as the nanocomposites was conducted by focusing on their dynamo-mechanical, shape memory and antimicrobial properties. To that end, their dynamic thermo-mechanical and shape memory behavior, as well as antifungal properties, were studied. The plasticizing effect of OLA notably lowers the T_g value of the neat PLA by at least 30 °C, according to DMTA measurements. In addition, a significant decrease also occurs in the storage and loss modulus at temperatures close to the physiological one of interest. PLA/OLA based nanocomposites showed a complex thermo-mechanical behavior, where the presence of the AgCH-NPs, especially high loads (3 wt.%), sufficiently affected the crystallinity and thermo-mechanical properties. Importantly, the addition of 0.5–1 wt.% of nanoparticles into the PLA/OLA matrix slightly reduces the glass transition temperature, while loads of 3 wt.% hinder the molecular motions of the chains, increasing the T_g to values closer to the neat material. Although an increase in the T_g and activation energy occurs in the nanocomposite with the highest load, those values are still below the ones of neat PLA/OLA. The presence of AgCH-NPs is responsible for the enhanced crystallinity, the lower glass transition and activation energies, as well as the reinforcement of the storage modulus in the rubbery region (45–100 °C) of the nanocomposites in comparison to the neat matrix. All these facts together are the driving forces for the better overall performance of the nanocomposites in the shape memory tests at the temperature of interest, resulting in higher and faster recovery ratios. In this sense, AgCH-NPs seem to provide an extra force to the PLA/OLA matrix to return the original configuration. In addition, the nanocomposites showed excellent antimicrobial effects, which, combined with the enhancement of their shape memory properties, give these materials great potential for performing as substrates with multi-functionality in the biomedical field.

Author Contributions: Conceptualization, D.L., A.S. and S.M.; methodology, D.L., A.S., G.R., M.F.-G., C.E., S.M., A.M.-B., L.P., A.L.; validation, D.L., A.S., L.P. and M.F.-G.; investigation, D.L., A.S., G.R., M.F.-G., C.E., S.M., A.M.-B., L.P.; writing—original draft preparation, A.S.; writing—review and editing, A.S., D.L., M.F.-G., A.M.-B., L.P.; supervision, D.L.; project administration, D.L. and M.F.-G.; funding acquisition, D.L., L.P. and M.F.-G. All authors have read and agreed to the published version of the manuscript.

Funding: This research was funded by Spanish Ministry of Science and Innovation (AEIMICINN/FEDER); Projects MAT2016-78437-R, MAT2017-88123-P and PCIN-2017-036.

Acknowledgments: A. S. acknowledges her "APOSTD/2018/228" and "PAID-10-19" postdoctoral contracts from the Education, Research, Culture and Sport Council from the Government of Valencia and from the Polytechnic University of Valencia, respectively. C.E. acknowledges the IJCI-2015-26432 contract from MICINN. L.P. acknowledges the Ramon y Cajal (RYC-2014-15595) contract from MICINN.

Conflicts of Interest: The authors declare no conflict of interest.

References

1. Yakacki, C.M.; Gall, K. Shape-Memory Polymers for Biomedical Applications. In *Shape-Memory Polymers. Advances in Polymer Science Series*; Lendlein, A., Ed.; Springer: Berlín/Heidelberg, Germany, 2009; Volume 226, pp. 147–175.
2. Lendlein, A.; Behl, M.; Hiebl, B.; Wischke, C. Shape-memory polymers as a technology platform for biomedical applications. *Expert Rev. Med. Devices* **2010**, *7*, 357–379. [CrossRef] [PubMed]
3. Serrano, M.C.; Ameer, G.A. Recent insights into the biomedical applications of shape-memory polymers. *Macromol. Biosci.* **2012**, *12*, 1156–1171. [CrossRef] [PubMed]
4. Sabir, M.; Xu, X.; Li, L. A review on biodegradable polymeric materials for bone tissue engineering applications. *J. Mater. Sci.* **2009**, *44*, 5713–5724. [CrossRef]
5. Feng, Y.; Guo, J. Biodegradable polydepsipeptides. *Int. J. Mol. Sci.* **2009**, *10*, 589–615. [CrossRef]
6. Grijpma, D.W.; Pennings, A.J. (Co)polymers of L-lactide, 2. Mechanical properties. *Macromol. Chem. Phys.* **1994**, *195*, 1649–1663. [CrossRef]
7. Liu, H.; Zhang, J. Toughening modification of poly(lactic acid) via melt blending. *ACS Symp. Ser.* **2012**, *1105*, 27–46.
8. Choi, N.-Y.; Kelch, S.; Lendlein, A. Synthesis, Shape-Memory Functionality and Hydrolytical Degradation Studies on Polymer Networks from Poly(rac-lactide)-b-poly(propylene oxide)-b-poly(rac-lactide) dimethacrylates. *Adv. Eng. Mater.* **2006**, *8*, 439–445. [CrossRef]
9. Kelch, S.; Choi, N.Y.; Wang, Z.; Lendlein, A. Amorphous, elastic AB copolymer networks from acrylates and poly[(L-lactide)-ran-glycolide]dimetfiacrylates. *Adv. Eng. Mater.* **2008**, *10*, 494–502. [CrossRef]
10. Min, C.; Cui, W.; Bei, J.; Wang, S. Effect of comonomer on thermal/mechanical and shape memory property of L-lactide-based shape-memory copolymers. *Polym. Adv. Technol.* **2007**, *18*, 299–305. [CrossRef]
11. Yang, J.; Liu, F.; Yang, L.; Li, S.M. Hydrolytic and enzymatic degradation of poly(trimethylene carbonate-co-D,L-lactide) random copolymers with shape memory behavior. *Eur. Polym. J.* **2010**, *46*, 783–791. [CrossRef]
12. Liu, C.; Mather, P.T. Thermomechanical characterization of blends of poly(vinyl acetate) with semicrystalline polymers for shape memory applications. In Proceedings of the 61st Annual Technical Conference ANTEC 2003, Nashville, TN, USA, 4–8 May 2003; pp. 1962–1966.
13. Wang, L.S.; Chen, H.C.; Xiong, Z.C.; Pang, X.B.; Xiong, C.D. Novel degradable compound shape-memory-polymer blend: Mechanical and shape-memory properties. *Mater. Lett.* **2010**, *64*, 284–286. [CrossRef]
14. Sessini, V.; Navarro-Baena, I.; Arrieta, M.P.; Dominici, F.; López, D.; Torre, L.; Kenny, J.M.; Dubois, P.; Raquez, J.M.; Peponi, L. Effect of the addition of polyester-grafted-cellulose nanocrystals on the shape memory properties of biodegradable PLA/PCL nanocomposites. *Polym. Degrad. Stab.* **2018**, *152*, 126–138. [CrossRef]
15. Leonés, A.; Sonseca, A.; López, D.; Fiori, S.; Peponi, L. Shape memory effect on electrospun PLA-based fibers tailoring their thermal response. *Eur. Polym. J.* **2019**, *117*, 217–226. [CrossRef]
16. Choi, A.H.; Ben-Nissan, B.; Matinlinna, J.P.; Conway, R.C. Current perspectives: Calcium phosphate nanocoatings and nanocomposite coatings in dentistry. *J. Dent. Res.* **2013**, *92*, 853–859. [CrossRef]
17. Beyth, N.; Farah, S.; Domb, A.J.; Weiss, E.I. Antibacterial dental resin composites. *React. Funct. Polym.* **2014**, *75*, 81–88. [CrossRef]
18. Hook, E.R.; Owen, O.J.; Bellis, C.A.; Holder, J.A.; O'Sullivan, D.J.; Barbour, M.E. Development of a novel antimicrobial-releasing glass ionomer cement functionalized with chlorhexidine hexametaphosphate nanoparticles. *J. Nanobiotechnol.* **2014**, *12*, 1–9. [CrossRef]
19. Chien, C.Y.; Liu, T.Y.; Kuo, W.H.; Wang, M.J.; Tsai, W.B. Dopamine-assisted immobilization of hydroxyapatite nanoparticles and RGD peptides to improve the osteoconductivity of titanium. *J. Biomed. Mater. Res. Part A* **2013**, *101 A*, 740–747. [CrossRef]
20. Sonseca, A.; Madani, S.; Rodríguez, G.; Hevilla, V.; Echeverría, C.; Fernández-García, M.; Muñoz-Bonilla, A.; Charef, N.; López, D. Multifunctional PLA blends containing chitosan mediated silver nanoparticles: Thermal, mechanical, antibacterial, and degradation properties. *Nanomaterials* **2020**, *10*, 22. [CrossRef]
21. Wei, D.; Sun, W.; Qian, W.; Ye, Y.; Ma, X. The synthesis of chitosan-based silver nanoparticles and their antibacterial activity. *Carbohydr. Res.* **2009**, *344*, 2375–2382. [CrossRef]

22. Kalaivani, R.; Maruthupandy, M.; Muneeswaran, T.; Hameedha Beevi, A.; Anand, M.; Ramakritinan, C.M.; Kumaraguru, A.K. Synthesis of chitosan mediated silver nanoparticles (Ag NPs) for potential antimicrobial applications. *Front. Lab. Med.* **2018**, *2*, 30–35. [CrossRef]
23. Behl, M.; Lendlein, A. Shape-memory polymers. *Mater. Today* **2007**, *10*, 20–28. [CrossRef]
24. Wagermaier, W.; Kratz, K.; Heuchel, M.; Lendlein, A. Characterization Methods for Shape-Memory Polymers. *Adv. Polym. Sci.* **2010**, *226*, 97–145.
25. Ratna, D.; Karger-Kocsis, J. Recent advances in shape memory polymers and composites: A review. *J. Mater. Sci.* **2008**, *43*, 254–269. [CrossRef]
26. ASTM E2149-13a. *Standard Test Method for Determining the Antimicrobial Activity of Antimicrobial Agents under Dynamic Contact Conditions*; ASTM International: West Conshohocken, PA, USA, 2013.
27. Shmool, T.A.; Zeitler, J.A. Insights into the structural dynamics of poly lactic-: Co-glycolic acid at terahertz frequencies. *Polym. Chem.* **2019**, *10*, 351–361. [CrossRef]
28. Frequency Dependence of Glass Transition Temperatures. Applications Notes Library, TA Instruments, Thermal Analysis, Rheology, TA423. Available online: https://www.tainstruments.com/applications-library-search/ (accessed on 10 May 2020).
29. Kuljanin-Jakovljević, J.; Stojanović, Z.; Nedeljković, J.M. Influence of CdS-filler on the thermal properties of poly(methyl methacrylate). *J. Mater. Sci.* **2006**, *41*, 5014–5016. [CrossRef]
30. Liu, X.; Wang, T.; Chow, L.C.; Yang, M.; Mitchell, J.W. Effects of Inorganic Fillers on the Thermal and Mechanical Properties of Poly(lactic acid). *Int. J. Polym. Sci.* **2014**, *2014*, 827028. [CrossRef]
31. Xu, J.; Shi, W.; Pang, W. Synthesis and shape memory effects of Si–O–Si cross-linked hybrid polyurethanes. *Polymer (Guildf)* **2006**, *47*, 457–465. [CrossRef]
32. Kalichevsky, M.T.; Jaroszkiewicz, E.M.; Ablett, S.; Blanshard, J.M.V.; Lillford, P.J. The glass transition of amylopectin measured by DSC, DMTA and NMR. *Carbohydr. Polym.* **1992**, *18*, 77–88. [CrossRef]
33. Seo, M.K.; Lee, J.R.; Park, S.J. Crystallization kinetics and interfacial behaviors of polypropylene composites reinforced with multi-walled carbon nanotubes. *Mater. Sci. Eng. A* **2005**, *404*, 79–84. [CrossRef]
34. Sheth, M.; Kumar, R.A.; Dave, V. Biodegradable Polymer Blends of Poly (lactic acid) and Poly (ethylene glycol). *J. Appl. Polymer Sci.* **2008**, *66*, 1495–1505. [CrossRef]
35. Patidar, D.; Agrawal, S.; Saxena, N.S. Glass transition activation energy of CdS/PMMA nano-composite and its dependence on composition of CdS nano-particles. *J. Therm. Anal. Calorim.* **2011**, *106*, 921–925. [CrossRef]
36. Zhang, W.; Chen, L.; Zhang, Y. Surprising shape-memory effect of polylactide resulted from toughening by polyamide elastomer. *Polymer (Guildf)* **2009**, *50*, 1311–1315. [CrossRef]
37. Zhang, H.; Wang, H.; Zhong, W.; Du, Q. A novel type of shape memory polymer blend and the shape memory mechanism. *Polymer (Guildf)* **2009**, *50*, 1596–1601. [CrossRef]
38. Lai, S.M.; Lan, Y.C. Shape memory properties of melt-blended polylactic acid (PLA)/thermoplastic polyurethane (TPU) bio-based blends. *J. Polym. Res.* **2013**, *20*, 2–9. [CrossRef]
39. Peponi, L.; Navarro-Baena, I.; Sonseca, A.; Gimenez, E.; Marcos-Fernandez, A.; Kenny, J.M. Synthesis and characterization of PCL–PLLA polyurethane with shape memory behavior. *Eur. Polym. J.* **2013**, *49*, 893–903. [CrossRef]
40. Peponi, L.; Sessini, V.; Arrieta, M.P.; Navarro-Baena, I.; Sonseca, A.; Dominici, F.; Gimenez, E.; Torre, L.; Tercjak, A.; López, D.; et al. Thermally-activated shape memory effect on biodegradable nanocomposites based on PLA/PCL blend reinforced with hydroxyapatite. *Polym. Degrad. Stab.* **2018**, *151*, 36–51. [CrossRef]

© 2020 by the authors. Licensee MDPI, Basel, Switzerland. This article is an open access article distributed under the terms and conditions of the Creative Commons Attribution (CC BY) license (http://creativecommons.org/licenses/by/4.0/).

Article

Highly Stretchable and Flexible Melt Spun Thermoplastic Conductive Yarns for Smart Textiles

G. M. Nazmul Islam [1], Stewart Collie [2], Muhammad Qasim [1] and M. Azam Ali [1,*]

[1] Centre for Bioengineering & Nanomedicine, Department of Food Science, Division of Sciences, University of Otago, P.O. Box 56, Dunedin 9054, New Zealand; nazmul.islam@postgrad.otago.ac.nz (G.M.N.I.); muhammadqasim.qasim@otago.ac.nz (M.Q.)
[2] Bioproduct & Fiber Technology, AgResearch, Christchurch 8140, New Zealand; stewart.collie@agresearch.co.nz
* Correspondence: azam.ali@otago.ac.nz

Received: 16 October 2020; Accepted: 20 November 2020; Published: 24 November 2020

Abstract: This study demonstrates a scalable fabrication process for producing biodegradable, highly stretchable and wearable melt spun thermoplastic polypropylene (PP), poly(lactic) acid (PLA), and composite (PP:PLA = 50:50) conductive yarns through a dip coating process. Polydopamine (PDA) treated and poly(3,4-ethylenedioxythiophene):poly(styrenesulfonate) (PEDOT:PSS) coated conductive PP, PLA, and PP/PLA yarns generated electric conductivity of 0.75 S/cm, 0.36 S/cm and 0.67 S/cm respectively. Fourier Transform Infrared Spectroscopy (FTIR) confirmed the interactions among the functional groups of PP, PLA, PP/PLA, PDA, and PEDOT:PSS. The surface morphology of thermoplastic yarns was characterized by optical microscope and Scanning Electron Microscope (SEM). The mechanical properties of yarns were also assessed, which include tensile strength (TS), Young's modulus and elongation at break (%). These highly stretchable and flexible conductive PP, PLA, and PP/PLA yarns showed elasticity of 667%, 121% and 315% respectively. The thermal behavior of yarns was evaluated by differential scanning calorimetry (DSC) and thermo-gravimetric analysis (TGA). Wash stability of conductive yarns was also measured. Furthermore, ageing effect was determined to predict the shelf life of the conductive yarns. We believe that these highly stretchable and flexible PEDOT:PSS coated conductive PP, PLA, and PP/PLA composite yarns fabricated by this process can be integrated into textiles for strain sensing to monitor the tiny movement of human motion.

Keywords: thermoplastic polymer; melt spinning; thermoplastic yarn; electric conductivity; wearable textile

---

1. Introduction

Smart textiles have drawn increased attention from the academic researchers and industry people due to their high sensitivity, high flexibility, breathability, multitasking capability, availability, low cost, deformability and comfort [1–5]. Textiles can be conductive applying various methods including spinning [6–9], knitting [10], coating [11–15], screen printing [16], inkjet printing [17,18] and 3D printing [19–22]. Electrically conductive yarn is one of the most basic and essential components of smart textiles due to its light weight, high stretchability, elasticity, flexibility, and comfort [23,24]. Integrating electronic mechanisms into textile structures can impart various smart functionalities including sensing, monitoring tiny body movement and information processing to conventional clothing [25].

Poly(lactic acid) (PLA) is one of the most promising and cheapest bio-based materials among the various biodegradable polyesters available in the market such as polyglycolic acid (PGA), polyhydroxybutyrate (PHB), and polycaprolactone (PCL) due to easy process characteristics [26,27].

PLA has been used in the field of biomedical, medical textiles, agricultural textiles, geo-textiles, food industry, filters, towels, home furnishings, industrial fabrics, and personal belongings due to their natural origin, adequate mechanical properties, permeability, low flammability, and excellent UV resistance [28]. Polypropylene (PP), an outstanding semi crystalline and non-polar thermoplastic polymer has been used in a wide range of applications such as protective textiles, geo-textiles, automotive interior, filaments, furniture, antistatic materials, medical devices, soft tissue replacement, plastic, piping systems, and other consumer food packaging due to its low cost, availability, gas barrier properties, adequate mechanical properties, and thermo-plasticity [29–31]. However, the use of PLA is restricted to the biomedical and packaging applications due to slow degradation, high processing cost and low shelf life though PLA shows high rigidity and good biocompatibility [30]. The main limitations of PLA, including low toughness and high brittleness, limit its application in stressful conditions [32,33]. It is a great challenge to achieve high levels of toughness of PLA film. The properties of PLA can be modified by blending, plasticization, and/or by reactive processing [34]. Blending polymer with other nanoparticles or polyolefin polymers is a simple method to potentially improve the property of pure polymer [35]. To minimize the above limitations of PLA, the blending of PLA with PP can lead to the desired properties such as high productivity and quick formulation changes at low price. Blending of PLA, starch, polyethylene glycol, polyethylene oxide, and polycaprolactone (PCL) with PP improves the degradability of PP. This polymer blending and composite have been widely used in biomedical textile, medical packaging, energy storage, plastic industries, and food packaging industries due to its enhanced mechanical, thermal, electrical, and biodegradation properties [36,37]. Introduction of PP with PLA decreases the stiffness property of PLA and thus enhance the mechanical properties of composite yarn. In this present investigation, melt spinning was applied to produce PP, PLA, and blend PP/PLA thermoplastic yarns due to low investment cost, solvent free simple spinning process, high production rate, and no environmental pollution. Melt spun PP, PLA, and blend PP/PLA yarns are hydrophobic and act as insulators. PP, PLA, and blend PP/PLA yarns do not absorb any chemical during coating due to hydrophobicity. Surface modifications play a vital role in a variety of application domains from electronics to medicine including interfacing with cells, bio-sensing, and drug delivery [38]. Polydopamine (PDA), a dopamine derived synthetic eumelanin polymer, can modify many kinds of substrates [12,39]. Polydopamine acts as a universal surface modification agent for different applications such as nanotechnology [40], biotechnology [41]. Here all thermoplastic yarns were treated with dopamine and Tris HCl solution. This PDA treatment converted hydrophobic yarns into hydrophilic which was proved by contact angle (CA) analysis.

Inherently conducting polymers (ICPs) such as polypyrrole, PEDOT, and polyaniline have become popular choices for producing multi-functional fibers, films, and fabrics because of their high conductivity, excellent electrochemical properties, promising catalytic activity, ease of handling, and excellent solution processability [42–50]. PEDOT:PSS is an automatic choice for researchers due to its high conductivity and high stability for developing highly conductive and flexible sensors for various applications such as biomedical and limb motion sensing [51], pH sensing [52,53], flexible heating element on textiles [54], strain sensing [14,55,56], temperature sensing [57], and wearable e-textiles [58]. Martin et al. [59] developed a multi-walled carbon nanotube (MWCNT, 4%)/polyethylene (PE) conducting polymeric composite by melt spinning technique and achieved conductivity only 0.1 S/cm. Soroudi et al. [60] also demonstrated filaments of blends of polypropylene (PP)/polyaniline (PANI) (20 wt%)/MWCNT (7.5 wt%) by melt spinning and this blend filaments showed maximum conductivity about 0.16 S/cm. Wang et al. [61] developed a PEDOT:PSS/PVA composite fiber via wet-spinning process for increasing the electrical conductivity and thermal stability but no information of washing is available. In this experiment, these conductive thermoplastic yarns were rinsed and the effect of rinsing on electric conductivity was assessed. These developed thermoplastic yarns also showed better electrical conductivity, thermal, and mechanical stability compared to others, which is suitable for strain sensing.

The target of this experiment is to fabricate PEDOT:PSS coated stable conductive thermoplastic PP, PLA and blend PP/PLA yarn which is free of metal, carbon, and silica nanoparticles. For predicting and analyzing the aging properties of conductive yarns, the aging behavior was assessed. Figure 1a,b show the key possible chemical reaction steps for producing PDA treated and PEDOT:PSS coated conductive PP and PLA yarns respectively.

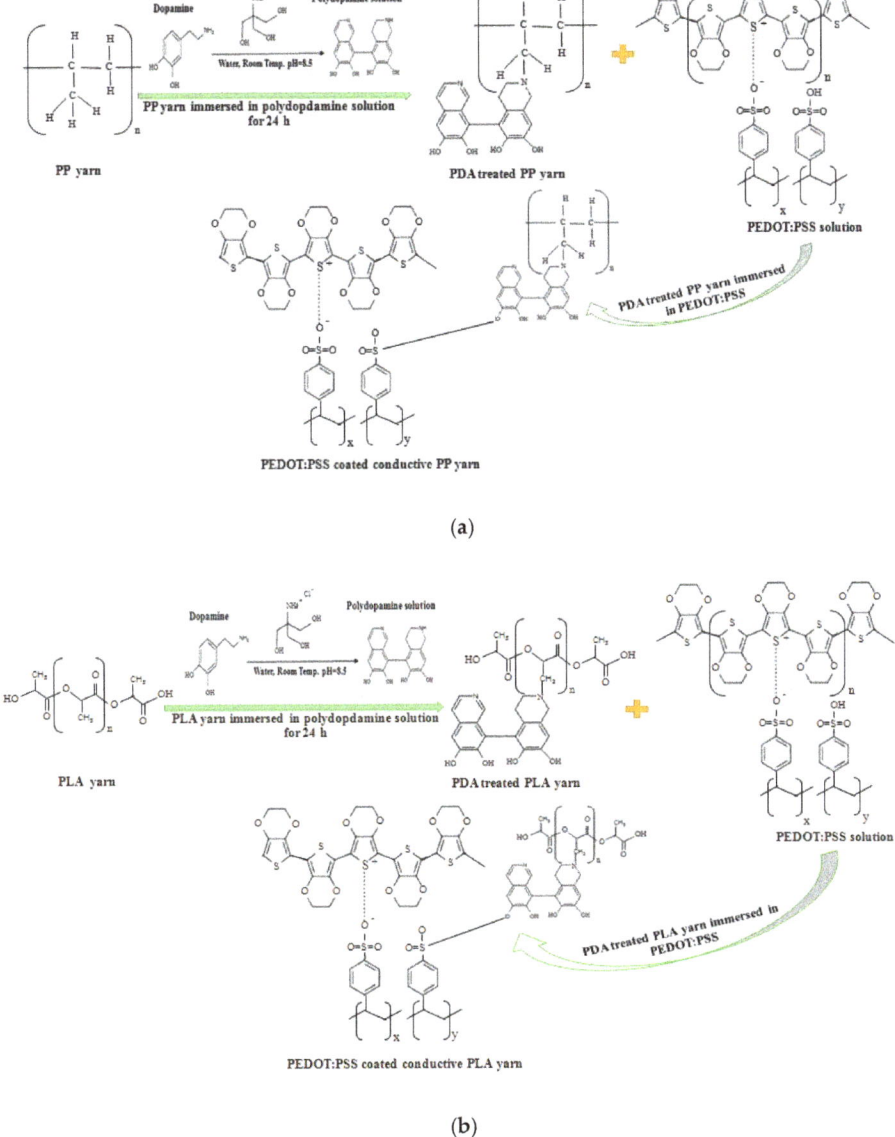

Figure 1. Possible key chemical reactions involved for producing PEDOT:PSS coated conductive (a) PP and (b) PLA yarn.

2. Materials and Methods

2.1. Materials

A melt spinner (LE-075 Mixing Extruder, CSI, USA) shown in Figure 2 was used for producing thermoplastic melt spun yarns. Poly (lactic acid) (PLA) was provided by Imagin Plastics Ltd., Auckland, New Zealand (average molecular weight, Mw~2.08×10^5, melt flow index, MFI~210 °C/2.16 kg of 15–25 g/10 min). Polypropylene (PP) was procured from Lyondell Basell, New Zealand (Mw~2.20×10^5, MFI~230 °C/2.16 kg of 25 g/10 min). PEDOT:PSS dispersion was purchased from Sigma-Aldrich, New Zealand with a ratio of PSS:PEDOT = 1:1.5, pH = 2–3.5 used as conducting material. Tris hydrochloride (Tris HCl) (Bio-Froxx, GmbH, Germany) was used as buffer agent and dopamine hydrochloride (98%, Sigma-Aldrich, New Zealand) was used as binding agent for surface modification of thermoplastic yarn. Hydrochloric acid (HCl) (Sigma-Aldrich, New Zealand) was used to maintain pH = 8.5 for dopamine and Tris HCl solution.

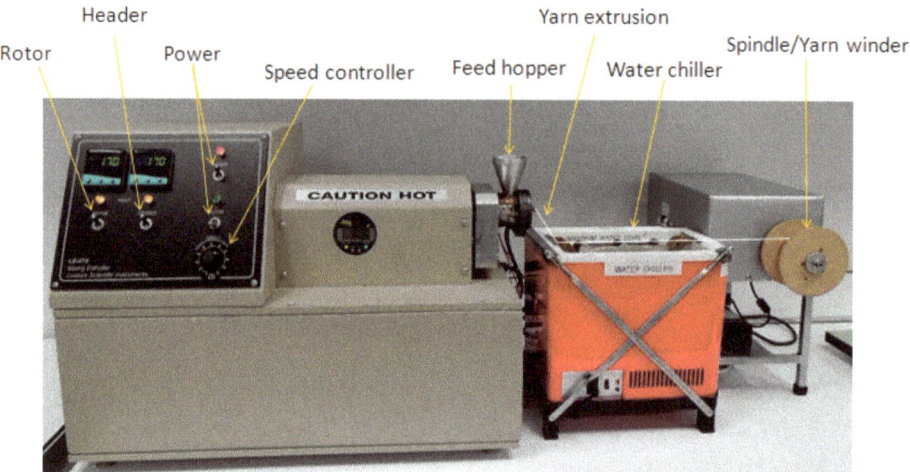

Figure 2. Schematic diagram of LE-075 Mixing Extruder.

2.2. Preparation of Melt Spun Conductive Composite Yarns

Before thermoplastic yarn extrusion, the most important parameters such as melting point of filler and matrix, resident time, rotation per minute (rpm) for the extruder were identified. Speed/output voltage of the extruder plays a vital role for maintaining the same diameter of the fine filament. Thermoplastic yarns were produced by identifying and applying the melting point of the fillers and rpm of extrusion. In this experiment, the resident time, rpm, and speed/output voltage for the extruder were 3 min, 90 and 50 V respectively. In this present investigation, three (3) types of thermoplastic yarns were developed maintaining the residence time, voltage percentage, and rpm of extruder. Figure 3a–c illustrate the schematic diagram of PDA treated and PEDOT:PSS coated melt spun PP, PLA, and blend PP/PLA conductive yarns respectively. At first PP and PLA thermoplastic melt spun yarns were produced by a melt extruder at 170 °C and 155 °C respectively. Then a mixture (50% PP and 50% PLA) of thermoplastic polymers was manually measured and mixed and also put into the hopper of the extruder. Considering the melting point of PP and PLA, the composite thermoplastic melt spun yarn was extruded at 170 °C. For uniform blending, the produced composite yarn was cut into small sections using a scissor and again put into the feed hopper. Maintaining the same temperature, rpm, and voltage percentage of the extruder, the final composite (PP/PLA) yarn was produced by repeating this process for two times. It is mentioned that considering the residence time (3 min), rpm (90) and output voltage (50 V) of the melt extruder, the take up speed of uniform yarn production is approximately 1 m/min.

The take up speed of yarn production can be increased by increasing the speed/output voltage of the melt extruder.

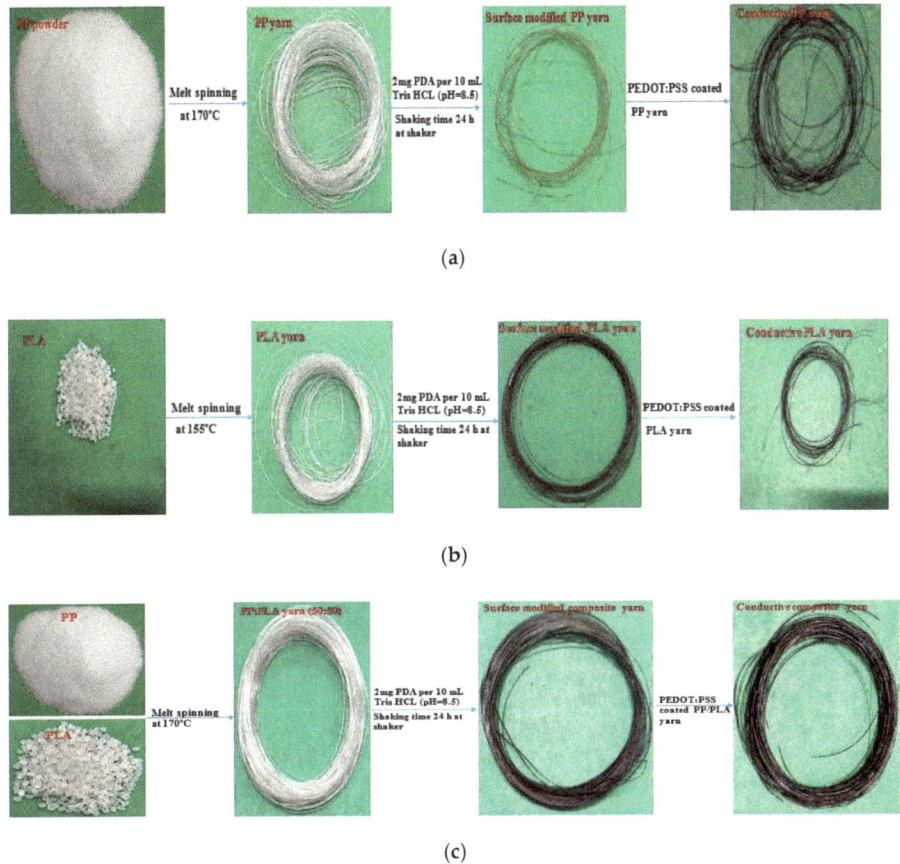

Figure 3. Schematic diagram of PEDOT:PSS coated conductive (**a**) PP, (**b**) PLA and (**c**) blend PP/PLA yarn.

These hydrophobic yarns were chemically modified to increase the hydrophilicity with PDA and Tris HCl. 12.11 g Tris HCl was added in 80 mL distilled water and pH = 8.5 was maintained by gently adding 1 µM of hydrochloric acid (HCl) for producing Tris HCl solution. 2 mg of dopamine hydrochloride per 10 mM Tris HCl was added to produce the aqueous solution. PP, PLA, and blend PP/PLA based hydrophobic yarns were immersed in this aqueous solution and kept in a shaker for 24 h with 55 rpm at room temperature. The surface modified yarns were rinsed with distilled water for 1 min and line dried at room temperature. These PDA treated hydrophilic PP, PLA and blend PP/PLA yarns were immersed in PEDOT:PSS dispersion for 5 min. Then the coated yarns were dried at room temperature for 4 h by hanging them on a clothes line using wooden clip hangers shown in Figure 4. This coating process was repeated for two dip coating cycles. All the thermoplastic yarns were coated for two times considering the flexibility, stiffness, and rigidity of conductive yarns. Though more coating cycles increased the conductivity but made the yarns stiff and rigid. Stiff and rigid yarns are not suitable for integrating into textiles for wearable applications. The above fabrication process can be described as: production of thermoplastic yarns by melt spinning > surface modification by polydopamine > PEDOT:PSS coated conductive yarns by dip coating.

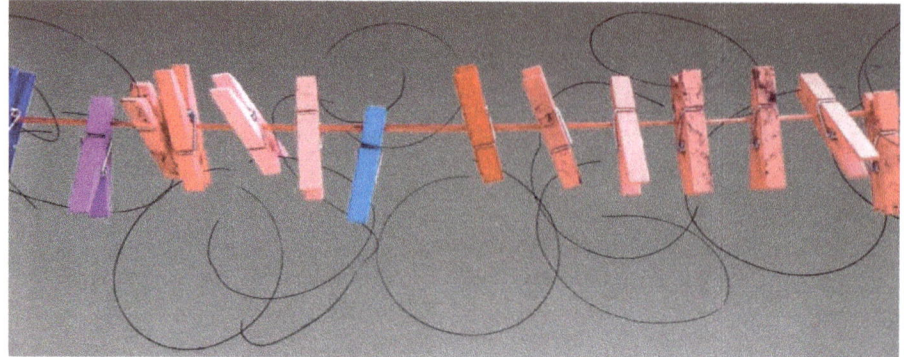

Figure 4. Line drying system for drying PEDOT:PSS coated conductive yarns.

2.3. Characterizations of Thermoplastic Yarns

The chemical interactions among different components such as pure PP, pure PLA, blend PP/PLA, PDA, and PEDOT:PSS were studied using FTIR. This measurement was performed with a total of 24 scans/sample over the range of 4000–400 cm^{-1} at resolution of 4 cm^{-1}.

The contact angle (CA) was analyzed using FTA200 Dynamic Contact Angle Analyzer (First Ten Angstroms, Portsmouth, VA, USA) with the static sessile drop method. A droplet (1 µL) of distilled water was placed on the surface of the raw thermoplastic and PDA treated yarns by a syringe. A video camera (Sony ICX274 CCD) was used to record the water contact angles of the raw thermoplastic and PDA treated yarns at room temperature.

The DC electrical resistance of 20 cm lengths of conductive yarns was measured by FLUKE 114 TRUE RMS Multimeter (Everett, WA, USA) before and after rinsing. Electric resistance was measured three times for each conductive yarn every after each dip coating cycle and averaged. Then electric conductivity (σ) was measured applying the following formula [62]:

$$\sigma = L/RA \qquad (1)$$

where R is the electrical resistance (Ω), A is the total cross-sectional surface area (cm^2) and L is the distance between electrodes (cm).

Optical microscopy was performed on several pure PP, pure PLA, blend PP/PLA, surface modified, and PEDOT:PSS coated yarns to determine the presence of PDA and conductive polymer on pure yarns. Each yarn was attached at both sides on a glass slide using clear scotch tape exposing 30 mm of yarn. Optical microscopy was carried out by an optical microscope (OLYMPUS, Tokyo, Japan). Optical images were captured with a HUWAEI Y9 camera (Shenzhen, China) and images were cropped using Photoshop software. An image of the all yarns at 100 times magnification was captured in all stages in the same position.

For analyzing the yarn surface morphology, PDA treated and PEDOT:PSS coated samples were attached to the scanning electron microscope (SEM) specimen stub using double sided carbon tape. Before SEM analysis, they were sputter-coated with 5 nm gold-palladium using a Q150T sputter coater (Quorum Technologies Ltd., East Sussex, UK) in order to prevent the surface charging effect which gives a blurred picture and to promote the emission of secondary electrons for providing a homogeneous surface for analysis. This morphological analysis of yarns was characterized using Tabletop Microscope TM3030 (Hitachi, Japan) with voltage of 15 kV at different magnifications. The thickness of the coating was measured using ImageJ software by taking three measurements of six different samples of the coatings. Data are expressed as mean ± SD.

The washing stability of conductive yarns of each dip cycle was assessed. The coated conductive yarns of each dip cycle were rinsed for 1 min and line dried for 2 h at room temperature. Then their electrical resistance was measured. This rinsing process was carried out five times.

Differential Scanning Calorimetry (DSC) was performed by TA analyzer (TA) Q1000 instrument (TA Instruments, New Castle, DE, USA) to measure the glass transition (T_g) and melting (T_m) temperature characteristics of thermoplastic PP, PLA, and blend PP/PLA conductive yarns to determine the thermal stability. Samples were weighed (10–15 mg) into a pan (Tzero pan; TA Instruments Ltd., New Castle, DE, USA). Each sample was heated over the temperature range from 20 to 200 °C at the rate of 5 °C/min under nitrogen atmosphere (50 mL/min).

The thermal stability and degradation of pure PP, pure PLA, PP/PLA, PDA treated and PEDOT:PSS coated yarns were analyzed by Q50 TGA analyzer (TA instruments, New Castle, DE, USA). The weight of samples was 20–35 mg. These stability analyses were performed over the temperature range 200 to 600 °C at a heating rate 20 °C/min under the nitrogen atmosphere (20 mL/min).

The mechanical properties of all yarns were investigated by a TA.HD plusC Texture Analyzer (UK) applying 5 kg load cell, gauge length of 25 mm and tensile speed 20 mm/min at room temperature.

To evaluate the aging effect, the conductive yarns of each dipping cycle were stored in a ambient room conditions for five weeks in separate polythene bags with minimal exposure to air and moisture. The loss of electrical resistance during aging was measured every week in order to determine the shelf life of conductive yarn.

3. Results and Discussion

3.1. Fourier Transform Infrared Spectroscopy (FTIR) Analysis

The functional groups of all the samples such as pure PP, pure PLA, blend PP/PLA, PDA and PEDOT:PSS were confirmed by interpretation of the FTIR spectra. Figure 5a, Figure 5b, and Figure 5c represent the FTIR spectra of PDA treated and PEDOT:PSS coated conductive PP, PLA and PP/PLA yarns respectively. All the transmittance bands are also listed in Table 1. Here Figure 5a depicts the transmittance bands corresponding to PP at 2950–2850 cm^{-1}, 1454 cm^{-1} and 1377 cm^{-1} were assigned to C–H stretching, –CH$_3$ bending and C–H bending respectively [30]. Here Figure 5b depicts the FTIR spectra of PLA, transmittance bands at 2995–2945 cm^{-1}, 1749 cm^{-1}, 1182–1045 cm^{-1} and 1453 cm^{-1} referred to CH and CH$_3$ group, C=O stretching, symmetric C–O–C stretching and asymmetric bending absorption of CH$_3$ respectively [63]. Ploypeetchara et al. [64] analyzed the spectra of different PP/PLA ratios and found the transmittance bands that represent PP and PLA were observed in the PP/PLA blend around 2952–2848cm^{-1}, 1456–1454 cm^{-1}, 1376 cm^{-1}, 1183–1182 cm^{-1} and 1086–1184 cm^{-1}. A specific peak for all PP/PLA blends appeared at 1749 cm^{-1} is corresponded to the stretching of the ester group (–COO) where the chemical interaction of the anhydride group of PP with the carbonyl group of PLA formed a new linkage which indicates the PP/PLA blends [37]. From Figure 5c, the transmittance bands that represent PP and PLA were observed in the PP/PLA blends around 2950–2848 cm^{-1}, 1743 cm^{-1}, 1454 cm^{-1}, 1376 cm^{-1} and 1182 cm^{-1}. After surface modification of PP, PLA and PP/PLA by PDA, the transmittance bands at 3186–3345 cm^{-1}; 3184–3345 cm^{-1} and 3184–3345 cm^{-1} respectively corresponded to stretching vibrations of O–H and N–H groups of PDA [65].

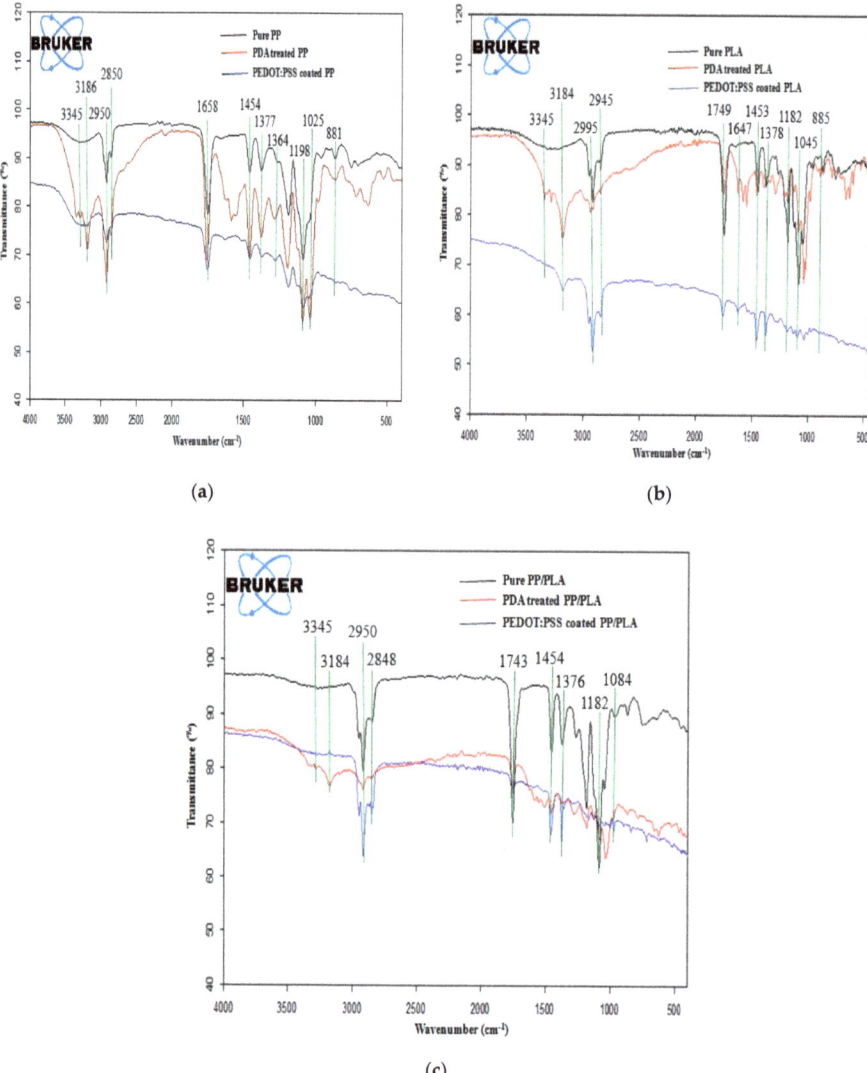

Figure 5. FTIR spectra of PEDOT:PSS coated (**a**) PP, (**b**) PLA and (**c**) blend PP/PLA yarn.

Table 1. FTIR transmittance bonds of thermoplastic yarns.

IR Absorption Bands (cm^{-1})	Description
3184–3345	stretching vibrations of O–H and N–H groups
2945–2850	C–H stretching
1743–1454	–CH$_3$ stretching
1647–1658	C=C stretching
1045–1182	C=O stretching
1376	C–H bending
1183 and 1025–881	Stretching C–S
1453	Symmetric C–O–C stretching

From Figure 5a–c, it is seen that the PDA was coated successfully onto the surfaces of PP, PLA and PP/PLA yarns. From Figure 5a it is seen that the absorption spectra of PEDOT:PSS coating on PDA treated conductive polypropylene yarn displayed the polymeric interactions in the thiophene backbone including C=C, C–C and C–S bonds at 1658 cm^{-1}, 1364 cm^{-1}, 1198 cm^{-1} and 1025–881 cm^{-1} respectively [63]. Similarly Figure 5b shows that the FTIR spectra of PEDOT:PSS coated and PDA treated conductive PLA yarn displayed the polymeric interactions in the thiophene back bone, including C=C, C–C and C–S bonds at 1647 cm^{-1}, 1378 cm^{-1}, 1199 cm^{-1} and 1025–885 cm^{-1} respectively [63]. Figure 5c also indicates that the absorption spectra of PDA treated and PEDOT:PSS coated conductive PP/PLA yarn displayed the transmittance bands at 1647 cm^{-1}, 1376 cm^{-1}, 1183 cm^{-1} and 1084–1042 cm^{-1} corresponded to C=C, C–C and C–S bonds respectively [66]. From Figure 5a–c, it is confirmed that after two coating layers of PEDOT:PSS on PDA treated PP, PLA, and PP/PLA yarns, all transmittance bands were found to be almost similar due to low PSS adsorption.

3.2. Contact Angle (CA) Analysis

Wettability, an important phenomena of substrates which is related to the surface roughness and surface charge. Here the melt spun thermoplastic yarns do not absorb any chemicals due to their hydrophobicity. Using the general method of contact angle measurement, it is hard to analyze a tiny fiber and yarn. Therefore, we modified the procedure and used adhesive tape to put yarn on contact angle machine stage. For dropping water on yarn surface, we did not use the machine connected syringe pump. However, we manually placed the drop of water on yarn surface using micro-pipette volume (1 µL). The contact angles (θ) of raw thermoplastic polypropylene yarn and PDA treated polypropylene yarn were measured and shown in Figure 6. From Figure 6, it is seen that the contact angle (CA) of raw thermoplastic polypropylene yarn is θ = 135°. It has a CA value of 135° in all groups before surface modification. As this raw thermoplastic polypropylene yarn is hydrophobic, the surface of this yarn was modified by polydopamine. After polydopamine treatment, the CA of treated polypropylene yarn decreased which is θ = 60°. From the CA of PDA treated yarn, it is confirmed that PDA converted hydrophobic thermoplastic polypropylene yarns into hydrophilic.

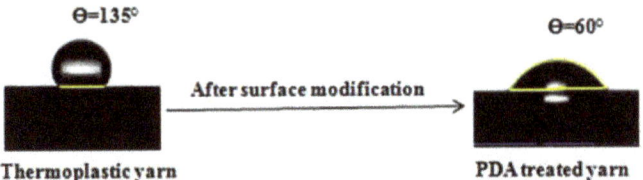

Figure 6. The images of water contact angle of untreated and PDA treated thermoplastic yarns.

3.3. Electrical Conductivity before Rinsing

Electrical conductivity is one of the most important key aspect and requirements for wearable conductive yarns. Table 2 shows the electric conductivity of PDA treated and PEDOT:PSS coated 20 cm long conductive PP, PLA, and blend PP/PLA yarns before rinsing.

Table 2. Electrical conductivity (S/cm) of PEDOT:PSS coated PP, PLA, and blend PP/PLA yarn before rinsing.

Yarn Type	Radius (cm)	Area (cm²)	Coating Cycle 1			Coating Cycle 2		
			Mean Electrical Resistance (kΩ)	SD	Conductivity (S/cm)	Mean Electrical Resistance (kΩ)	SD	Conductivity (S/cm)
PP	0.014	0.000616	131.00	3.42	0.25	43.04	2.20	0.75
PLA	0.013	0.000531	223.33	7.77	0.17	122.67	5.47	0.36
PP/PLA	0.012	0.000452	181.13	3.62	0.24	65.93	1.23	0.67

SD = Standard Deviation.

After the first and second dip coating cycles, the electric conductivity of conductive PP yarn is 0.25 S/cm and 0.75 S/cm, respectively. Similarly, the electric conductivity of conductive PLA yarn is 0.17 S/cm and 0.36 S/cm respectively. In addition, the electrical conductivity of blend PP/PLA yarn is 0.24 and 0.67 S/cm respectively before rinsing. The number of dip coating cycle increases the electrical conductivity of the coated yarns. After the second coating cycles, the electrical conductivity of each yarn increases at least two times compared to the first coating cycle. In Table 2, it is seen that the PDA treatment has converted the hydrophobic yarns into hydrophilic yarns successfully which was proved by contact angle analysis and the number of coating cycles increased the PEDOT:PSS pick up% which increased the electrical conductivity.

3.4. Tensile Properties Analysis

The mechanical properties (tensile strength, Young's modulus, and elongation at break %) of pure PP, pure PLA, pure PP/PLA, PDA treated and PEDOT:PSS coated conductive PP, PLA, and PP/PLA yarns were investigated. The role of the PDA introduction and PEDOT:PSS coating on melt spun PP, PLA and PP/PLA yarns was characterized by their mechanical properties. Figure 7 illustrates the stress-strain curves to analyze the mechanical properties of the various types of yarns. Three (3) replicates were tested for each yarn and the average values of tensile strength, Young's modulus and elongation at break (%) were reported in Table 3.

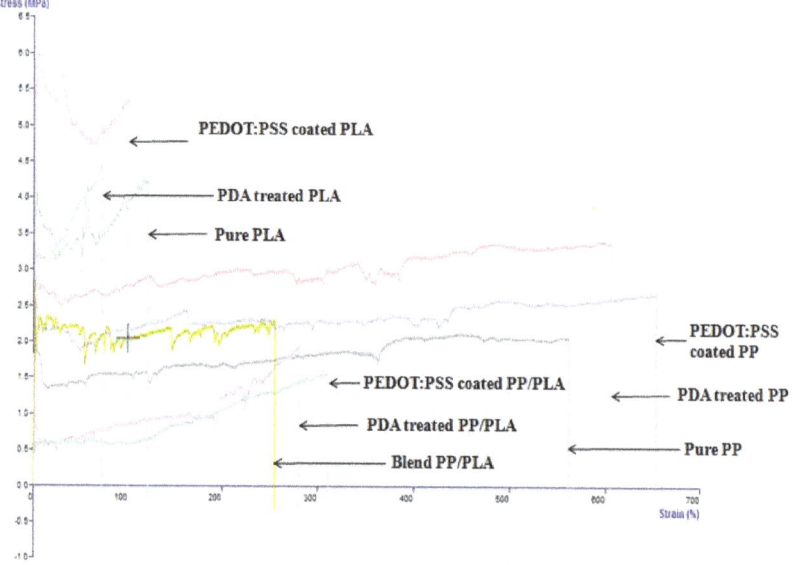

Figure 7. The stress-strain curves of various thermoplastic yarns.

Table 3. Mechanical properties of thermoplastic yarns.

Types of Yarn	Tensile Strength (MPa)	SD (MPa)	Tensile/Young's Modulus (MPa)	SD (MPa)	Elongation at Break (%)	SD
PP	1.22	0.14	76.98	4.99	594.53	5.76
Modified PP	1.81	0.09	87.92	5.47	636.51	7.98
PEDOT:PSS coated PP	1.97	0.02	116.39	8.94	667.47	5.92
PLA	2.99	0.18	230.70	7.67	42.30	0.39
Modified PLA	3.41	0.40	291.87	5.78	76.89	0.63
PEDOT:PSS coated PLA	3.57	0.23	309.29	7.51	121.35	0.48
PP/PLA	1.35	0.15	96.164	7.71	227.17	1.34
Modified PP/PLA	2.08	0.49	172.11	5.33	263.64	1.01
PEDOT:PSS coated PP/PLA	2.56	0.08	188.40	5.76	315.33	0.98

SD = Standard Deviation.

Figure 7 shows all full stress-strain curves for analyzing the tensile strength, Young's modulus, and elongation at break%. From Table 3, it is seen that PLA yarns have better tensile strength compared to PP and PP/PLA yarns. The tensile strength of pure PP, pure PLA, blend PP/PLA, polydopamine treated and PEDOT:PSS coated PP, PLA and PP/PLA is 1.22 MPa, 1.81 MPa, 1.97 MPa; 2.99 MPa, 3.41 MPa, 3.57 MPa; and 1.35 MPa, 2.08 MPa, 2.56 MPa respectively. Moreover, the tensile strength of PP/PLA blends improved with the addition of the PLA content due to the higher Young's modulus of the PLA yarn compared to the PP yarn. The bridged two immiscible PP and PLA polymers have formed a strong chemical bond which was confirmed by the FTIR analysis. However, polydopamine treatment and PEDOT:PSS coating also increased the mechanical properties of these treated and coated yarns. The mechanical properties of PP/PLA blends are strongly influenced with greater physical properties of PLA including the degree of crystallinity, melting point, density, heat capacity hardness, Young's modulus, tensile strength, glass transition temperature, and mechanical properties.

From Table 3 it is seen that the Young's modulus of pure PP, pure PLA, blend PP/PLA, polydopamine treated and PEDOT:PSS coated PP, PLA and PP/PLA is 76.98 MPa, 87.92 MPa, 116.39 MPa; 230.70 MPa, 291.87 MPa, 309.29 MPa; 96.16 MPa, 172.11 MPa and 188.40 MPa respectively. The elongation at break of pure PP, pure PLA, blend PP/PLA, polydopamine treated and PEDOT:PSS coated PP, PLA and PP/PLA is 594.53%, 636.51%, 667.47%; 42.30%, 76.89%, 121.35% and 227%, 264%, 315% respectively. So it is obvious that the polydopamine treatment and PEDOT:PSS coating have played vital a role for improving the mechanical properties of blend PP/PLA yarn.

However, it is clearly exhibited that introducing of PDA treatment and PEDOT:PSS coating illustrated a good improvement of mechanical properties of the treated and coated PP, PLA and PP/PLA yarns due to the $-NH_2$ functional group of dopamine and C-S bonds reaction happened among PDA treated thermoplastic yarns and PEDOT:PSS. This developed conductive yarns showed higher elongation at break% compared to others development. For example, Luo et al. [67] developed PEDOT:PSS/PDMS blend conductive polymer films which showed elongation at break of about 82%. Azizi et al. [37] also developed PP, PLA, and PP/PLA nanocomposite and the elongation at break of PP, PLA and PP/PLA are 210%, 20% and 25–150% respectively. So this high stretchability feature from this present investigation was a good upshot for this study which may also be an intelligent aspects of these new yards to be applied for strain sensing application.

3.5. Optical Microscopy Images Analysis

Optical microscopy images were used to analyze the coating thickness of the PP, PLA and PP/PLA yarns at different stages to analyze the changes of coating thickness of PP, PLA and PP/PLA yarns after PDA treatment and PEDOT:PSS coating. Optical microscope images captured of several single yarns in a original state before PDA treatment, after PDA treatment and PEDOT:PSS coating show that there are significant changes in the thickness of boundary layers shown in Figure 8. It is mentioned that an image of the all yarns at 100 times magnification was captured in all stages.

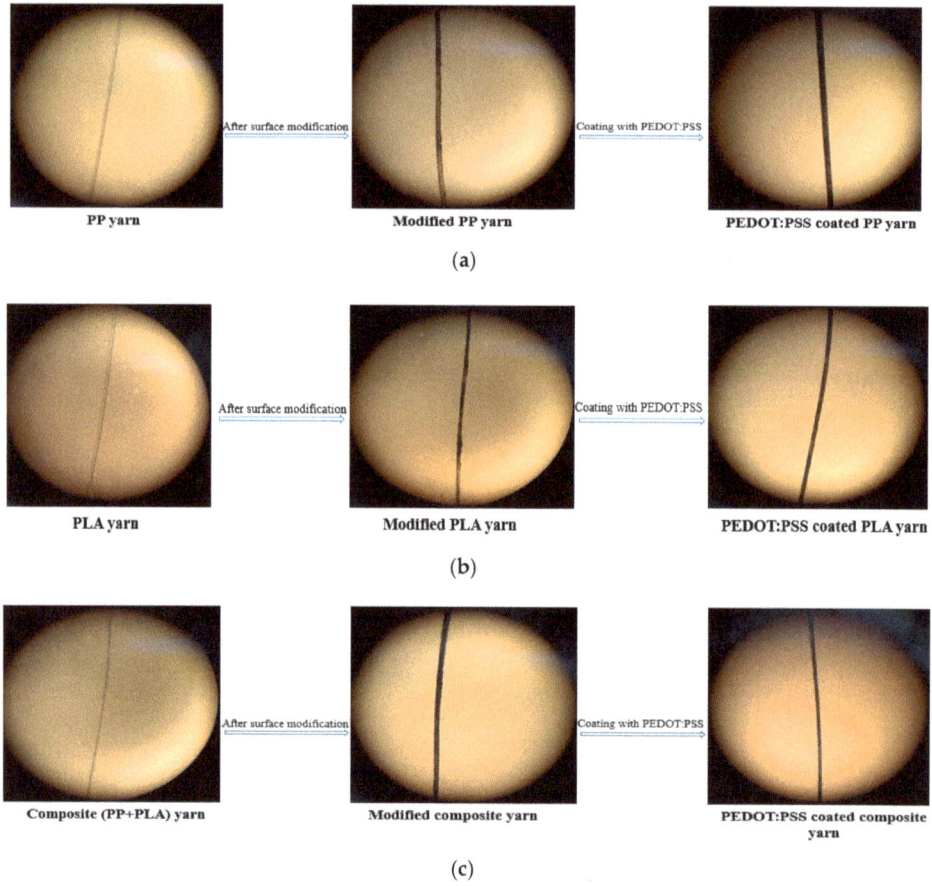

Figure 8. Optical microscopic images of conductive (a) PP, (b) PLA and (c) blend PP/PLA yarn.

From Figure 8, it is seen that PDA treatment converts the white color of pure PP, PLA, and blend PP/PLA yarn into black which confirms the successful coating on thermoplastic yarns. So it can be assumed that PDA treatment has a great impact in the increased thickness of boundary layers of yarns and PEDOT:PSS pickup%.

3.6. Scanning Electron Microscope (SEM) Analysis

The surface morphology of pure PP, pure PLA, blend PP/PLA, PDA treated, and PEDOT:PSS coated conductive PP, PLA, and PP/PLA yarns were also analyzed using SEM as shown in Figure 9, Figure 10, and Figure 11 respectively. A smooth surface morphology was observed without PDA coating while rough surface was observed in yarns which had been coated with PDA. This fact is evident from Figures 9a, 10a and 11a as a flat, smooth, and featureless surface of pure PP, pure PLA, and blend PP/PLA yarn can be observed.

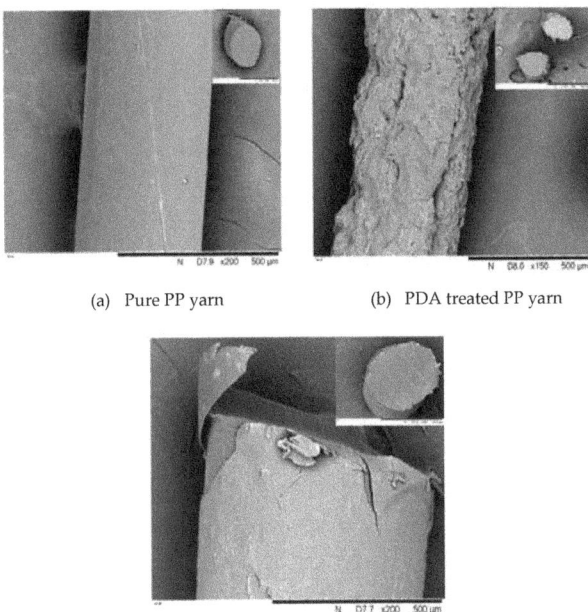

(a) Pure PP yarn (b) PDA treated PP yarn

(c) PEDOT:PSS coated conductive PP yarn

Figure 9. SEM images (surface and cross-section) of PP yarn at various stages.

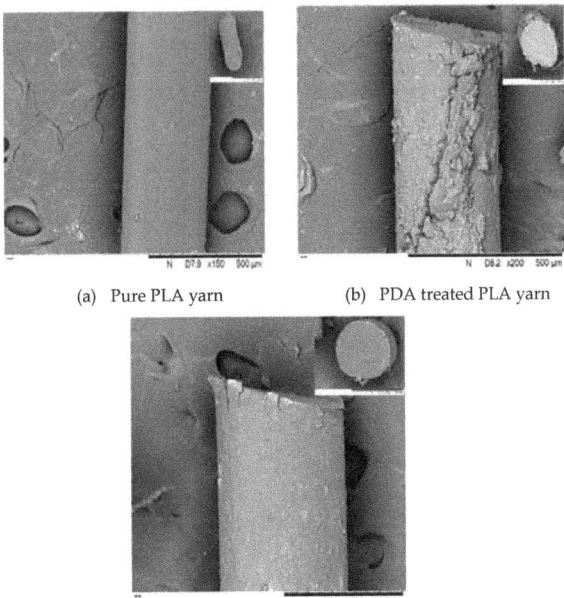

(a) Pure PLA yarn (b) PDA treated PLA yarn

(c) PEDOT:PSS coated conductive PLA yarn

Figure 10. SEM images (surface and cross-section) of PLA yarn at various stages.

(a) Blend PP/PLA yarn

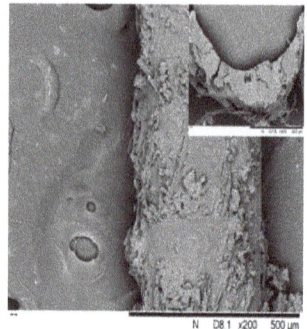
(b) PDA treated PP/PLA yarn

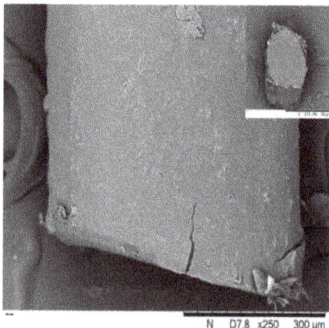
(c) PEDOT:PSS coated conductive PP/PLA yarn

Figure 11. SEM images (surface and cross-section) of PP/PLA yarn at various stages.

Figure 9b, Figure 10b, and Figure 11b revealed densely rough, more intact features and granular morphology of PDA treated PP, PLA and PP/PLA yarns. After surface modification of thermoplastic yarns by PDA, the results showed that the thickness of the modified PP, PLA and PP/PLA coating were 3.96 ± 1.45 µm, 3.52 ± 5.12 µm and 6.71 ± 3.9 µm respectively. Though PDA coating layer was observed over surfaces of PP, PLA and PP/PLA surface but it was not smooth and cracks are visible. However, still this roughness created a hydrophilic base for further coating and increased the tensile strength of yarns.

Figure 9c, Figure 10c, and Figure 11c displayed the PEDOT:PSS coated conductive PP, PLA and PP/PLA yarns respectively. A clear wrapping of PEDOT:PSS can be seen in Figures 9c, 10c and 11c over the yarns of PP, PLA, and blend PP/PLA. After PEDOT:PSS coating, the thickness of coating layer of the coated PP, PLA and PP/PLA were 5.79 ± 1.44 µm, 5.62 ± 1.0 µm and 8.3 ± 2.3 µm respectively. This smooth morphology is critical for conductivity of materials as brittle surface can act as a barrier to flow of charges. The possible reason of achieving smooth surfaces of PEDOT:PSS coated thermoplastic yarns may be attributed to higher wettability, absorption and better linkage between the conductive polymer dispersion and the flexible substrates [68].

3.7. Thermal Behavior Analysis

3.7.1. Thermo-Gravimetric Analysis (TGA)

The thermal stability of pure PP, pure PLA, blend PP/PLA, PDA modified and PEDOT:PSS polymer coated conductive yarns were analyzed by thermos-gravimetric analysis under nitrogen atmosphere are shown in Figure 12a–c respectively. Here 5% and 50% mass loss occurring was investigated to maintain the accuracy of the thermal degradation temperatures characteristics. The two degradation temperatures $T_{5\%}$ and $T_{50\%}$ correspond to 5% and 50% mass loss of the samples respectively. The remaining ash (%) at 500 °C was also measured to determine the stability of various yarns. The mass loss (5%, 50%) and the remaining ash (%) were summarized in Table 4.

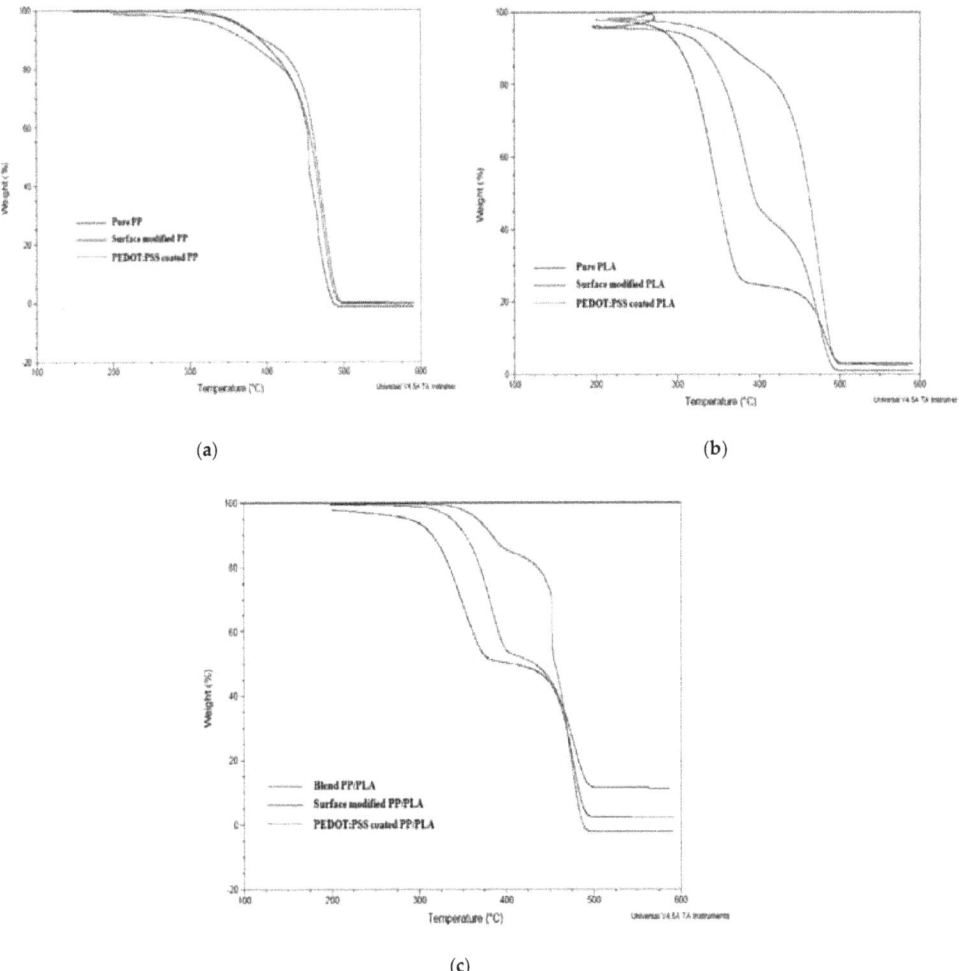

Figure 12. Thermogravimetric analysis (TGA) of PDA treated and PEDOT:PSS coated (a) PP, (b) PLA and (c) PP/PLA yarn.

Table 4. TGA results of conductive PP, PLA and PP/PLA: $T_{5\%}$, $T_{50\%}$, ΔT and the remaining ash (%) at 500 °C.

Yarn Type	$T_{5\%}$ (°C)	$T_{50\%}$ (°C)	ΔT (°C)	Remaining Ash (%) at 500°C
PP	332	455	-	0.93
Modified PP	344	461	6	2.16
PEDOT:PSS coated PP	364	465	11	2.72
PLA	296	354	-	0.20
Modified PLA	307	372	18	0.30
PEDOT:PSS coated PLA	333	393	39	0.39
PP/PLA	315	386	-	1.16
Modified PP/PLA	317	412	26	2.46
PEDOT:PSS coated PP/PLA	336	430	44	4.84

ΔT = temperature difference at 50% mass loss among the cross-linked samples and the neat polymers.

The thermal degradation curves of pure PP and pure PLA are also shown for comparison with modified and coated yarns. Figure 12a,b illustrate that the pure PP, PLA, modified and coated yarns experience single stage mass loss. Remaining ash (%) at 500 °C indicates that the introduction of PDA treatment and PEDOT:PSS coating improve the thermal stability of polymers with increase onset thermal degradation temperature and high molecular chain interaction with thermoplastic polymer.

From Figure 12c, it is seen that the thermo-grams of blend PP/PLA polymers reveal two-step degradation processes which indicate two mass loss. The first weight loss is due to the vanishing of the ester groups in the PLA polymer structure [37]. The second weight loss observed at ~380 °C which indicates the decomposition of PP polymer. The addition of PLA in PP polymer to produce blend PP/PLA decreases the initial degradation temperature to 315 °C due to the incompatibility between PP and PLA polymers. However, the introduction of PDA and PEDOT:PSS coating increase the interfacial adhesion between PP and PLA. From remaining ash (%) at 500 °C, it is confirmed that the thermal stability of thermoplastic yarns has been enhanced by addition of PDA and PEDOT:PSS coating.

3.7.2. Differential Scanning Calorimetry (DSC) Analysis

To determine the thermal properties of pure PP, pure PLA, blend PP/PLA, PDA treated and PEDOT:PSS coated conductive yarns, DSC analysis was carried out and the thermo-grams are shown in Figure 13a–c respectively. The glass transition temperature (T_g) and melting temperature (T_m) are summarized in Table 5. From Figure 13a, it is seen that no glass transition temperature is detected. The melting temperature (Tm) of PP yarn is detected at 130.92 °C [69]. After PDA treatment and PEDOT:PSS coating, melting point temperature (T_m) of PP is revealed at 131.98 °C and 132.61 °C respectively. So it is noted that PDA treatment and PEDOT:PSS coating have increased the melting point (T_m) of PP polymer. Moreover, from Figure 13b, it is seen that the T_g of the pure PLA shows a hysteresis peak. The T_g and T_m of PLA are 56.44 °C [64] and 131.55 °C [70] respectively.

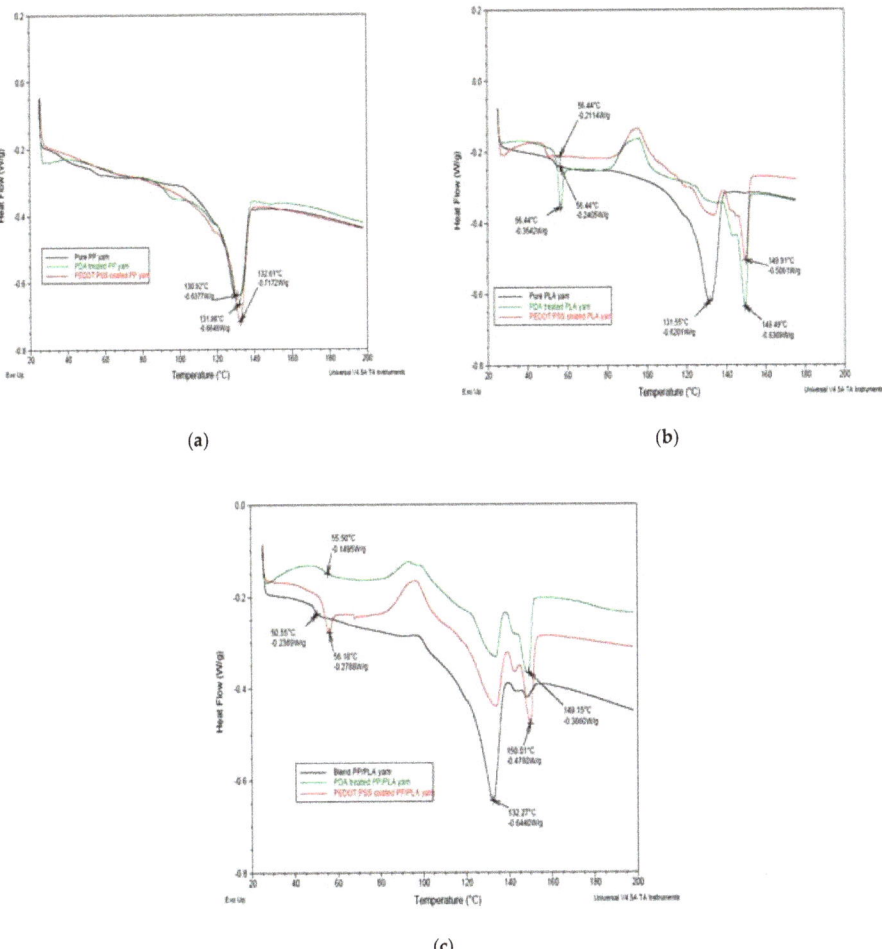

Figure 13. Differential scanning calorimetry (DSC) thermograms of PDA treated and PEDOT:PSS coated (**a**) PP, (**b**) PLA and (**c**) blend PP/PLA yarn.

Table 5. DSC results of glass transition temperature (T_g) and melting point (T_m) of PDA treated and PEDOT:PSS coated PP, PLA and PP/PLA yarns.

Yarn Type	T_g (°C)	T_m (°C)
PP	-	130.92
Modified PP	-	131.98
PEDOT:PSS coated PP	-	132.61
PLA	56.44	131.55
Modified PLA	56.44	149.49
PEDOT:PSS coated PLA	56.44	149.91
PP/PLA	50.55	132.27
Modified PP/PLA	55.50	149.15
PEDOT:PSS coated PP/PLA	56.18	150.01

The T_g of the PLA remains same due to the reduction of the mobility of the amorphous character in the PLA polymer and the physical cross links with lower addition of PDA and PEDOT:PSS.

A considerable increase is found in the T_m of PLA after PDA treatment and coating due to the physical crosslink with PLA polymer. From Figure 13c, it is seen that the glass transition temperature of blend PP/PLA polymer is 50.55 °C [64]. After polydopamine treatment and PEDOT:PSS coating, the glass transition temperature of blend PP/PLA polymers have increased at 55.50 °C and 56.18 °C respectively.

Polymer blending of PLA with PP decrease the melting point of PP. The melting point of blend PP/PLA is 132.27 °C. The melting points of polydopamine treated and PEDOT:PSS coated blend PP/PLA yarn are 149.15 °C and 150.01 °C respectively. So it is obvious that the addition of PEDOT:PSS coating increases the melting point of blend PP/PLA polymer because of the increased interfacial adhesion and interaction between the two polymer chains.

3.8. Aging Effect on Electrical Conductivity under Different Processing Conditions

For predicting and improving the shelf life of the developed conductive yarn, it is essential to analyze the degradation of the conducting material under end-use conditions during aging. Textile sensors will be used in several times. Consumers will wear this type of sensors and go outside. Then this sensors will be exposed to moisture, oxygen, and sunlight. So the developed conductive yarns were stored in a real conditioning room maintaining the parameters (temp. 20 °C and R.H. 65 ± 4%) for five weeks to analyze and measure the effect of oxygen, sunlight, and moisture content on the electric conductivity of conductive yarns. Table 6 represents the aging effect on electrical resistance of PEDOT:PSS coated conductive PP, PLA and PP/PLA yarn.

Table 6. Aging effect of PEDOT:PSS coated conductive yarns.

Conductive Yarn Type	Aging Duration (Week)	Electrical Resistance (KΩ)							
		For 1st Coating				For 2nd Coating			
		Mean	SD	Increased Electrical Resistance Every Week (%)	Total Increased Electrical Resistance (%) from Week 0–Week 5 (%)	Mean	SD	Increased Electrical Resistance (%)	Total Increased Electrical Resistance (%) from Week 0–Week 5 (%)
PP	0	131.00	3.41	-	23.43	43.00	2.08	-	20.98
	1	143.07	1.63	9.21		44.70	0.46	3.95	
	2	149.67	0.57	4.61		46.93	0.57	4.99	
	3	154.70	2.03	3.36		48.73	0.75	3.84	
	4	159.30	1.91	2.97		51.37	1.32	5.42	
	5	164.53	0.83	3.28		52.80	0.30	2.78	
PLA	0	223.00	7.76	-	26.85	122.00	5.47	-	23.00
	1	251.10	1.11	12.60		134.13	1.05	9.94	
	2	262.40	2.80	4.50		139.40	1.20	3.92	
	3	269.10	1.74	2.55		143.33	0.86	2.82	
	4	279.10	2.16	3.72		148.50	0.86	3.61	
	5	288.80	2.35	3.48		152.53	0.83	2.71	
PP/PLA	0	181.00	3.63	-	25.45	66.00	1.23	-	21.96
	1	199.50	0.79	10.22		73.43	0.65	11.26	
	2	210.50	1.12	5.51		75.67	0.57	3.05	
	3	215.33	0.96	2.29		78.50	0.87	3.74	
	4	223.40	0.79	3.75		80.80	0.36	2.93	
	5	231.63	0.42	3.68		81.63	0.87	1.03	

Here the gain of electrical resistance on aging was evaluated in every week. By calculating this total increased electrical resistance (%), the effect of atmospheric storage was analyzed. Considering the dip coating cycle 1 and 2, electrical resistance of conductive PP, PLA and PP/PLA yarn increased by ~23.43%, ~20.98%; ~26.85%, ~23% and ~25.45%, ~21.96% respectively in five weeks due to aging under storage conditions. Considering the increase of electric resistance, it is confirmed that aging has enormous effect on the shelf life of these conductive materials. The reason for increase in the electrical resistance of PEDOT:PSS coated conductive yarns may be the oxidative degradation by oxygen and the degradation of the conductive material by the atmospheric moisture.

3.9. Electrical Conductivity after Rinsing

Conductive yarns must be sufficiently robust to be suitable for daily use particularly in respect of bending, abrasion, and cleaning. Wash durable conductive yarns production is a great technical challenge for repeat use. Conductive tracks typically cannot survive machine washing due to the mechanical stresses, reaction between detergent and water. The cleaning stability and washing performance of both PEDOT:PSS coated conductive yarns were analyzed. Figure 14 shows the relationship between electrical conductivity (S/cm) and rinsing cycles of PEDOT:PSS coated PP, PLA, and PP/PLA yarns after five rinsing cycles. Electrical conductivity of the conductive yarns was considerably decreased every after rinsing cycle.

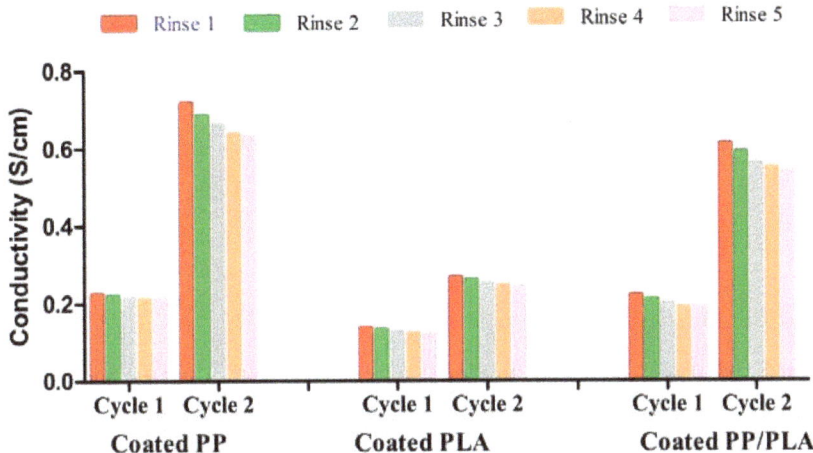

Figure 14. Conductivity (S/cm) of PEDOT:PSS coated PP, PLA and PP/PLA yarns after rinsing.

Before rinsing the conductivity for dip coating cycle 1 and cycle 2 of conductive PP yarn was 0.25 S/cm and 0.75 S/cm respectively. After 5th rinsing of conductive PP yarn, the conductivity for dip coating cycle 1 and cycle 2 is 0.213 S/cm and 0.632 S/cm respectively. The decreased conductivity of PP yarn for cycle 1 and cycle 2 is 14.8% and 15.73% respectively. Similarly before rinsing the conductivity for dip coating cycle 1 and cycle 2 of conductive PLA yarn was 0.17 S/cm and 0.36 S/cm respectively. After 5th rinsing of conductive PLA yarn, the conductivity for dip coating cycle 1 and cycle 2 is 0.12 S/cm and 0.241 S/cm, respectively. Therefore the decreased conductivity of PLA yarn for dip coating cycle 1 and cycle 2 is 29.41% and 33.06% respectively. Before rinsing the conductivity for dip coating cycle 1 and cycle 2 of conductive composite (PP/PLA)yarn was 0.24 S/cm and 0.67 S/cm respectively. After 5th rinsing of conductive PP/PLA yarn, the conductivity for dip coating cycle 1 and cycle 2 of conductive PP/PLA yarn is 0.19 S/cm and 0.54 S/cm respectively. The decreased conductivity of blend PP/PLA yarn for dip coating cycle 1 and cycle 2 is 20.83% and 19.40%, respectively. This decreased electrical conductivity could be due to the removal of excess unfix PEDOT:PSS on the yarn surface. Here conductive PP yarn showed better cleaning stability compared to conductive PLA and composite PP/PLA yarn.

4. Conclusions

A new class of smart interactive textiles (*i*-textiles) is being designed to develop new strategies toward smart materials for innovative applications in the various fields including public safety, healthcare, artificial muscles, military, strain sensing, space exploration, stretchable displays, sports, and consumer fitness. This manuscript detailing of it study results has demonstrated to construct highly stretchable, cost effective, durable, and environmentally friendly melt spun thermoplastic conductive yarns with excellent thermal and mechanical properties. Here we have introduced mussel-inspired polydodapine (PDA) treatment to modify the surface of the melt spun thermoplastic yarns. This PDA treatment acts not only as a coupling or bonding agent but also as plasticizer. This dual characteristics illustrate significant improvement of surface properties of the thermoplastic yarns. These PDA treated thermoplastic yarns consisting of PP, PLA, and PP/PLA that were effectively coated with PEDOT:PSS toward increasing the efficacy of wearable textile sensors. Mechanically the conductive PP, PLA, and PP/PLA yarns were highly stretchable and flexible. These highly stretchable and flexible conductive yarns can be used for producing conductive textiles by knitting and also can be integrated into any textile substrates/fabrics by sewing. The usage of these conductive yarns might be applied to analyze sporting performance and heart beat of a sportsman, tiny joint movement of human body, the health record of patients, speaking, swallowing and breathing.

Author Contributions: Conceptualization, G.M.N.I.; methodology, G.M.N.I.; software, G.M.N.I.; validation, G.M.N.I.; formal analysis, G.M.N.I.; investigation, G.M.N.I. and M.Q.; resources, G.M.N.I.; M.A.A. and S.C.; data curation, G.M.N.I. and M.Q.; writing—original draft preparation, G.M.N.I.; writing—review and editing, all authors; visualization, G.M.N.I.; supervision, M.A.A. and S.C.; project administration, M.A.A. and S.C. All authors have read and agreed to the published version of the manuscript.

Funding: This research received no external funding.

Acknowledgments: The authors would like to acknowledge the financial support from Home Science Alumnae/Todhunter/Carpenter Scholarship, University of Otago, Dunedin, New Zealand.

Conflicts of Interest: The authors declare no conflict of interest.

References

1. Islam, G.M.N.; Ali, A.; Collie, S. Textile sensors for wearable applications: A comprehensive review. *Cellulose* **2020**, *27*, 6103–6131. [CrossRef]
2. Cheng, Y.; Zhu, T.; Li, S.; Huang, J.; Mao, J.; Yang, H.; Gao, S.; Chen, Z.; Lai, Y. A novel strategy for fabricating robust superhydrophobic fabrics by environmentally-friendly enzyme etching. *Chem. Eng. J.* **2019**, *355*, 290–298. [CrossRef]
3. Zhang, Y.; Lin, Z.; Huang, X.; You, X.; Ye, J.; Wu, H. A Large area, stretchable, textile based tactile sensor. *Adv. Mater. Technol.* **2020**, *5*, 1901060–1901069. [CrossRef]
4. Fan, X.; Zhan, Y.; Xu, H.; Hou, Z.; Liu, X.; Riedel, R. Highly flexible, light-weight and mechanically enhanced (Mo2C/PyC)f fabrics for efficient electromagnetic interference shielding. *Compos. Part A Appl. Sci. Manuf.* **2020**, *136*, 105955–105963. [CrossRef]
5. Islam, G.M.N.; Ke, G. Ultrasonic effects on the kinetics and thermodynamics of dyeing wool fiber with reactive dye. *Fibers Polym.* **2020**, *21*, 1071–1077. [CrossRef]
6. Xu, T.; Zhang, Z.; Qu, L. Graphene based fibers: Recent advances in preparation and application. *Adv. Mater.* **2020**, *32*, 1901979–1901994. [CrossRef] [PubMed]
7. Cui, Y.; Zhang, M.; Li, J.; Luo, H.; Zhang, X.; Fu, Z. WSMS: Wearable stress monitoring system based on IoT multi-sensor platform for living sheep transportation. *Electronics* **2019**, *8*, 441. [CrossRef]
8. Kayser, L.V.; Lipomi, D.J. Stretchable conductive polymers and composites based on PEDOT and PEDOT:PSS. *Adv. Mater.* **2019**, *31*, 1806133–1806145. [CrossRef]
9. Mohan Bhasney, S.; Kumar, A.; Katiyar, V. Microcrystalline cellulose, polylactic acid and polypropylene biocomposites and its morphological, mechanical, thermal and rheological properties. *Compos. Part B Eng.* **2020**, *184*, 107717–107731. [CrossRef]
10. Schwarz, P.A.; Obermann, M.; Weber, M.; Ehrmann, A. Smarten up garments through knitting. In Proceedings of the 48th Conference of the International Federation of Knitting Technologists (IFKT), IOP Conference Series, Moenchengladbach, Germany, 8–11 June 2016; pp. 1–8.
11. Kong, X.; Zhu, C.; Lv, J.; Zhang, J.; Feng, J. Robust fluorine-free superhydrophobic coating on polyester fabrics by spraying commercial adhesive and hydrophobic fumed SiO_2 nanoparticles. *Prog. Org. Coat.* **2020**, *138*, 105342–105351. [CrossRef]
12. Sadi, M.S.; Pan, J.; Xu, A.; Cheng, D.; Cai, G.; Wang, X. Direct dip-coating of carbon nanotubes onto polydopamine-templated cotton fabrics for wearable applications. *Cellulose* **2019**, *26*, 7569–7579. [CrossRef]

13. Baribina, N.; Baltina, I.; Oks, A. Application of additional coating for conductive yarns protection against washing. *Key Eng. Mater.* **2018**, *762*, 396–401. [CrossRef]
14. Zhang, C.; Zhou, G.; Rao, W.; Fan, L.; Xu, W.; Xu, J. A simple method of fabricating nickel-coated cotton fabrics for wearable strain sensor. *Cellulose* **2018**, *25*, 4859–4870. [CrossRef]
15. Li, W.; Xu, F.; Liu, W.; Gao, Y.; Zhang, K.; Zhang, X.; Qiu, Y. Flexible strain sensor based on aerogel spun carbon nanotube yarn with a core-sheath structure. *Compos. Part A Appl. Sci. Manuf.* **2018**, *108*, 107–113. [CrossRef]
16. Memarian, F.; Rahmani, S.; Yousefzadeh, M.; Latifi, M. Wearable Technologies in Sportswear. In *Materials in Sports Equipment*; Elsevier: Amsterdam, The Netherlands, 2019; pp. 123–160. [CrossRef]
17. Karim, N.; Afroj, S.; Malandraki, A.; Butterworth, S.; Beach, C.; Rigout, M.; Novoselov, K.S.; Casson, A.J.; Yeates, S.G. All inkjet-printed graphene based conductive patterns for wearable e-textile applications. *J. Mater. Chem. C* **2017**, *5*, 11640–11648. [CrossRef]
18. Aslam, S.; Bokhari, T.H.; Anwar, T.; Khan, U.; Nairan, A.; Khan, K. Graphene oxide coated graphene foam based chemical sensor. *Mater. Lett.* **2019**, *235*, 66–70. [CrossRef]
19. Wu, B.; Xu, P.; Yang, W.; Hoch, M.; Dong, W.; Chen, M.; Bai, H.; Ma, P. Super toughened heat resistant poly(lactic acid) alloys by tailoring the phase morphology and the crystallization behaviors. *J. Polym. Sci.* **2020**, *58*, 500–509. [CrossRef]
20. Bouchart, F.; Vidal, O.; Lacroix, J.M.; Spriet, C.; Chamary, S.; Brutel, A.; Hornez, J.C. 3D printed bioceramic for phage therapy against bone nosocomial infections. *Mater. Sci. Eng. C Mater. Biol. Appl.* **2020**, *111*, 110840–110852. [CrossRef]
21. Chen, W.; Xu, Y.; Li, Y.; Jia, L.; Mo, X.; Jiang, G.; Zhou, G. 3D printing electrospinning fiber-reinforced decellularized extracellular matrix for cartilage regeneration. *Chem. Eng. J.* **2020**, *382*, 122986–122995. [CrossRef]
22. Ueda, M.; Kishimoto, S.; Yamawaki, M.; Matsuzaki, R.; Todoroki, A.; Hirano, Y.; Le Duigou, A. 3D compaction printing of a continuous carbon fiber reinforced thermoplastic. *Compos. Part A Appl. Sci. Manuf.* **2020**, *137*, 105985–105993. [CrossRef]
23. Sarker, F.; Potluri, P.; Afroj, S.; Koncherry, V.; Novoselov, K.S.; Karim, N. Ultrahigh performance of nanoengineered graphene based natural jute fiber composites. *ACS Appl. Mater. Interfaces* **2019**, *11*, 21166–21176. [CrossRef]
24. Afroj, S.; Tan, S.; Abdelkader, A.M.; Novoselov, K.S.; Karim, N. Highly conductive, scalable and machine washable graphene-based E-textiles for multifunctional wearable electronic applications. *Adv. Funct. Mater.* **2020**, *30*, 2000293–2000302. [CrossRef]
25. Afroj, S.; Karim, N.; Wang, Z.; Tan, S.; He, P.; Holwill, M.; Ghazaryan, D.; Fernando, A.; Novoselov, K.S. Engineering graphene flakes for wearable textile sensors via highly scalable and ultrafast yarn dyeing technique. *ACS Nano* **2019**, *13*, 3847–3857. [CrossRef] [PubMed]
26. Su, S.; Kang, P.M. Systemic review of biodegradable nanomaterials in nanomedicine. *Nanomaterials* **2020**, *10*, 656. [CrossRef] [PubMed]
27. Liu, H.; Jian, R.; Chen, H.; Tian, X.; Sun, C.; Zhu, J.; Yang, Z.; Sun, J.; Wang, C. Application of biodegradable and biocompatible nanocomposites in electronics: Current status and future directions. *Nanomaterials* **2019**, *9*, 950. [CrossRef]
28. Sonseca, A.; Madani, S.; Muñoz Bonilla, A.; Fernández García, M.; Peponi, L.; Leonés, A.; Rodríguez, G.; Echeverría, C.; López, D. Biodegradable and antimicrobial PLA–OLA blends containing chitosan-mediated silver nanoparticles with shape memory properties for potential medical applications. *Nanomaterials* **2020**, *10*, 1065. [CrossRef]
29. Huang, T.; Kwan, I.; Li, K.D.; Ek, M. Effect of cellulose oxalate as cellulosic reinforcement in ternary composites of polypropylene/maleated polypropylene/cellulose. *Compos. Part A Appl. Sci. Manuf.* **2020**, *134*, 105894–105904. [CrossRef]
30. Mandal, D.K.; Bhunia, H.; Bajpai, P.K. Thermal degradation kinetics of PP/PLA nanocomposite blends. *J. Thermoplast. Compos. Mater.* **2019**, *32*, 1714–1730. [CrossRef]
31. Zakaria, M.; Nakane, K. Fabrication of polypropylene nanofibers from polypropylene/polyvinyl butyral blend films using laser assisted melt-electrospinning. *Polym. Eng. Sci.* **2020**, *60*, 362–370. [CrossRef]

32. Zengwen, C.; Lu, Z.; Pan, H.; Bian, J.; Han, L.; Zhang, H.; Dong, L.; Yang, Y. Structuring poly (lactic acid) film with excellent tensile toughness through extrusion blow molding. *Polymer* **2020**, *187*, 122091–122098. [CrossRef]
33. Muthuraj, R.; Misra, M.; Mohanty, A.K. Biodegradable compatibilized polymer blends for packaging applications: A literature review. *J. Appl. Polym. Sci.* **2018**, *135*, 45726–45760. [CrossRef]
34. Cui, X.; Honda, T.; Asoh, T.A.; Uyama, H. Cellulose modified by citric acid reinforced polypropylene resin as fillers. *Carbohydr. Polym.* **2020**, *230*, 115662–115670. [CrossRef] [PubMed]
35. Pivsa Art, S.; Kord Sa Ard, J.; Pivsa Art, W.; Wongpajan, R.; O-Charoen, N.; Pavasupree, S.; Hamada, H. Effect of Compatibilizer on PLA/PP Blend for Injection Molding. *Energy Procedia* **2016**, *89*, 353–360. [CrossRef]
36. Mohamad, I.N.; Rohani, R.; Masdar, M.S.; Nor, M.; Tusirin, M.; Jamaliah, M.J. Permeation properties of polymeric membranes for biohydrogen purification. *Int. J. Hydrogen Energy* **2016**, *41*, 4474–4488. [CrossRef]
37. Azizi, S.; Azizi, M.; Sabetzadeh, M. The role of multiwalled carbon nanotubes in the mechanical, thermal, rheological and electrical properties of PP/PLA/MWCNTs nanocomposites. *J. Compos. Sci.* **2019**, *3*, 64. [CrossRef]
38. Lynge, M.E.; van der Westen, R.; Postma, A.; Städler, B. Polydopamine—A nature-inspired polymer coating for biomedical science. *Nanoscale* **2011**, *3*, 4916–4928. [CrossRef]
39. El Yakhlifi, S.; Ball, V. Polydopamine as a stable and functional nanomaterial. *Colloids Surf. B Biointerfaces* **2020**, *186*, 110719–110726. [CrossRef]
40. Hu, H.; Yu, B.; Ye, Q.; Gu, Y.; Zhou, F. Modification of carbon nanotubes with a nanothin polydopamine layer and polydimethylamino-ethyl methacrylate brushes. *Carbon* **2010**, *48*, 2347–2353. [CrossRef]
41. Cherenack, K.; Zysset, C.; Kinkeldei, T.; Münzenrieder, N.; Tröster, G. Woven electronic fibers with sensing and display functions for smart textiles. *Adv. Mater.* **2010**, *22*, 5178–5182. [CrossRef]
42. Bontapalle, S.; Varughese, S. Understanding the mechanism of ageing and a method to improve the ageing resistance of conducting PEDOT:PSS films. *Polym. Degrad. Stab.* **2020**, *171*, 1–11. [CrossRef]
43. Lund, A.; van der Velden, N.M.; Persson, N.-K.; Hamedi, M.M.; Müller, C. Electrically conducting fibres for e-textiles: An open playground for conjugated polymers and carbon nanomaterials. *Mater. Sci. Eng. R Rep.* **2018**, *126*, 1–29. [CrossRef]
44. Kim, S.M.; Kim, C.H.; Kim, Y.; Kim, N.; Lee, W.J.; Lee, E.H.; Kim, D.; Park, S.; Lee, K.; Rivnay, J.; et al. Influence of PEDOT:PSS crystallinity and composition on electrochemical transistor performance and long-term stability. *Nat. Commun.* **2018**, *9*, 3858–3866. [CrossRef] [PubMed]
45. Yin, F.; Ye, D.; Zhu, C.; Qiu, L.; Huang, Y. Stretchable, highly durable ternary nanocomposite strain sensor for structural health monitoring of flexible aircraft. *Sensors* **2017**, *17*, 2677. [CrossRef] [PubMed]
46. Wang, J.; Wang, J.; Kong, Z.; Lv, K.; Teng, C.; Zhu, Y. Conducting polymer based materials for electrochemical energy conversion and storage. *Adv. Mater.* **2017**, *29*, 1703044–1703054. [CrossRef]
47. Naveen, M.H.; Gurudatt, N.G.; Shim, Y.-B. Applications of conducting polymer composites to electrochemical sensors: A review. *Appl. Mater. Today* **2017**, *9*, 419–433. [CrossRef]
48. Le, T.H.; Kim, Y.; Yoon, H. Electrical and electrochemical properties of conducting polymers. *Polymers* **2017**, *9*, 150. [CrossRef]
49. Eom, J.; Jaisutti, R.; Lee, H.; Lee, W.; Heo, J.S.; Lee, J.Y.; Park, S.K.; Kim, Y.H. Highly sensitive textile strain sensors and wireless user interface devices using all polymeric conducting fibers. *ACS Appl. Mater. Interfaces* **2017**, *9*, 10190–10197. [CrossRef]
50. Allison, L.; Hoxie, S.; Andrew, T.L. Towards seamlessly-integrated textile electronics: Methods to coat fabrics and fibers with conducting polymers for electronic applications. *Chem. Commun.* **2017**, *53*, 7182–7193. [CrossRef]
51. Gao, Q.; Wang, M.; Kang, X.; Zhu, C.; Ge, M. Continuous wet-spinning of flexible and water-stable conductive PEDOT: PSS/PVA composite fibers for wearable sensors. *Compos. Commun.* **2020**, *17*, 134–140. [CrossRef]
52. Reid, D.O.; Smith, R.E.; Garcia Torres, J.; Watts, J.F.; Crean, C. Solvent Treatment of wet-spun PEDOT: PSS fibers for fiber-based wearable pH sensing. *Sensors* **2019**, *19*, 4213. [CrossRef]
53. Shan, Y.; Xu, C.; Zhang, H.; Chen, H.; Bilal, M.; Niu, S.; Cao, L.; Huang, Q. Polydopamine modified metal–organic frameworks, NH2-Fe-MIL-101, as pH-sensitive nanocarriers for controlled pesticide release. *Nanomaterials* **2020**, *10*, 2000. [CrossRef] [PubMed]

54. Moraes, M.R.; Alves, A.C.; Toptan, F.; Martins, M.S.; Vieira, E.M.F.; Paleo, A.J.; Souto, A.P.; Santos, W.L.F.; Esteves, M.F.; Zille, A. Glycerol/PEDOT:PSS coated woven fabric as a flexible heating element on textiles. *J. Mater. Chem. C* **2017**, *5*, 3807–3822. [CrossRef]
55. Cai, G.; Yang, M.; Xu, Z.; Liu, J.; Tang, B.; Wang, X. Flexible and wearable strain sensing fabrics. *Chem. Eng. J.* **2017**, *325*, 396–403. [CrossRef]
56. Chen, H.; Lv, L.; Zhang, J.; Zhang, S.; Xu, P.; Li, C.; Zhang, Z.; Li, Y.; Xu, Y.; Wang, J. Enhanced stretchable and sensitive strain sensor via controlled strain distribution. *Nanomaterials* **2020**, *10*, 218. [CrossRef] [PubMed]
57. Chen, X.; An, J.; Cai, G.; Zhang, J.; Chen, W.; Dong, X.; Zhu, L.; Tang, B.; Wang, J.; Wang, X. Environmentally friendly flexible strain sensor from waste cotton fabrics and natural rubber latex. *Polymers* **2019**, *11*, 404. [CrossRef]
58. Tadesse, M.G.; Mengistie, D.A.; Chen, Y.; Wang, L.; Loghin, C.; Nierstrasz, V. Electrically conductive highly elastic polyamide/lycra fabric treated with PEDOT:PSS and polyurethane. *J. Mater. Sci.* **2019**, *54*, 9591–9602. [CrossRef]
59. Strååt, M.; Toll, S.; Boldizar, A.; Rigdahl, M.; Hagström, B. Melt spinning of conducting polymeric composites containing carbonaceous fillers. *J. Appl. Polym. Sci.* **2011**, *119*, 3264–3272. [CrossRef]
60. Soroudi, A.; Skrifvars, M. Melt blending of carbon nanotubes/polyaniline/polypropylene compounds and their melt spinning to conductive fibres. *Synth. Met.* **2010**, *160*, 1143–1147. [CrossRef]
61. Wang, X.Y.; Feng, G.Y.; Li, M.J.; Ge, M.Q. Effect of PEDOT:PSS content on structure and properties of PEDOT:PSS/poly(vinyl alcohol) composite fiber. *Polym. Bull.* **2018**, *76*, 2097–2111. [CrossRef]
62. Hassan, M.M.; Tucker, N.; Le Guen, M.J. Thermal, mechanical and viscoelastic properties of citric acid-crosslinked starch/cellulose composite foams. *Carbohydr. Polym.* **2020**, *230*, 115676–115685. [CrossRef]
63. Li, L.; Li, X.; Du, M.; Guo, Y.; Li, Y.; Li, H.; Yang, Y.; Alam, F.E.; Lin, C.-T.; Fang, Y. Solid-phase coalescence of electrochemically exfoliated graphene flakes into a continuous film on copper. *Chem. Mater.* **2016**, *28*, 3360–3366. [CrossRef]
64. Ploypetchara, N.; Suppakul, P.; Atong, D.; Pechyen, C. Blend of polypropylene/poly(lactic acid) for medical packaging application: Physicochemical, thermal, mechanical, and barrier properties. *Energy Procedia* **2014**, *56*, 201–210. [CrossRef]
65. Boroumand, Y.; Razmjou, A.; Moazzam, P.; Mohagheghian, F.; Eshaghi, G.; Etemadifar, Z.; Asadnia, M.; Shafiei, R. Mussel inspired bacterial denitrification of water using fractal patterns of polydopamine. *J. Water Process Eng.* **2020**, *33*, 101105–101115. [CrossRef]
66. Chang, H.C.; Sun, T.; Sultana, N.; Lim, M.M.; Khan, T.H.; Ismail, A.F. Conductive PEDOT:PSS coated polylactide (PLA) and poly(3-hydroxybutyrate-co-3-hydroxyvalerate) (PHBV) electrospun membranes: Fabrication and characterization. *Mater. Sci. Eng. C Mater. Biol. Appl.* **2016**, *61*, 396–410. [CrossRef]
67. Luo, R.; Li, H.; Du, B.; Zhou, S.; Zhu, Y. A simple strategy for high stretchable, flexible and conductive polymer films based on PEDOT:PSS-PDMS blends. *Org. Electron.* **2020**, *76*, 105451–105458. [CrossRef]
68. Gholampour, N.; Brian, D.; Eslamian, M. Tailoring characteristics of PEDOT:PSS coated on glass and plastics by ultrasonic substrate vibration post treatment. *Coatings* **2018**, *8*, 337. [CrossRef]
69. Mofokeng, J.P.; Luyt, A.S.; Tábi, T.; Kovács, J. Comparison of injection moulded, natural fibre-reinforced composites with PP and PLA as matrices. *J. Thermoplast. Compos. Mater.* **2011**, *25*, 927–948. [CrossRef]
70. Kowalczyk, M.; Piorkowska, E.; Kulpinski, P.; Pracella, M. Mechanical and thermal properties of PLA composites with cellulose nanofibers and standard size fibers. *Compos. Part A Appl. Sci. Manuf.* **2011**, *42*, 1509–1514. [CrossRef]

Publisher's Note: MDPI stays neutral with regard to jurisdictional claims in published maps and institutional affiliations.

© 2020 by the authors. Licensee MDPI, Basel, Switzerland. This article is an open access article distributed under the terms and conditions of the Creative Commons Attribution (CC BY) license (http://creativecommons.org/licenses/by/4.0/).

Article

A Way to Predict Gold Nanoparticles/Polymer Hybrid Microgel Agglomeration Based on Rheological Studies

Coro Echeverría * and Carmen Mijangos

Institute of Polymer Science and Technology (ICTP-CSIC), C/Juan de la Cierva 3, 28006 Madrid, Spain; cmijangos@ictp.csic.es
* Correspondence: cecheverria@ictp.csic.es

Received: 10 September 2019; Accepted: 11 October 2019; Published: 21 October 2019

Abstract: In this work, a detailed rheological study of hybrid poly(acrylamide-*co*-acrylic acid) P(AAm-*co*-AAc) aqueous microgel dispersions is performed. Our intention is to understand how the presence of gold nanoparticles, AuNP, embedded within the microgel matrix, affects the viscoelastic properties, the colloidal gel structure formation, and the structure recovery after cessation of the deformation of the aqueous microgel dispersions. Frequency sweep experiments confirmed that hybrid microgel dispersions present a gel-like behavior and that the presence of AuNP content within microgel matrix contributes to the elasticity of the microgel dispersions. Strain sweep test confirmed that hybrid microgels aqueous dispersion also form colloidal gel structures that break upon deformation but that can be recovered when the deformation decreases. The fractal analysis performed to hybrid microgels, by applying Shih et al. and Wu and Morbidelli's scaling theories, evidenced that AuNP significantly affects the colloidal gel structure configuration ending up with the formation of agglomerates or microgel clusters with closer structures in comparison to the reference P(AAm-*co*-AAc) aqueous microgel dispersions.

Keywords: polymer microgels; hybrid microgels; thermoresponsive; rheology; scaling theory; fractal analysis

---

1. Introduction

Among all colloidal systems, the sub micrometer-sized hydrogel (microgels) particles are of special interest [1,2]. After decades of research, polymeric microgels have revealed their versatility from both the functionality (responsiveness) and applications perspective. These smart materials have received much attention owing to their environmentally tunable sizes and potential applications, such as chemical separation, catalysis, sensors, enzyme immobilization, drug delivery systems, biomimicking artificial synovial fluids, tissue mimicking, and injectable 3D cell scaffolds, among others [3–8]. Besides the mentioned advanced applications, microgels are used as building blocks to create structures such as colloidal crystals, films, and gels in the macroscopic scale [9–12], and more recently as active sites confined within electrospun polymer fibers toward the design of tailored multifunctional stimuli-responsive advanced materials [13–16]. In our previous works, we first developed poly(acrylamide-*co*-acrylic acid) microgels featured with the ability to swell upon heating, thus showing a positive thermosensitivity and an upper critical solution temperature-like (UCST) volume phase transition temperature [17,18]. This thermo-responsiveness derived from the presence of acrylic acid that forms hydrogen bonds with acrylamide. The obtained UCST-like microgels were characterized in terms size, shape, and thermoresponsiveness, so that the effect of the composition–crosslinking degree and acrylic acid comonomer content could be understood. Moreover, rheological behavior of aqueous poly (acrylamide-acrylic acid) microgel dispersions was also studied from the perspective of a colloidal system. This study outlined their macroscopic elasticity showing that the material behaves as a colloidal gel [17,18].

As deduced from the literature, the evolution of microgel systems advanced toward the development of hybrid systems (organic/inorganic systems). In fact, since the first work of Antonietti et al. [19], in which microgels were used as microreactor and "exo-templates" for the controlled growth of gold nanoparticles, the development of hybrid microgels has increased significantly [20–28]. This interest is related to the fact that the incorporation of inorganic nanoparticles into polymeric microgels provided additional functionalities to the final system [29,30]. For instance, the incorporation of gold nanoparticles to thermoresponsive microgels would provide optical properties and thus dual-stimuli responsiveness; the system could be remotely activated (swollen) via light [23].

The intention with this work, besides understanding the effect of adding a nanoparticle into a nano/microgel in the flow behavior, is to propose our approach as a tool to control/predict the formation, breakage, and reformation of agglomerates in colloidal dispersion that might influence their final applications, in particular as drug delivery systems, carriers, or similar. For such purpose, in the present work a detailed rheological study of hybrid P(AAm-co-AAc) microgel dispersions is presented. Thus, the aim of this research is to understand how the presence of gold nanoparticles AuNP, that are embedded within the P(AAm-co-AAc) microgel matrix, could modify the viscoelastic properties of the aqueous microgel dispersions and their macroscopic elasticity. In previous studies, we could determine that P(AAm-co-AAc) aqueous microgels dispersion present a gel-like behavior associated not only to the gel nature of the polymeric microgels themselves, but also to the formation of certain structure due to interactions occurring between microgel particles within the dispersion. Therefore, in this work we will determine if the presence of AuNP within microgels could modify such interactions and hence the described gel-like behavior. To do so, we used two scaling models (Shih et al. [31] and Wu and Morbidelli [32]) to perform a fractal analysis of the hybrid aqueous microgel dispersions.

2. Materials and Methods

2.1. Materials

As monomer and comonomer we used acrylamide (AA, 99% pure, Sigma-Aldrich, St. Quentin Fallavier, France) and acrylic acid (AAc, Sigma-Aldrich), respectively. To obtain crosslinked microgels N,N'-methylenebisacrylamide (MBA, Sigma-Aldrich, 99.5% pure) was used as crosslinking agent. The reaction was initiated using 2,2'-azobis(2-methylpropionamidine)dihydrochloride (AMPA-d, Sigma-Aldrich, 97% pure). As part of the organic phase, span 80 (sorbitan monooleate) (Fluka, Saint Louis, MO, USA) and dodecane (Fluka, 99% pure) were used as surfactant and organic solvent, respectively. Gold nanoparticles (AuNP) used for their encapsulation into microgel matrix were purchased from Nanogap, which are already covered with poly(n-vinyl-2-pyrrolidone) so that they can be stable in water dispersion. According to Nanogap, AuNP contain 16% gold and the size is approximately 5 ± 1 nm (See Figure S2). All the water used in the preparation and characterization of microgels was Millipore Milli Q grade.

2.2. Synthesis of Hybrid P(AAm-AAc)-AuNP Microgels

Microgel synthesis was performed by means of inverse emulsion polymerization (w/o) method as described in [27]. The aqueous phase of the emulsion is formed by acrylamide monomer, acrylic acid comonomer, crosslinker, water dispersed gold nanoparticles, and distilled water (See Table 1 for details). For the oil phase, emulsifier (SPAN 80) and the organic solvent (dodecane) were mixed. Prior to the reaction, both solutions were purged with nitrogen during 30 min. Then, we incorporated the organic phase into a three-necked round bottom flask. Over this organic phase we added the aqueous phase solution by means of a peristaltic pump with a feeding rate of 1.5 mL/min while the forming emulsion was mechanically stirred at 475 r.p.m. This procedure derived in the formation of aqueous phase droplets dispersed in the organic phase. These droplets act as reservoir where the radical polymerization reaction occurs. Finally, the polymerization was thermally initiated using a 2,2′Azo-bis-(2-methylpropionamidine-dihydrochloride (AMPA-d) solution. (The reaction took place at 50 °C, which is the decomposition temperature of the initiator.) Immediately after adding the initiator, the emulsion became turbid. From this stage, the polymerization reaction was allowed to continue for 3 h under nitrogen atmosphere. After this period of time, the reaction was cooled down to room temperature while the stirring and nitrogen flow was maintained to avoid aggregation. Finally, all the prepared emulsions were purified by removal of organic phase by decantation and the remaining aqueous phase was further precipitated in ethanol with subsequent washing by centrifugation at 4500 r.p.m. All samples were redispersed in deionized water and placed in dialysis bags (molecular weight cut off = 3500) for 1 week to remove any unreacted materials. For the sake of clarity, in Table 1 the recipe and reaction conditions for the hybrid microgel synthesis are described.

Table 1. Recipe for the synthesis of the hybrid microgels

	AQUEOUS PHASE				ORGANIC PHASE	
	Acrylamide (Mol)	Acrylic Acid (Mol)	MBA (Mol)	AuNP (%wt)	Dodecane (mL)	Span 80 (g)
P(AAm-co-AAc)	0.056	2.77×10^{-4}	1.29×10^{-3}	0	30	0.5056
P(AAm-co-AAc)–5% AuNP	0.056	2.77×10^{-4}	1.29×10^{-3}	5	30	0.5056
P(AAm-co-AAc)–10% AuNP	0.056	2.77×10^{-4}	1.29×10^{-3}	10	30	0.5056

The volume of water in the aqueous phase is 10 mL. 1 wt% of AMPA-d intitiator was used.

2.3. Characterization Methods

The morphological analysis of the microgels was performed by scanning electron microscopy, SEM (ESEM, XL30, Philips, North Billerica, MA, USA) to determine the shape, size, and dispersion of microgels. Transmission Electron Microscopy, TEM (JEM 3000F, 300kv, JEOL, Tokyo, Japan) was used to confirm the presence and distribution of AuNP within microgel matrix. For this analysis, dried microgels were further dispersed in acetone (We used acetone in order to promote a rapid evaporation of the solvent.). A drop of the dispersion was deposited in a glass wafer, waited till the solvent evaporates, and sputter-coated with gold to minimize charging at fixed conditions for SEM analysis. For TEM analysis a drop of the dispersion was deposited in formvar/carbon-coated grids.

For the rheological study of the hybrid microgel dispersions ARG2 (TA Instrument, New Castle, DE, USA) stress-controlled rheometer was used. First, we determined the linear viscoelastic range by performing strain sweep tests at a constant and non-destructive frequency of 0.5 Hz for hybrid microgel aqueous dispersion with concentrations of 1, 2, and 5 wt%. Then, we carried out frequency sweep tests at a constant strain of 1%, within the linear viscoelastic region. All measurements were performed at 20 °C, the temperature at which microgels are in their collapsed state [27] and using a 60 mm acrylic parallel plate. The temperature sweep tests were performed, from 1 to 40 °C, at a non-destructive frequency of 0.5 Hz, and at a constant strain of 3%. The protocol used for loading the sample were the same for all the aliquots. We used a 1 mL micropipette to add the same volume. The plastic tip used for the loading was previously cut in order to avoid any unwanted deformation prior to the measurement. Microgel dispersions were squeezed the same, by selecting the same gap for all the microgel dispersions and controlling the normal force exerted to each dispersion, so that the initial state is the same for all the samples. In order to preserve or not affect structural equilibrium of the microgels dispersions, we did not perform any preshear prior to the strain sweep test. To ensure a reproducibility and to obtain an average of the determined parameters we performed a minimum of 5 measurements of each microgel dispersion. To determine the critical strain from the stain sweep tests, we selected the second point that comes out of the linearity that coincides with the >5% deviation rule. Fractal analysis was carried out by the application of two scaling theories: Shih et al. [31] and Wu and Morbidelli [32].

3. Results and Discussion

3.1. Morphology

Figure 1 shows SEM (A, B, C) micrographs corresponding to P(AAm-co-AAc) microgels and hybrid microgels containing 5% and 10% of AuNP. As observed, the synthesized P(AAm-co-AAc) microgels are of spherical shape with diameters in the range of 200 to 500 nm (Figure 1A), which is in agreement with the definition of microgel: intramolecularly crosslinked polymer particle with diameter size in the range of 100 nm to 1 µm. The incorporation of AuNP did not affect the morphology of P(AAm-co-AAc) microgels as seen in both Figure 1B,C and already envisaged in previous work [27]. The hybrid microgels are also of spherical shape with similar diameter sizes. Therefore, in terms of morphology, microgels shape and size were not affected by the addition of AuNP, probably due to the small quantity of added nanoparticles and their small size (5 nm) as well. Nevertheless, SEM technique does not allow to detect AuNP.

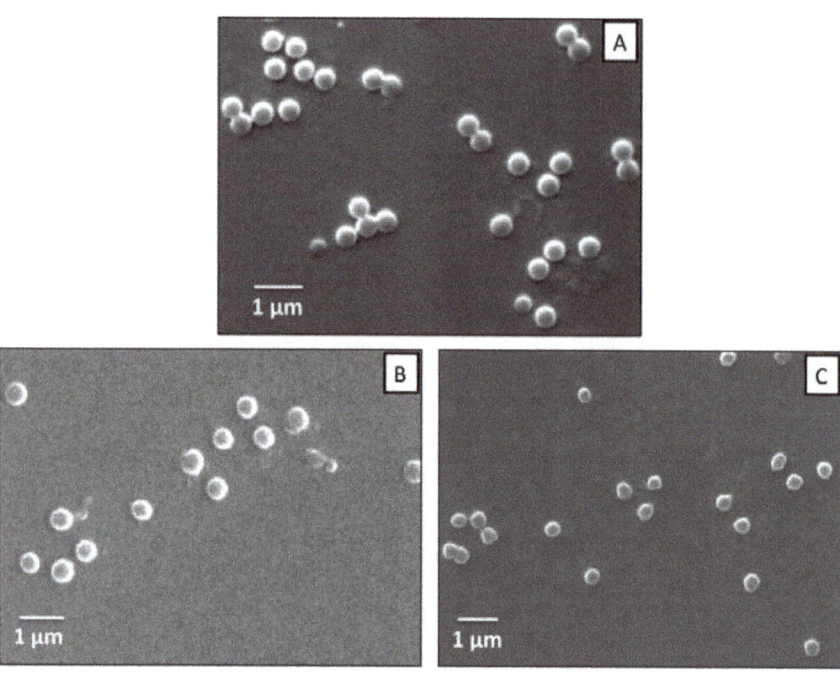

Figure 1. Representative SEM micrographs corresponding to the samples (**A**) P(AAm-co-AAc), (**B**) P(AAm-co-AAc)-5% AuNP, and (**C**) P(AAm-co-AAc)-10% AuNP.

In order to confirm the successful encapsulation of AuNP, TEM micrographs corresponding to P(AAm-co-AAc)-10% AuNP hybrid microgel sample (Figure 1D,E) are depicted in Figure 2A–D. Those micrographs, reprinted with permission of Elsevier, were taken from our previous work [27]. As shown in Figure 2A, AuNP appear as black spheres in an organic matrix. In particular, in Figure 2A it is possible to observe the atomic planes of the Au nanoparticles, besides confirming that the particles diameter is approximately 5 ± 1 nm. In Figure 2B the image of a single hybrid microgel with black spheres can be observed. Such image is not clear enough to identify whether AuNP are located inside the microgel matrix or on the surface. But if we focus on the Figure 2C,D (magnified from C), near the edge of a single microgel AuNP particles (black spheres) clearly embedded inside the microgel matrix can be observed. This result confirmed the successful encapsulation of AuNP [27]. In order to determine the final AuNP content we also performed thermogravimetric analysis. From TGA experiments we could conclude that the samples P(AAm-*co*-AAc)-5% AuNP and P(AAm-*co*-AAc)-10% AuNP have a final AuNP content of 3% and 8%, respectively. (See Figure S1 and Table S1 of supporting information).

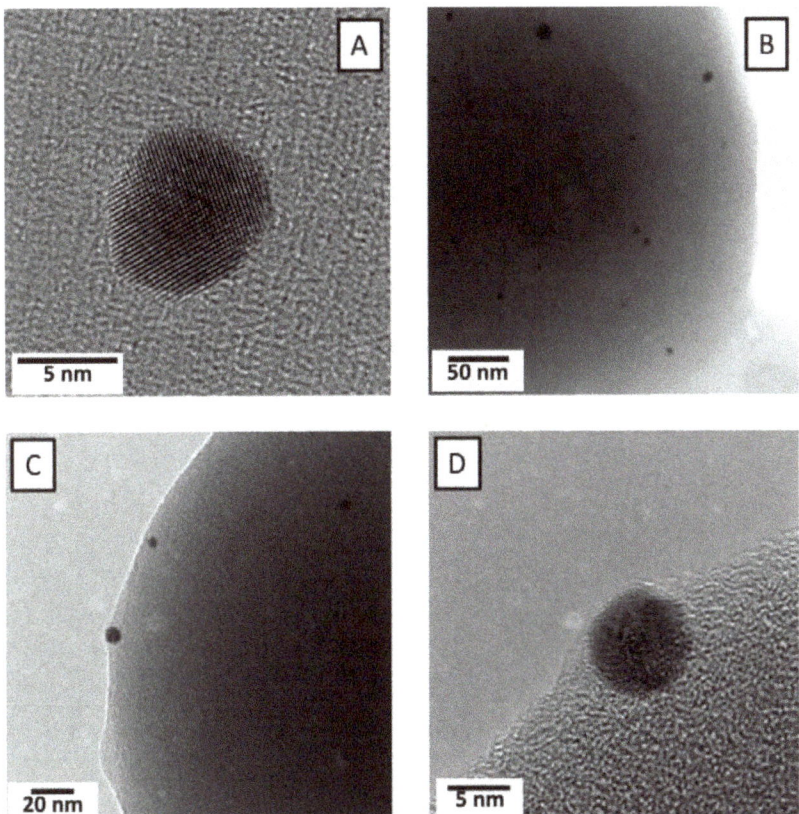

Figure 2. A to D figures show TEM images corresponding to the sample P(AAm-*co*-AAc)-10% AuNP (Adapted from [27] with permission from, Wiley-VCH, Copyright 2010).

Regarding the influence of AuNP in the swelling ability and thermo-responsiveness, in a previous work we demonstrated that the incorporation of AuNP shifted the volume phase transition temperature of P(AAm-*co*-AAc) towards temperatures close to 37 °C [27].

As mentioned in the introductory section of the manuscript, the versatility of polymeric microgels make them very useful for a wide and diverse spectrum of applications. In this particular case, we developed this hybrid dual responsive microgel system so that it could be potentially used as drug carriers for a further controlled drug release [18]. Therefore, it is crucial to understand the rheology of the microgel aqueous dispersions so that their application could not be conditioned due to undesirable or uncontrollable agglomerations that could occur when applied by injection, for instance, or when circulating in the blood stream so that blood clot could be induced. Taking this into consideration, in this work we put effort in determining the rheological properties of these hybrid microgels so that we could better understand their structure-properties-applications relationship.

3.2. Effect of the Incorporation of AuNP in the Viscoelastic Properties of Aqueous Hybrid Microgel Dispersions

Our first step in the rheological characterization of microgel dispersion was to determine their viscoelastic behavior. Figure 3A shows representative frequency sweep tests corresponding to

P(AAm-co-AAc), P(AAm-co-AAc)-5% AuNP, and P(AAm-co-AAc)-10% AuNP aqueous microgel dispersion (C = 5 wt%). As general behavior, the three microgel samples possess the elastic modulus (full symbols) higher than the viscous modulus (empty symbols) at all the frequency ranges studied. Having G' > G" indicates that microgel dispersion presents a solid-like behavior. Moreover, microgel dispersions also present an elastic modulus G' which is constant and independent of the frequency (showing a finite value at zero frequency). These are the two conditions that define a polymer gel rheologically. Therefore, these two facts together serve to affirm that microgel dispersions present a gel-like behavior as it was also the case of similar systems [17,25]. In addition to these two conditions, the three microgel samples described a minimum in the frequency dependence of G" (Figure 3A), which is characteristic of colloidal gels, as stated by Mewis and Wagner [33]. This minimum is better observed in the graph corresponding to the P(AAm-co-AAc) microgel (without AuNP).

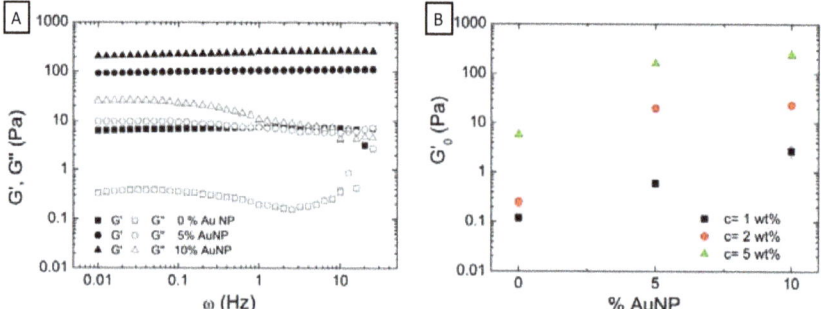

Figure 3. (**A**) Evolution of the elastic (G') and viscous modulus (G") with frequency for P(AAm-co-AAc), P(AAm-co-AAc)-5% AuNP and P(AAm-co-AAc)-10% AuNP microgel aqueous dispersions at a single concentration of 5 wt%. (From 10 to 1 Hz measurement was performed under a constant strain of 3% and from 1 to 0.01 Hz under a constant value of 1% strain). (**B**) Elastic modulus plateau (G'0) represented as a function of AuNP content, for the three different concentration of microgel dispersion: 1, 2, and 5 wt%.

For a deeper analysis of the results, in Figure 3B we have collected the elastic modulus plateau G'0 (extrapolated at zero frequency) obtained from the frequency sweep tests performed for hybrid microgel dispersions at three different concentrations (1, 2, and 5 wt%), and represented as a function of AuNP content within the microgel matrix. Two results can be drawn: First, that the elastic modulus increases as the concentration of the aqueous microgel dispersions increases, and second and probably more relevant, that the encapsulation of AuNP within the microgel matrix contributes to increase the elastic modulus of the dispersion. For instance, if we analyze the effect of AuNP content in the dispersions containing 5 wt% of microgel, it is observed that the encapsulation of 5% of AuNP within microgel matrix provokes an increase of the G' modulus one order of magnitude. When the encapsulated AuNP increases from 5% to 10%, the elastic modulus increases also the double, that is, it evolves from approximately 150 Pa (5% AuNP) to 260 Pa (10% AuNP). This result clearly indicates that AuNP act as a filler reinforcing the microgel matrix. The aqueous microgel dispersions' elastic character is increased giving rise to a stronger gel-like system with improved viscoelastic properties.

We have confirmed that the encapsulation of AuNP contributes to the elasticity of the system deriving in a stronger gel-like behavior compared to pure microgel dispersions. In previous studies we could determine that the gel-like behavior was associated not only to the gel nature of the polymeric microgels but also to the formation of certain structure (a colloidal gel) due to interactions occurring between microgel particles within the aqueous dispersion [17,25]. In the present work we have studied how hybrid microgel dispersions behave under strain (deformation) and evaluated the influence of AuNP embedded in the microgel matrix. Accordingly, in Figure 4A the evolution of G' and G" with increasing applied strain and for decreasing applied strain is evaluated for P(AAm-co-AAc) (5 wt% microgel concentration).

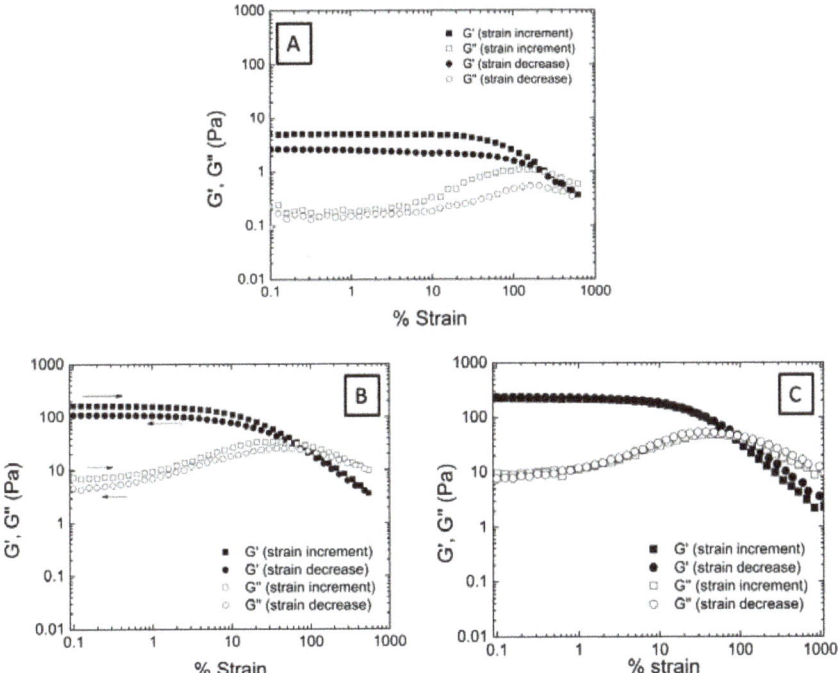

Figure 4. Evolution of G′ (full symbols) and G″ (empty symbols) as a function of % strain determined as the strain increases (square) and as the applied strain decreases to the initial state (circle) for (**A**) P(AAm-co-AAc), (**B**) P(AAm-co-AAc)-5% AuNP and (**C**) P(AAm-co-AAc)-10% AuNP microgel aqueous dispersions at 5 wt% concentration.

If we analyze the obtained graphs, at low strains both G′ and G″ keep constant being G′ > G″. This first stage defines the linear viscoelastic range (LVR). But as the strain increases, both G′ and G″ reach the limit of linearity at a critical strain value (γ0), becoming dependent on the strain. At certain strain value the modulus crossover occurs (G′ = G″). From this point on, viscous modulus is higher than the elastic modulus, which is indicative of the liquid-like behavior of the microgel dispersion. When the applied strain is reduced up to the initial value, both G′ and G″ start to increase again until recovering their independency against the strain ending up in the gel-like behavior described before. The fact that aqueous microgel dispersions are able to recover their initial gel-like behavior implies the formation of some kind of structure that breaks upon deformation that restructure when the deformation disappears [34–36]. However, the obtained absolute values of G′ and G″ are slightly lower in the upturn measurement, showing an incomplete recovery or hysteresis.

When analyzing the strain dependent behavior of the hybrid microgels shown in Figure 4B,C, a similar trend is observed. Hybrid microgel dispersions described a region in the curve at which G′ and G″ are independent of the applied strain, but as the strain increases both G′ and G″ become dependent, the cross-over point is achieved and G′, G″ end up decreasing. When the strain is reduced, hybrid microgel dispersions recover the initial G′, G″ values, and therefore, the initial gel-like behavior. Interestingly, the encapsulation of 5% of AuNP within microgel matrix derived in weakened hysteresis. But even more remarkable is the behavior described for the P(AAm-co-AAc)-10% AuNP microgel dispersions where no hysteresis is observed. This means that, P(AAm-co-AAc)-10% AuNP hybrid microgel dispersion is capable to totally recover the initial colloidal gel structure, and therefore the interactions that held the structure (cluster).

In fact, strain dependence behavior (increasing and further decreasing the applied strain) corresponding to P(AAm-co-AAc)-10% AuNP shows two superimposed curves indicating that there is no loss of elasticity; G′ and G″ recover from the imposed deformation. Therefore, hybrid microgel dispersions are capable to totally recover the interactions between microgels and thus the initial colloidal gel structure.

3.3. Fractal Analysis of Aqueous Hybrid Microgel Dispersions by Means of Shi et al. and Wu and Morbidelli Scaling Theory: Effect of AuNP in the Microgel Interactions

At this stage, from the evaluation of the viscoelastic properties we have confirmed the reinforcement role that AuNP have in the microgel. Additionally, the analysis of the hybrid microgel dispersions behavior under strain put in evidence the formation of some kind of structure that breaks upon deformation and with the ability to recover as the applied deformation decreases gradually. Previous research regarding the rheological behavior of aqueous dispersions of poly (acrylamide-acrylic acid) microgels already outlined their macroscopic elasticity showing that the material behaves as a colloidal gel [17,18]. These two results are relevant for further potential applications that might imply flow of aqueous dispersions. Indeed, shear-thinning of a colloidal suspension could enable a more homogeneous and easy delivery of the material in the case of injectable materials [37]. And if this behavior is complemented with the total recovery of the elastic properties immediately after injection/deformation, this may prevent the flow of the colloidal solution and facilitate that the material remains on the target site.

There are several studies regarding the interplay of microgel inter-particle interaction modifications by changing particle size, surface charge (functionalization), and crosslinking degree [38–42] (most of them used poly(N-isopropylacrylamide) PNIPAM microgels), but reports aimed to evaluate, control, and predict its influence on the fractal structure and cluster formation through rheology are scarce. For instance, Liao et al. [8,43] generated an in-situ formed hydrogel, using PNIPAM microgels as the building blocks to construct injectable thermal gelling scaffold for 3D cell culture, in the presence of Ca^{2+} to induce changes in the inter-particle interaction. They studied their fractal structure and concluded that both salt concentration and temperature modify the interactions among microgels [8,43]. Recently, we also attempt to modify microgels colloidal behavior through quaternization but no relevant differences were obtained [44]. Therefore, to determine the flow behavior of the hybrid microgel dispersions and to understand how hybrid microgels interact within the aqueous dispersion could be as relevant as the characterization of their responsiveness. Being so, we took advantage of the scaling theories developed by Shih et al. [31] and Wu and Morbidelli [32] and used them as a tool to perform a fractal analysis. These two scaling theories, which are an extension of the computer model proposed by Brown and Ball [45] are based on the fact that microgel dispersions are a collection of flocs–fractal objects closely packed throughout the sample-. These models were recently used to study the interactions occurring among poly(N-Isopropylacrylamide) polycationic microgels [44].

Shih et al.'s [31] models differentiated two extreme situations and unravel the intra- and inter-floc interactions in two separate regimes: strong-link regime where inter-floc interactions are stronger than intra-floc (among particles) interactions and weak-link regime where the elasticity is driven by the mechanically weaker part of the system, that is the weak links between flocs. Shih et al. described each regime based on the scaling relationship of both the elastic constant (elastic modulus plateau) and critical deformation with concentration as shown in the following equations:

(i) Strong-link regime:

$$G'_0 \sim \varphi^{(d+x)/(d-D_f)} \tag{1}$$

$$\gamma_0 \sim \varphi^{-(1+x)/(d-D_f)} \tag{2}$$

being φ the concentration, d the Euclidean dimension of the system (d = 3), D_f the fractal dimension, and x is the fractal dimension of the aggregated backbone that has to be lower than the fractal dimension and positive ($D_f > x > 0$) [31,46], being a reasonable value 1–1.3 [31].

(ii) Weak-link regime:

$$G'_0 \sim \varphi^{(d-2)/(d-D_f)} \tag{3}$$

$$\gamma_0 \sim \varphi^{1/(d-D_f)} \tag{4}$$

Wu and Morbidelli extended the Shih et al. model to the case where samples belong to a transition regime between the strong and weak link regimes. Therefore, in order to gather these two regimes in a single model, they introduced a new parameter, microscopic elastic constant, α. This parameter describes a range of values from 0 to 1, where $\alpha = 0$, corresponds to Shih et al.'s strong-link regime whereas $\alpha = 1$ matches the weak-link regime. Wu and Morbidelli defined the model with the following equations:

$$G'_0 \sim \varphi^{\beta/(d-D_f)} \tag{5}$$

$$\gamma_0 \sim \varphi^{(d-\beta-1)/(d-D_f)} \tag{6}$$

$$\beta = (d-2) + (2+x)(1-\alpha) \tag{7}$$

being β an auxiliary parameter that relates α with x.

In order to perform the fractal analysis, strain sweep tests of hybrid microgels aqueous dispersions were carried out at three different concentrations (1, 2, and 5 wt%) as shown in Figure 5: P(AAm-co-AAc) (Figure 5A), P(AAm-co-AAc)-5% AuNP (Figure 5B), and P(AAm-co-AAc)-10% AuNP (Figure 5C). As observed from the figure, the three samples describe an increase of G' and G" with the microgel concentration. However, the increase of microgel concentration lead to a decrease of the linear viscoelastic range; less strain is necessary to break the formed structure within the dispersion.

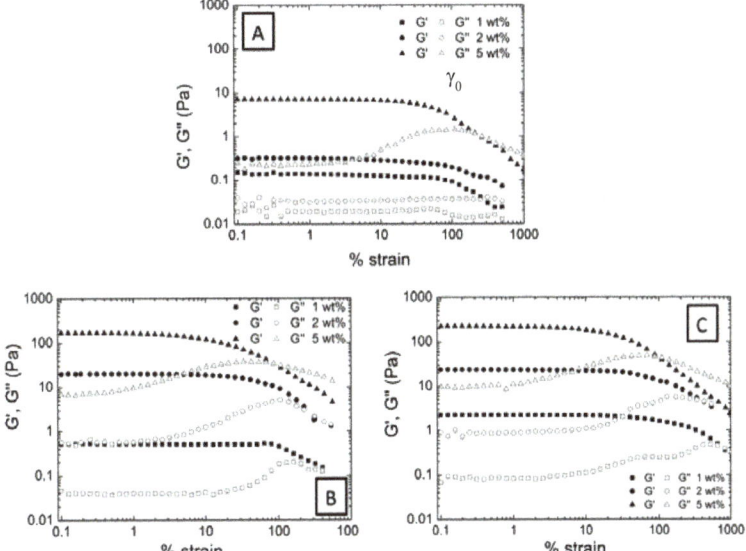

Figure 5. Strain sweep test corresponding to (**A**) P(AAm-co-AAc), (**B**) P(AAm-co-AAc)-5% AuNP, and (**C**) P(AAm-co-AAc)-10% AuNP microgel aqueous dispersions at three different concentrations: 1, 2, and 5 wt%.

We have extracted the average elastic modulus plateau G'_0 and the average critical deformation γ_0, for each aqueous microgel dispersions at each studied concentration, and represented as a function of the concentration in Figure 6A,B. As it was expected, both G'_0 and γ_0, exhibit a power law relationship with the concentration that can be fitted to the form: $G'_0 \sim C^A$ and $\gamma_0 \sim C^B$. When evaluating the concentration dependence of G'_0, positive slopes are observed for the studied microgel dispersions (as collected in Table 2). In contrast, the evolution of γ_0 exhibit negative slopes.

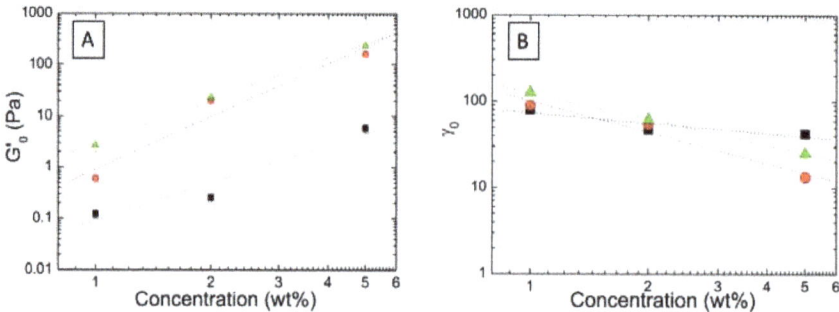

Figure 6. Double-logarithmic plot of (**A**) the elastic modulus plateau, G'_0, and (**B**) critical strain, γ_0, as a function of the aqueous microgel concentration for the samples: P(AAm-*co*-AAc) (black square), P(AAm-*co*-AAc)-5% AuNP (red circle) and P(AAm-*co0*-AAc)-10% AuNP (green triangle).

Table 2. Summary of the results obtained by applying Shi et al., and Wu and Morbidelli's fractal models.

Samples	Slopes (Figure 6A,B)		Shi et al.		Wu and Morbidelli			Regime
	A	B	Df	x	Df	β	α	
P(AAm-*co*-AAc)	2.56 ± 0.16	−0.22 ± 0.17	—	<0	2.14 ± 0.28	2.2 ± 0.5	0.63 ± 0.17	Transition (weak)
P(AAm-*co*-AAc)-5% AuNP	3.82 ± 0.78	−0.62 ± 0.22	—	<0	2.3 ± 0.7	2.3 ± 1.2	0.57 ± 0.38	Transition (weak)
P(AAm-*co*-AAc)-10% AuNP	2.29 ± 0.06	−1.18 ± 0.13	1.19	1.13	1.2 ± 0.4	4.14 ± 0.1	0.04 ± 0.03	Strong-link

According to the Shi et al. scaling model [31], the fact that γ_0 exhibit a negative slope indicates that the three systems fall into the strong-link regime, in which inter-floc interactions are stronger than intra-floc. In the case of P(AAm-*co*-AAc)-5% AuNP microgel dispersion, this strong-link regime is also evidenced in the strain hardening depicted in Figure 5B for the dispersion containing 1 wt% of microgels. Thus, fractal dimension is obtained by applying Equations (1) and (2) of the Shi et al. model and collected in Table 2. But, as shown in the table, results obtained with the Shi et al. model are only valid for the P(AAm-*co*-AAc)-10% AuNP microgel dispersion, being this sample the one that fulfill the requirements mentioned above (having an *x* value lower than the fractal dimension and positive). In the case of P(AAm-*co*-AAc)-5% AuNP microgel dispersion, although 1 wt% dispersion clearly described a strain hardening behavior, thus supporting the strong-link regime, it is observed that upon addition of more microgels, 2 and 5 wt %, dispersions turn to strain thinning (see Figure 5B), which evidenced a transition between the two regimes, strong- and weak-link. Consequently, in order to evaluate the type of interactions Wu and Morbidelli's scaling theory needed to be applied. From the applications of Equations (5)–(7) we could obtain the fractal dimension (Df) and microscopic elastic interaction regime (α) of the three studied microgel dispersions (Table 2). Some of the parameters showed significant experimental errors, in particular for P(AAm-*co*-AAc)-5% AuNP; these results would be more accurate with the study of more concentrations however, we have to indicate that our intention with this experiment was to obtain semiquantitative results for comparative purposes.

The estimation of α parameter resulted in values that evolves from 0.63 for P(AAm-*co*-AAc) to 0.04 for P(AAm-*co*-AAc)-10% AuNP, indicating an evolution from a transitioning regime toward a strong-link regime. It is worth mentioning that there is a lack of agreement between the semi-quantitative strong-link regime assessed for P(AAm-co-AAc)-10% AuNP microgel dispersion through Wu and Morbidelli's model, and the qualitative shear thinning described in Figure 5C.

The encapsulation of AuNP within the microgel matrix contributes to increase the inter-floc interactions, or in other words, hampered the formation of cluster or agglomerates. This result is also in agreement with the estimation for the fractal dimension Df that varies from 2.14 to 1.2 with the incorporation of AuNP. This remarkable decrease of the fractal dimension also indicates that the incorporation of AuNP has changed the growth/aggregation mechanism of the aqueous dispersions. Fractal dimension is an indication of the growth mechanism and stacking density of the particles forming the floc. As stated in the literature two growing mechanisms are identified: reaction-limited and diffusion-limited aggregation mechanisms. Each mechanism gives rise to different agglomerates structures and thus, to different fractal dimensions. In the case of the reaction-limited mechanism, this ends up in agglomerates with denser structures (Df ~ 2.0–2.2) [47,48]. For the diffusion-limited mechanism, agglomerates form looser structures (Df ~ 1.7–1.8) [49]. Taking this into account, we can conclude that the incorporation of AuNP has modified the growth mechanism of P(AAm-co-AAc) microgels, ending up in the formation of agglomerates with looser structures. It is worth mentioning that the obtained results could also be affected by the spatial distribution of the AuNP within microgel matrix that could be different as the content of AuNP increases [46]. This issue will be studied in further work.

4. Conclusions

The detailed rheological study performed in this work resolves that hybrid microgel dispersions present a gel-like behavior, besides confirming the elasticity reinforcement role of AuNP, as it was expected. By analyzing the behavior under deformation (strain sweep tests), we confirmed the colloidal gel structure which can break upon deformation but completely restructure as the deformation gradually disappears. Metal nanoparticles (AuNP) also play a major role in the recovery of the structure. From the application of the scaling theories, we also determined that the incorporation of metal nanoparticles (AuNP) affects significantly the colloidal structure formation to the point of modifying the growth mechanism from a reaction-limited to a diffusion-limited aggregation mechanism with the incorporation of 10 wt% AuNP. In conclusion, the incorporation of AuNP modifies the agglomerate growth mechanism giving rise to agglomerates with looser structures at rest that are easily deformable.

The strategy used in this work serves as a tool to control the formation, breakage, and reformation of agglomerates under stress through rheological characterization. Moreover, it opens a new way to predict metal nanoparticle effect in the microgel dispersions behavior through simple stress sweep tests, being the targeted objective. In addition, this facilitates the preparation of a more homogeneous and easy applicable system, for instance when applied for injectable systems [37]. In addition, the recovery of the elastic properties immediately after injection (deformation) may prevent the flow of the colloidal solution and facilitate that the material remains on the target site [50]. In summary, the approach presented herein is straightforward and simple, besides potentially inducing additional functionalities to the microgel system derived from the nanoparticles responsiveness itself.

Being able to control and program the flow behavior of microgel dispersion would be extremely helpful in order to ensure the success of their final application, in particular as drug delivery systems, carriers, or similar. Being able to tailor how stimuli responsive microgels should behave under deformation, and at rest, would avoid undesirable or unexpected agglomeration problems that could limit their responsiveness or have dramatic consequences as blood clots when referring to real in-vivo use of the smart dispersions. In conclusion, this work puts in evidence the importance of understanding the rheological behavior and fractal structure of stimuli responsive microgels for further potential applications that might be implied in flow/deformation of aqueous dispersions.

Supplementary Materials: The following are available online at http://www.mdpi.com/2079-4991/9/10/1499/s1, Figure S1: TGA Thermograms corresponding to P(AAm-co-AAc) (black), P(AAm-co-AAc)-5% AuNP (red) and P(AAm-*co*-AAc)-10% AuNP (green). Table S1: Results corresponding to the residue obtained from TGA experiments.

Author Contributions: Investigation, C.E. and C.M.; methodology, C.E.; supervision, C.M.; writing—original draft, C.E.; writing—review and editing, C.E. and C.M.

Acknowledgments: CE acknowledges IJCI-2015-26432 contract from MINEICO.

Conflicts of Interest: The authors declare no conflict of interest.

References

1. Gennes, P.-G. *Nobel Lecture—Soft Matter*; 4930838533; Angewandte Chemie International Edition: Paris, France, 1991.
2. De Gennes, P.G. Ultradivided matter. *Nature* **2001**, *412*, 385. [CrossRef] [PubMed]
3. Thorne, J.B.; Vine, G.J.; Snowden, M.J. Microgel applications and commercial considerations. *Colloid Polym. Sci.* **2011**, *289*, 625–646. [CrossRef]
4. Muratalin, M.; Luckham, P.F.; Esimova, A.; Aidarova, S.; Mutaliyeva, B.; Madybekova, G.; Sharipova, A.; Issayeva, A. Study of N-isopropylacrylamide-based microgel particles as a potential drug delivery agents. *Colloids Surf. A Physicochem. Eng. Asp.* **2017**, *532*, 8–17. [CrossRef]
5. Madrigal, J.L.; Sharma, S.N.; Campbell, K.T.; Stilhano, R.S.; Gijsbers, R.; Silva, E.A. Microgels produced using microfluidic on-chip polymer blending for controlled released of VEGF encoding lentivectors. *Acta Biomater.* **2018**, *69*, 265–276. [CrossRef]
6. Liu, G.; Liu, Z.; Li, N.; Wang, X.; Zhou, F.; Liu, W. Hairy Polyelectrolyte Brushes-Grafted Thermosensitive Microgels as Artificial Synovial Fluid for Simultaneous Biomimetic Lubrication and Arthritis Treatment. *ACS Appl. Mater. Interfaces* **2014**, *6*, 20452–20463. [CrossRef]
7. Joshi, A.; Nandi, S.; Chester, D.; Brown, A.C.; Muller, M. Study of Poly(N-isopropylacrylamide-*co*-acrylic acid) (pNIPAM) Microgel Particle Induced Deformations of Tissue-Mimicking Phantom by Ultrasound Stimulation. *Langmuir* **2018**, *34*, 1457–1465. [CrossRef]
8. Gan, T.; Zhang, Y.; Guan, Y. In Situ Gelation of P(NIPAM-HEMA) Microgel Dispersion and Its Applications as Injectable 3D Cell Scaffold. *Biomacromolecules* **2009**, *10*, 1410–1415. [CrossRef]
9. Gelissen, A.P.; Schmid, A.J.; Plamper, F.A.; Pergushov, D.V.; Richtering, W. Quaternized microgels as soft templates for polyelectrolyte layer-by-layer assemblies. *Polymer* **2014**, *55*, 1991–1999. [CrossRef]
10. Jia, S.; Tang, Z.; Guan, Y.; Zhang, Y. Order–Disorder Transition in Doped Microgel Colloidal Crystals and Its Application for Optical Sensing. *ACS Appl. Mater. Interfaces* **2018**, *10*, 14254–14258. [CrossRef]
11. Park, C.W.; South, A.B.; Hu, X.; Verdes, C.; Kim, J.-D.; Lyon, L.A. Gold nanoparticles reinforce self-healing microgel multilayers. *Colloid Polym. Sci.* **2010**, *289*, 583–590. [CrossRef]
12. Krüger, A.J.D.; Köhler, J.; Cichosz, S.; Rose, J.C.; Gehlen, D.B.; Haraszti, T.; Möller, M.; De Laporte, L.; Krüger, A.J.D.; Koehler, J. A catalyst-free, temperature controlled gelation system for in-mold fabrication of microgels. *Chem. Commun.* **2018**, *54*, 6943–6946. [CrossRef] [PubMed]
13. Faria, J.; Echeverria, C.; Borges, J.P.; Godinho, M.H.; Soares, P.I.P. Towards the development of multifunctional hybrid fibrillary gels: Production and optimization by colloidal electrospinning. *RSC Adv.* **2017**, *7*, 48972–48979. [CrossRef]
14. Marques, S.C.S.; Soares, P.I.P.; Zabala, C.E.; Godinho, M.H.; Borges, J.P. Confinement of thermoresponsive microgels into fibres via colloidal electrospinning: Experimental and statistical analysis. *RSC Adv.* **2016**, *6*, 76370–76380. [CrossRef]
15. Diaz, J.E.; Barrero, A.; Marquez, M.; Fernandez-Nieves, A.; Loscertales, I.G. Absorption properties of microgel-pvp composite nanofibers made by electrospinning. *Macromol. Rapid Commun.* **2010**, *31*, 183–189. [PubMed]
16. Leon, A.M.; Aguilera, J.M.; Park, D.J. Mechanical, rheological and structural properties of fiber-containing microgels based on whey protein and alginate. *Carbohydr. Polym.* **2019**, *207*, 571–579. [CrossRef] [PubMed]
17. Echeverria, C.; López, D.; Mijangos, C.; Zabala, C.E. UCST Responsive Microgels of Poly(acrylamide–acrylic acid) Copolymers: Structure and Viscoelastic Properties. *Macromolecules* **2009**, *42*, 9118–9123. [CrossRef]
18. Zabala, C.E.; Peppas, N.A.; Mijangos, C. Novel strategy for the determination of UCST-like microgels network structure: Effect on swelling behavior and rheology. *Soft Matter* **2012**, *8*, 337–346.

19. Antonietti, M.; Gröhn, F.; Hartmann, J.; Bronstein, L. Nonclassical Shapes of Noble-Metal Colloids by Synthesis in Microgel Nanoreactors. *Angew. Chem. Int. Ed.* **1997**, *36*, 2080–2083. [CrossRef]
20. Yoshida, A.; Kitayama, Y.; Kiguchi, K.; Yamada, T.; Akasaka, H.; Sasaki, R.; Takeuchi, T. Gold Nanoparticle-Incorporated Molecularly Imprinted Microgels as Radiation Sensitizers in Pancreatic Cancer. *ACS Appl. Bio Mater.* **2019**, *2*, 1177–1183. [CrossRef]
21. Peng, J.; Tang, D.; Jia, S.; Zhang, Y.; Sun, Z.; Yang, X.; Zou, H.; Lv, H. In situ thermal synthesis of molybdenum oxide nanocrystals in thermoresponsive microgels. *Colloids Surf. A Physicochem. Eng. Asp.* **2019**, *563*, 130–140. [CrossRef]
22. Brändel, T.; Sabadasch, V.; Hannappel, Y.; Hellweg, T. Improved Smart Microgel Carriers for Catalytic Silver Nanoparticles. *ACS Omega* **2019**, *4*, 4636–4649. [CrossRef] [PubMed]
23. Das, M.; Sanson, N.; Fava, D.; Kumacheva, E. Microgels Loaded with Gold Nanorods: Photothermally Triggered Volume Transitions under Physiological Conditions†. *Langmuir* **2007**, *23*, 196–201. [CrossRef] [PubMed]
24. Zhang, J.; Xu, S.; Kumacheva, E. Polymer Microgels: Reactors for Semiconductor, Metal, and Magnetic Nanoparticles. *J. Am. Chem. Soc.* **2004**, *126*, 7908–7914. [CrossRef] [PubMed]
25. Echeverria, C.; Mijangos, C.; Zabala, C.E. UCST-Like Hybrid PAAm-AA/Fe_3O_4 Microgels. Effect of Fe_3O_4 Nanoparticles on Morphology, Thermosensitivity and Elasticity. *Langmuir* **2011**, *27*, 8027–8035. [CrossRef] [PubMed]
26. Echeverria, C.; Soares, P.; Robalo, A.; Pereira, L.; Novo, C.M.; Ferreira, I.; Borges, J.P. One-pot synthesis of dual-stimuli responsive hybrid PNIPAAm-chitosan microgels. *Mater. Des.* **2015**, *86*, 745–751. [CrossRef]
27. Echeverria, C.; Mijangos, C. Effect of gold nanoparticles on the thermosensitivity, morphology, and optical properties of poly(acrylamide-acrylic acid) microgels. *Macromol. Rapid Commun.* **2010**, *31*, 54–58. [CrossRef] [PubMed]
28. Scarabelli, L.; Schumacher, M.; De Aberasturi, D.J.; Merkl, J.-P.; Henriksen-Lacey, M.; De Oliveira, T.M.; Janschel, M.; Schmidtke, C.; Bals, S.; Weller, H.; et al. Encapsulation of Noble Metal Nanoparticles through Seeded Emulsion Polymerization as Highly Stable Plasmonic Systems. *Adv. Funct. Mater.* **2019**, *29*, 29. [CrossRef]
29. Pérez-Juste, J.; Pastoriza-Santos, I.; Liz-Marzán, L.M. Multifunctionality in metal@microgel colloidal nanocomposites. *J. Mater. Chem. A* **2013**, *1*, 20–26. [CrossRef]
30. Strozyk, M.S.; Carregal-Romero, S.; Henriksen-Lacey, M.; Brust, M.; Liz-Marzán, L.M. Biocompatible, Multiresponsive Nanogel Composites for Codelivery of Antiangiogenic and Chemotherapeutic Agents. *Chem. Mater.* **2017**, *29*, 2303–2313. [CrossRef]
31. Shih, W.-H.; Shih, W.Y.; Kim, S.-I.; Liu, J.; Aksay, I.A. Scaling behavior of the elastic properties of colloidal gels. *Phys. Rev. A* **1990**, *42*, 4772–4779. [CrossRef]
32. Wu, H.; Morbidelli, M. A Model Relating Structure of Colloidal Gels to Their Elastic Properties. *Langmuir* **2001**, *17*, 1030–1036. [CrossRef]
33. Mewis, J.; Wagner, N.J. *Colloidal Suspension Rheology*; Cambridge University Press: Cambridge, UK, 2012.
34. Sprakel, J.; Lindström, S.B.; Kodger, T.E.; Weitz, D.A. Stress enhancement in the delayed yielding of colloidal gels. *Phys. Rev. Lett.* **2011**, *106*, 248303. [CrossRef]
35. Stokes, J.R.; Frith, W.J. Rheology of gelling and yielding soft matter systems. *Soft Matter* **2008**, *4*, 1133. [CrossRef]
36. Diba, M.; Wang, H.; Kodger, T.E.; Parsa, S.; Leeuwenburgh, S.C.G. Highly Elastic and Self-Healing Composite Colloidal Gels. *Adv. Mater.* **2017**, *29*, 1604672. [CrossRef] [PubMed]
37. Guvendiren, M.; Lu, H.D.; Burdick, J.A. Shear-thinning hydrogels for biomedical applications. *Soft Matter* **2012**, *8*, 260–272. [CrossRef]
38. Liao, W.; Zhang, Y.; Guan, Y.; Zhu, X.X. Gelation Kinetics of Thermosensitive PNIPAM Microgel Dispersions. *Macromol. Chem. Phys.* **2011**, *212*, 2052–2060. [CrossRef]
39. Gan, T.; Guan, Y.; Zhang, Y. Thermogelable PNIPAM microgel dispersion as 3D cell scaffold: Effect of syneresis. *J. Mater. Chem.* **2010**, *20*, 5937. [CrossRef]
40. Cheng, D.; Wu, Y.; Guan, Y.; Zhang, Y. Tuning properties of injectable hydrogel scaffold by PEG blending. *Polymer* **2012**, *53*, 5124–5131. [CrossRef]

41. Wang, T.; Jin, L.; Song, Y.; Li, J.; Gao, Y.; Shi, S. Rheological study on the thermoinduced gelation behavior of poly(N -isopropylacrylamide-co -acrylic acid) microgel suspensions. *J. Appl. Polym. Sci.* **2017**, *134*, 45259. [CrossRef]
42. Fraylich, M.R.; Liu, R.; Richardson, S.M.; Baird, P.; Hoyland, J.; Freemont, A.J.; Alexander, C.; Shakesheff, K.; Cellesi, F.; Saunders, B.R. Thermally-triggered gelation of PLGA dispersions: Towards an injectable colloidal cell delivery system. *J. Colloid Interface Sci.* **2010**, *344*, 61–69. [CrossRef]
43. Liao, W.; Zhang, Y.; Guan, Y.; Zhu, X.X. Fractal Structures of the Hydrogels Formed in Situ from Poly(N-isopropylacrylamide) Microgel Dispersions. *Langmuir* **2012**, *28*, 10873–10880. [CrossRef] [PubMed]
44. Echeverría, C.; Aragón-Gutiérrez, A.; Fernández-García, M.; Muñoz-Bonilla, A.; López, D. Thermoresponsive Poly(N-Isopropylacrylamide-*co*-Dimethylaminoethyl Methacrylate) Microgel Aqueous Dispersions with Potential Antimicrobial Properties. *Polymers* **2019**, *11*, 606. [CrossRef] [PubMed]
45. Brown, W.D.; Ball, R.C. Computer simulation of chemically limited aggregation. *J. Phys. A Math. Gen.* **1985**, *18*, L517–L521. [CrossRef]
46. Vermant, J.; Ceccia, S.; Dolgovskij, M.K.; Maffettone, P.L.; Macosko, C.W. Quantifying dispersion of layered nanocomposites via melt rheology. *J. Rheol.* **2007**, *51*, 429. [CrossRef]
47. Kolb, M.; Botet, R.; Jullien, R. Scaling of Kinetically Growing Clusters. *Phys. Rev. Lett.* **1983**, *51*, 1123–1126. [CrossRef]
48. Meakin, P. Formation of Fractal Clusters and Networks by Irreversible Diffusion-Limited Aggregation. *Phys. Rev. Lett.* **1983**, *51*, 1119–1122. [CrossRef]
49. Witten, T.A.; Sander, L.M. Diffusion-Limited Aggregation, a Kinetic Critical Phenomenon. *Phys. Rev. Lett.* **1981**, *47*, 1400–1403. [CrossRef]
50. Yan, C.; Mackay, M.E.; Czymmek, K.; Nagarkar, R.P.; Schneider, J.P.; Pochan, D.J. Injectable Solid Peptide Hydrogel as a Cell Carrier: Effects of Shear Flow on Hydrogels and Cell Payload. *Langmuir* **2012**, *28*, 6076–6087. [CrossRef]

© 2019 by the authors. Licensee MDPI, Basel, Switzerland. This article is an open access article distributed under the terms and conditions of the Creative Commons Attribution (CC BY) license (http://creativecommons.org/licenses/by/4.0/).

Review

Systemic Review of Biodegradable Nanomaterials in Nanomedicine

Shi Su and Peter M. Kang *

Cardiovascular Institute, Beth Israel Deaconess Medical Center and Harvard Medical School, 3 Blackfan Circle, CLS 910, Boston, MA 02215, USA; ssu@bidmc.harvard.edu
* Correspondence: pkang@bidmc.harvard.edu; Tel.: +1-(617)-735-4290; Fax: +1-(617)-735-4207

Received: 29 February 2020; Accepted: 25 March 2020; Published: 1 April 2020

Abstract: Background: Nanomedicine is a field of science that uses nanoscale materials for the diagnosis and treatment of human disease. It has emerged as an important aspect of the therapeutics, but at the same time, also raises concerns regarding the safety of the nanomaterials involved. Recent applications of functionalized biodegradable nanomaterials have significantly improved the safety profile of nanomedicine. Objective: Our goal is to evaluate different types of biodegradable nanomaterials that have been functionalized for their biomedical applications. Method: In this review, we used PubMed as our literature source and selected recently published studies on biodegradable nanomaterials and their applications in nanomedicine. Results: We found that biodegradable polymers are commonly functionalized for various purposes. Their property of being naturally degraded under biological conditions allows these biodegradable nanomaterials to be used for many biomedical purposes, including bio-imaging, targeted drug delivery, implantation and tissue engineering. The degradability of these nanoparticles can be utilized to control cargo release, by allowing efficient degradation of the nanomaterials at the target site while maintaining nanoparticle integrity at off-target sites. Conclusion: While each biodegradable nanomaterial has its advantages and disadvantages, with careful design and functionalization, biodegradable nanoparticles hold great future in nanomedicine.

Keywords: biodegradable; nanomaterials; nanomedicine

---

1. Introduction

Nanotechnology is being applied in many aspects of human life, including agriculture, transportation, electronics, communication, food industry and medicine [1–5]. Nanotechnology is the manipulation of matter at the nanoscale (1 to 100 nm) to create new particles and devices [5]. Nanotechnology assisted medicine, known as nanomedicine, is an interdisciplinary field of science and technology applying materials at the nanoscale for the diagnosis and treatment of human disease [6,7]. Nanomedicine has emerged as an important aspect of the therapeutic regimen for different types of diseases as it holds great potential for personalized medicine. Nanomedicine also has very diverse applications, including smart imaging, molecular detection, and targeted therapy [7]. Many unique properties of nanoparticles depend on the size and shape, the surface charge and modification, and the hydrophobicity of the nanoparticles [8]. The unique properties of nanoparticles could provide great advantages of nanomedicine. For example, the small size of nanoparticle can allow them cross biological barriers; different structures of nanoparticles can increase the bioavailability of non-soluble or unstable drugs; the modifiable surface of nanoparticles can allow desired targeting capacity to the diseased area for either imaging or specific drug delivery. Improved drug bioactivity, bioavailability and controlled delivery are being realized as drugs can be encapsulated into nanodrug delivery system. It is therefore deemed as a superior therapeutic approach compared to the conventional medicine. With the development of nanomedicine, concerns have also been raised regarding the safety of nanomaterials involved. In the notion of improving the safety profile of nanomedicine, biodegradable nanomaterials are gaining increasing attention in this field. Biodegradable nanomaterials are nano-scale materials that can be naturally degraded under biological conditions in the body [9]. The degradability of the nanoparticles can be a useful property to control cargo release as the ideal biodegradable nanoparticles require efficient degradation at the target site while remaining stable at off-target sites [7,9]. Biodegradable nanoparticles hold great promise in drug delivery system due to a number of reasons: they provide controlled releasing profile; they are stable in the circulation system; they are non-toxic and non-immunogenic; they are also capable of avoiding the reticuloendothelial system, part of the immune system in the body that takes up and clears foreign objects, thus prolonging their circulation time [9]. The rationale for this review is that while there are several reviews on one specific type of biodegradable nanomaterial, there is no recent systemic review on the biodegradable nanomaterials and their applications in nanomedicine [10–14]. In this review, our objective is to evaluate different types of biodegradable nanomaterials that are currently being investigated for their application in different diseases.

2. Literature Search Methods

We used PubMed as our source for literature research. The key words used were "biodegradable nanomaterial" and "nanomedicine" and the range of publication date was set within the past 10 years. Out of the 561 studies available at the time of writing, we selected the ones that we considered to be relevant to our review, which reported functionalized biodegradable nanomaterials and their application in nanomedicine. Subsequently, under each subsection, the specific name of the biodegradable nanomaterial were added as a key word to fine-tune the literature search.

3. Types of Biodegradable Nanoparticles

Similar to their nondegradable counterparts, biodegradable nanoparticles can be categorized based on their structure and arrangement of the nanomaterials, either by encapsulating the agents of interest as nanocapsule or incorporating those agents into a nanosphere [8]. The agents of interest can either be encapsulated in the nanoparticles or adsorbed on the surface of the nanomaterials [8]. Some examples of classic nanocapsules include micelles and liposomes, and dendrimers are an example of nanospheres. No matter what the structure is or how the payload is being incorporated, biodegradable nanoparticles remain the general advantages of nanoparticles in nanomedicine, such as slow and controlled release of the cargo and targeted delivery, which lead to enhanced therapeutic effects and decreased side-effects, especially for certain cytotoxic drugs, with one more advantage being that biodegradable nanomaterials decrease the cytotoxicity to the body. Further surface modifications can also be done to improve the drug release profile and targeting efficiency.

In making biodegradable nanoparticles, polymers have shown high biocompatibility and biosafety [15]. Polymer-based nanoparticles are solid colloidal particles with a size of 10–500 nm and can be used to carry therapeutic agents of interest by either embedding/encapsulating the agents within their polymeric matrix, or adsorbing/conjugating them onto the surface [15,16]. In addition, the particle surface and size can be modified to control drug release [15]. Based on the main materials used for the formation of the nanoparticles, polymer-based nanomaterials can be categorized into two main groups: synthesized materials such as poly-D-L-lactide-co-glycolide (PLGA), polyactic acid (PLA), and poly-e-caprolactone (PCL); and natural materials like chitosan. All these polymers can undergo degradation process to be degraded into products that can be safely processed in the body (Figure 1). The degradation rate of polymer-based nanoparticles are affected by many factors, including internal factors such as the size, structure and molecular weight of the nanoparticles, as well as external factors, such as pH and temperature, both of which influence the payload releasing profile [17]. Synthetic polymers have the general advantage of relatively long drug release period, compared to their natural polymer counterparts [18]. However, based on the type of materials being applied, certain disadvantages may also arise for certain nanoparticles, whether it is low drug-loading capacity, instability, or increased fragility. The existence of different advantages and disadvantages of different nanomaterials require careful consideration in order to choose proper nanomaterials when designing new nanoparticles. The detailed advantages and disadvantages of each nanomaterial will be summarized in the following sections (Table 1).

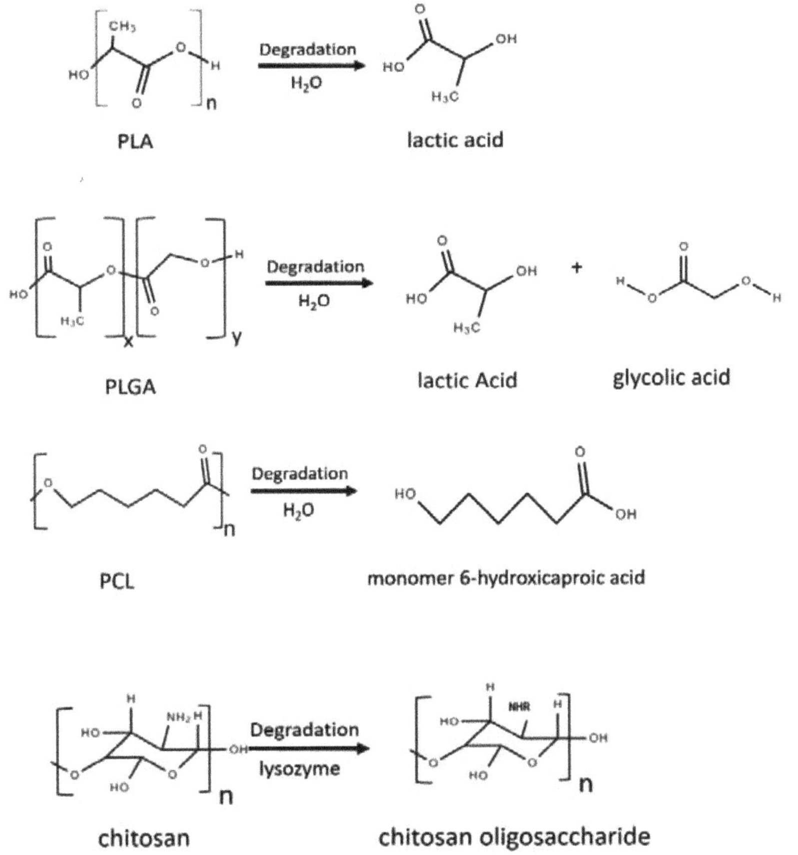

Figure 1. The biodegradation reaction of some commonly used biodegradable nanomaterials.

Table 1. Advantages and limitations of some biodegradable nanoparticles.

Biodegradable Nanoparticles		Advantages	Limitations
General		Biocompatibility Low immunogenicity Slow and controlled release of the cargo Targeted delivery Enhanced therapeutic effects Decreased side-effects and cytotoxicity Modifiable size and surface to improve drug release profile and targeting efficiency	Instable if not modified High cost of particle production Difficult to synthesize particles that are homogeneous in shape and size Relatively low drug encapsulation efficiency Difficulty in large scale production and sterilization
Polymer-based	PLA micelles	• Hydrophobicity useful for carrying low soluble drugs • Easily modifiable physical and chemical properties to obtain desirable pharmacokinetic and biodegradable properties	• Non-specific uptake by the reticuloendothelial systems • Low drug loading capacity • Low encapsulation efficiency • Particle size-dependent immunotoxicity • Easily affected degradation rate
	PLGA micelles	• Wide range of erosion times • Modifiable mechanical properties • Degradation rate can be changed by adjusting the ratio of PLA:PGA and their molecular weights	• The acidic nature of PLGA monomers are not suitable for certain drugs and bioactive molecules • Difficult to achieve optimal drug release profile • non-linear, dose-dependent, and easily altered biodistribution and pharmacokinetics
	PCL nanoparticles	• Degradation does not produce acidic byproducts • Slow degradation rate, ideal for long-term implantation device • Versatile mechanical properties	• Hydrophobicity that limits its production
	Chitosan nanoparticles	• Good absorbability, permeability, and moisture retention • Easily degradable • Low toxicity	• Very sensitive to environmental temperature • Degradation rate affected by environmental pH • Poor long-term stability
	Dendrimers	• High degree of branching and polyvalency, with very high surface-to-volume ratio, enabling high drug carrying efficiency • Capable of carrying drugs with poor solubility • High water solubility • Useful as MRI imaging agent	• Cytotoxicity may occur during the interactions with cell membrane depending on the charge and modifications made on the surface of the dendrimer
Lipid-based	Liposomes	• Self-assembly, enabling easy drug loading • High loading efficiency • Protect encapsulated drugs from early inactivation, degradation and dilution in the circulation • Can be formulated into different forms for various routes of administration • Easily functionalized with surface modification	• Fast clearance rate • Low stability • Complex production method

3.1. Micelles

A micelle is defined as a collection of amphiphilic molecules that can self-assemble in water into a spherical vesicle [19]. Micelles can be formed by either lipid- or polymer-based amphiphilic molecules [20–22]. Lipid-based micelles are composed of small molecules that have a hydrophilic head group and a hydrophobic tail, which is the hydrocarbon portion of long fatty acids [23]. Polymer-based micelles are formed with polymers of alternating hydrophilic, such as poly(ethylene oxide) (PEO), and hydrophobic blocks, such as poly(propylene oxide) (PPO), poly(lactic acid) (PLA), or other biocompatible and hydrophobic polyethers or polyesters [23] (Figure 2). Polymeric micelles range from 10 to 100nm in size and have a narrow size-distribution [17,24]. Self-assembly of the single chains of amphiphilic molecules occurs when they reach certain concentrations, which are defined as the critical micelle concentration (CMC) [17]. CMC is an important parameter to assess the stability of micelles as micelles with lower CMC are more thermodynamically stable [17]. The molecules of micelles are self-assembled in a way that the core is hydrophobic whereas the shell is hydrophilic [17]. The hydrophobic core allows encapsulation of poorly soluble drugs whereas the hydrophilic shell increases circulation time and structural stability to enable controlled and sustained release of the drugs, though their circulation time is shorter than that of liposomes due to their smaller size [20,22].

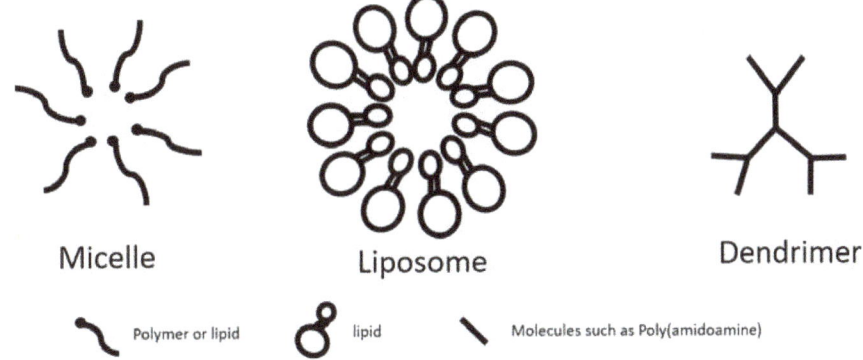

Figure 2. Examples of applications of nanomaterials in different nanoparticles.

3.1.1. Polylactic Acid (PLA) Micelles

Poly(D,L-lactic acid) (PLA) is a type of biodegradable nanomaterial that is widely used in nanomedicine. PLA is produced from the monomer of lactic acid (LA), which is obtained from glucose fermentation [25]. In the process of synthesizing PLA, LA is converted to lactide and eventually to PLA [25]. Under physiological conditions, PLA can be hydrolyzed into lactic acid, and eventually secreted out of the body [26,27].

PLA is relatively hydrophobic, and is therefore commonly used for implants (such as stents or screws for bone fixations), medical sutures, as well as drug delivery micelles as it improves oral bioavailability of hydrophobic drugs [10,11]. One of the advantages of using PLA to make micelles is that its physical and chemical properties, such as size, shape, molecular weight and liquid-to-gas ratio, can all be easily altered to obtain desirable pharmacokinetic and biodegradable properties [11]. However, limitations also exist for PLA nanoparticles, such as non-specific uptake by the reticuloendothelial systems, as well as low drug loading capacity and low encapsulation efficiency [11]. Although PLA generally elicits low immunotoxicity, it has been shown that the size of PLA nanoparticles affects their immunotoxicity- the smaller the nanoparticles, the more immunotoxic they are [28].

As PLA degrades primarily by hydrolysis while the polymer degradation rate is determined by its reactivity with water, any factors that can change the reactivity can affect the degradation rate [10]. Although the release of the drug in micelles is mainly controlled by the rate of diffusion of the drug from the micellar core and the rate of biodegradation of the micelles, other factors such as the compatibility between the drug and core forming block of copolymer, the amount of drug loaded, the molecular volume of the drug, and the length of the core forming block also affect the drug release profile [29]. A general consideration when choosing PLA as micelle block is to match the mechanical properties and the degradation rate to the need of the application [10]. Meanwhile modifications of the nanoparticles have also been made for better delivery efficiency. For example, micelle-templated Polylactic-co-glycolic acid (PLGA) nanoparticles have been developed for hydrophobic drug delivery with increased stability and loading capacity [21].

3.1.2. Polylactic-Co-Glycolic Acid (PLGA) Micelles

Polylactic-co-glycolic acid (PLGA) is one of the best characterized biodegradable polymers that is frequently used for drug delivery as it can be hydrolyzed in the body to produce metabolite monomers lactic acid and glycolic acid, and eventually degraded into non-toxic products (i.e., water and carbon dioxide) that can easily be eliminated from the body [8,30,31]. PLGA is a copolymer of hydrophobic polylactic acid (PLA) and hydrophilic polyglycolic acid (PGA) [32]. PGA is also a biodegradable material that can be degraded into glycolic acid, which is a natural metabolite [33]. For this reason, PGA has been most commonly used in the production of resorbable sutures [33].

As a copolymer, PLGA has a wide range of erosion time and modifiable mechanical properties [31,32]. The degradation rate of PLGA can be changed by adjusting the ratio of PLA:PGA and their molecular weights in order to control the release of incorporated drugs [31,32]. These characteristics make PLGA a very attractive type of material for drug delivery [31,32]. Different approaches of loading systems exist for PLGA-based nanoparticle therapeutics, including protein encapsulation, protein adsorption, and nucleic acid loading [12,34].

Despite the fact that more than three decades have passed since PLGA first received the approval from the Food and Drug Administration (FDA), there are still only 19 long releasing approved products containing PLGA [35,36]. The slow development of long releasing PLGA drug carriers are attributed to the challenge that regardless of its many modifiable properties, the acidic PLGA monomers are not suitable for certain drugs and bioactive molecules [8]. Another major challenge of formulating PLGA containing drugs still lies in the difficulty of achieving the desired drug release profile [36]. The biodistribution and pharmacokinetics of PLGA are non-linear, dose-dependent, and easily affected by different factors including the hydrophilicity, the inter-hydrolytic group chemical interactions, the crystallinity as well as the volume to surface ratio of PLGA [31].

3.1.3. Modification of Micelles

Poly-ethylene-glycol (PEG) is commonly used to modify the surface of nanoparticles to enable long-term circulation [8]. The process of incorporating PEG onto the surface of a nanoparticle is known as PEGylation [8]. PEGylation has been incorporated for the development of various block copolymers [31]. PEG-b-PLA micelles are used as a platform for the systemic multi-drug delivery of poorly water soluble anticancer agents because PEGylation stabilizes micelles and improves encapsulation capacity [37]. PEG-PLGA copolymers can form nanospheres, micelles and hydrogels, making them great biodegradable nanomaterials for the construction of nanodrug delivery system [38]. Depending on the structural differences, PEG-PLGA copolymers can have characteristics suitable for different loading agents, enabling multi-drug loading capacity, further improving their therapeutic efficacy [38]. For examples, PLGA-b-PEG-b-PLGA is a thermosensitive copolymer that can transition from solution into gel at body temperature for multi-drug delivery of both hydrophobic and hydrophilic anticancer agents [37]. Several such PEG-PLGA copolymer composed multi-drug delivery systems have been approved by the FDA as neoadjuvant therapy for cancer treatment [37].

In some micelles, functionalized PEG layers are added as the hydrophilic outer shell to attain receptor-mediated drug and gene delivery through PEG-conjugated ligands with a minimal non-specific interaction with other proteins [39]. Moreover, in order to meet the needs of delivering different types of drugs, other types of micelles have also been developed [40]. For example, in addition to conventional micelles with a hydrophobic core and a hydrophilic shell, there are also reverse micelles with a hydrophilic core and a hydrophobic shell to ensure sustained drug release through the hydrogen bond between the drug and the core [40]. In addition, since tumors have lower pH compared to healthy tissues, micelles can also be functionalized by adding peptides responsive to pH change for effective cancer imaging and therapy [41]. All these modifications enable pre-clinical evaluation and clinical translation of emerging agents [37].

3.2. Poly-ε-Caprolactone (PCL) Nanoparticles

Poly-ε-caprolactone (PCL) is a polymer member of the aliphatic polyester family that is typically obtained by polymerization processes using a monomer and an initiator [13]. PCL can also be biodegraded, by hydrolysis of its ester linkage, into 6-hydroxycaprioc acid and then into acetyl-CoA, which eventually becomes water and carbon dioxide via the citric acid cycle [42,43]. Particularly, unlike PLA and PLGA, the degradation of which produce acidic products that further catalyze the polymer degradation process, the degradation of PCL does not produce acidic byproducts, making PCL a more favorable nanomaterial for the development of long-term implantable devices for its slow degradation rate [8,17]. Cholic acid can functionalize branched PCL with different molecular weights to meet the need of different nanodrug delivery systems, as higher molecular weight of the polymer matrix results in a slower drug release rate [44].

Since PEGylation of nanoparticles can be used to reduce immunogenicity and toxicity, prolong circulation time, change bio-distribution and optimize nanoparticle activities, copolymers of hydrophilic PEG and hydrophobic PCL can yield high biocompatibility and biodegradability [45]. The high biocompatibility, biodegradability, long circulation time and easy modification of surface properties of micelles composed of PEG-PCL di-block copolymers make them favorable candidates as nanodrug delivery systems [45].

3.3. Chitosan Nanoparticles

Derived from natural biopolymer chitin, chitosan is a copolymer of D-glucosamine and N-acetylglucosamine bonded via the β(1–4) linkages [16,46]. It can be degraded in vivo by several enzymes, mainly by lysozyme, a protease that ubiquitously exists in mammalian tissues [46]. Lysozyme hydrolyses the β(1–4) linkages between N-acetylglucosamine and glucosamine in chitosan to produce oligosaccharides, which can then either be excreted or be part of glycosaminoglycans or glycoproteins [46,47].

Chitosan has good absorbability, permeability, moisture retention and are easily degradable [15]. It shows low toxicity in both in vitro and in vivo models [14]. However, it is very sensitive to environmental temperature and is recommended to be stored at low temperatures [47]. Generally, the drug delivery system of chitosan nanoparticles is similar to the PLGA system, but the chitosan system is more pH dependent [14]. Chitosan nanoparticles have been applied in many site specific drug delivery systems via different administrative routes, including oral, nasal, and pulmonary drug delivery systems [14]. The mucoadhesive properties of chitosan nanoparticles increase the absorption rate of the drugs in the intestine [14].

PEG can be incorporated to increase the stability of the chitosan nanoparticles [46]. Chitosan oligosaccharide can also be functionalized to be a "switch on" imaging agent by conjugating with aggregation induced emission active tetraphenylethene (TPE) and lipophilic–cationic triphenylphosphonium (TPP) molecules [48]. Once chitosan self-assembles, TPE provides self-assembly induced fluorescence and TPP helps the nanoparticle enter into the cell by lipid-raft endocytosis [48].

Since the environmental pH affects the degradation rate of chitosan, considerations need to be made when designing chitosan nanoparticles for their encapsulated drugs to be effectively released at the diseased site, especially when the microenvironment of the drug releasing site is acidic, for example, the acidic tumor site [46]. Other pH variables can also limit the development of chitosan nanoparticles. For example, the orientation of β(1–4) linkages of chitosan can change under physiological pH depending on the crosslinkers used in the formation of chitosan nanoparticles [46]. The change in orientation can decrease the accessibility of the β(1–4) linkages to lysozyme, which can then lead to limited degradation of nanoparticles, since the breaking of the β(1–4) linkages by lysozyme is essential to the degradation of chitosan nanoparticles (Figure 1) [46]. In addition, poor long-term stability is a major drawback for large manufacture of chitosan nanoparticles [46]. Therefore, further investigations are still needed to fully understand the mechanisms of the interactions between chitosan and lysozyme when designing chitosan nanoparticles.

3.4. Dendrimers

Dendrimers are the smallest of nanocarriers that present as sphere-shaped and are structurally similar to branching polymer chains [49]. They are between 1 nm to 100 nm of diameter in size [50]. These radially symmetric molecules with well-defined, homogeneous, and monodisperse structure consisting of tree-like branches are analogous to protein, enzymes, and viruses, and can be easily functionalized [51]. Because their amphiphilic copolymers have both hydrophilic and hydrophobic monomer units, they can be used to carry drugs with poor solubility [49]. Drugs can be incorporated into dendrimers by covalent binding, electrostatic interactions or encapsulation [50]. The branch structure gives dendrimers a very high surface-to-volume ratio, enabling them to increase drug carrying efficiency [49]. In addition, their high degree of branching, polyvalency, biocompatibility, high water solubility, and low immunogenicity make dendrimers excellent vehicles for safely and effectively transporting drugs, and they are particularly attractive as MRI imaging agents [50,52].

Dendrimers are transported into and across cells via endocytic pathways [52]. Depending on the charge and modifications made on the surface of dendrimers, cytotoxicity may occur during the interactions with cell membrane [50]. Poly(amidoamine) (PAMAM) dendrimers are the most commonly used dendrimer in nanomedicine [50]. PAMAM has been applied in many drug or gene delivery systems and can be administered through various routes [50]. It can also be applied in the treatment of inflammatory diseases such as atherosclerosis and rheumatoid arthritis because its structure exerts anti-inflammatory activity [50]. Furthermore, arginine functionalized peptide dendrimers can condense plasmid DNA and protect it from nuclease digestion, which can serve as potential gene delivery vehicles [53].

3.5. Lipid-Based Nanoparticles

Besides polymer-based nanoparticles, lipid-based nanoparticles such as liposomes have also been employed in the drug delivery system for decades. As lipid is generally considered safe in the human body, for example, lipoproteins are natural nanoparticles that are found inside the human body, lipid-based nanoparticles have been under development as drug delivery systems [24]. Liposomes are spherical vesicles ranging from 10 to 1000 nm that consist of one or more phospholipid bilayers [24]. The phospholipid can be naturally occurring or synthetic phospholipids such as phosphatidylcholine (PC), phosphatidylethanolamine (PE), phosphatidylserine, and phosphatidylglycerol [54]. Cholesterol is generally added to stabilize the lipid bilayers of liposomes [54]. The aqueous core of liposomes can load and hold hydrophilic agents while its lipid bilayers can load hydrophobic agents [55].

Due to the similarities between the composition of liposome and that of cell membranes, liposomes are considered more biocompatible than other synthetic materials [55]. They are non-hemolytic, non-toxic, non-immunogenic, biocompatible and biodegradable [49]. Liposomes can self-assemble, which enables easy drug loading. They can carry large drug payloads and protect their encapsulated drugs from early inactivation, degradation and dilution in the circulation and can be formulated into different forms for different routes of administration [49,56]. Liposomes can be applied in a wide range of areas. They can not only be utilized to carry low molecular weight drugs, imaging agents, peptides or nucleic acids, but also serve as part of surgical implants for tissue repair, or as biosensors [56–59].

In addition to PEGylation and size alteration, liposomes can also be functionalized by attaching certain chemistry functional groups, peptides, antibodies, or acids to their surface to improve cell targeting efficiency [55,60]. Similar to the antigen-antibody complex formed between the antibody conjugated to the nanoparticles and the antigen presented on the surface of the cell, chemistry functional groups conjugated to the surface of the nanoparticles can form strong covalent bonds with metabolically labelled cell surface glycans [60]. Such reaction is known as "click" chemistry [60]. Compared with antigen-antibody complex, the "click" chemistry requires fewer functional groups, and form stronger bonds to allow sufficient time for the encapsulated nano-drugs to be internalized into the cells [60]. Recently, Boyd's groups functionalized liposomes containing drug nanocrystals using PEGylation and attachment of azide functional groups to improve drug loading capacity and to achieve cell-targeted delivery [60].

However, while liposomes have high loading efficiency, their low stability, fast clearance rate and a complex method of fabrication limit their potential in industrial scale fabrication [24]. To overcome the challenges faced by liposomes, solid lipid nanoparticles (SLNs) and nanostructured lipid carriers (NLCs) have subsequently been developed [24]. SLNs and NLCs have higher stability and lower toxicity compared to polymeric nanoparticles due to their smaller size and natural materials [24]. However, their low encapsulation efficiency hinders their potential to be widely pursued in the biomedical field [61]. Since both lipid-based and polymer-based nanoparticles have their own limitations, lipid-polymer hybrid nanoparticles have recently been developed to provide wider opportunities for the biomedical applications [62]. New polymer liposomes such as electrostatically crosslinked polymer–liposomes have also been developed [63]. These pH-sensitive copolymer methoxy poly(ethylene glycol)-block-poly(methacrylic acid)-cholesterol (mPEG-b-P(MAAc)-chol) and crosslinking reagent poly(ethylene glycol) end-capped with lysine (PEG-Lys2) were crosslinked into polymer–liposomes through electrostatic interactions [63]. These polymer-liposomes are stable under physiological conditions, but breaks down under acidic condition- similar to tumor microenvironment- to rapidly release their payloads, offering a new approach for anti-cancer therapies [63].

3.6. Other Natural Materials

Similarly, other natural biodegradable materials such as gelatin are also widely used in the nanomedicine field. Gelatin is a denatured protein that can be obtained either by partial acid or alkaline hydrolysis or by thermal or enzymatic degradation of animal collagen protein [64]. Derived from collagen, the most abundant protein in animals, gelatin does not produce any harmful by-products upon enzymatic degradation in the body [64]. Gelatin is considered as GRAS (generally regarded as safe) by the FDA [65]. Because gelatin is stable, easily modifiable and biodegradable, it is involved in the development of many clinical applications, including drug delivery system and hydrogels [64]. Right now the challenge that limits gelatin nanomedicine production is to make commercial gelatin nanoparticles homogenous in size [64].

4. Applications of Biodegradable Nanoparticles

Depending on the size, shape and composition/structure of the nanoparticles, different nanoparticles have different encapsulation efficiency, pharmacokinetics, and releasing mechanism. For example, the administration and encapsulation efficiency, as well as stability are different for nanoparticles that are composed of PGLA and those of PCL [8]. With the add-on targeting properties of these biodegradable nanoparticles, biomedical applications such as imaging as well as targeted drug delivery have been greatly advanced. However, in the clinical application of the biodegradable nanoparticles, while the safety profile of these nanoparticles has improved due to the biodegradability of the nanomaterials used, for the very reason, general challenges persist regarding the circulation time as well as the drug incorporating and/or releasing efficiency as they need to compete with the degradation rate of the nanomaterials.

4.1. Imaging

One of the early applications of nanomedicine involves the use of nanomaterials as contrast agents in biomedical imaging for diagnosis purpose, as the easy modification and targeting property of nanoparticles enable them to localize the tissue of interest and visualize with high resolution [7]. These nanoparticles are applied in many imaging modalities such as computed tomography (CT), magnetic resonance imaging (MRI), positron emission tomography, fluorescence imaging and photoacoustic imaging [7]. Targeted delivery by nanocarriers can greatly reduce the concentration of contrast agents, reducing the risk of contrast-induced kidney injury [66]. Gold nanoparticles are central in the development of imaging contrast agents [7]. However, if the nanoparticle itself cannot be biodegraded, long-term safety concern still remains. For instance, although metal nanoparticles with surface plasmon resonance in the near-infrared region (NIR) were of great interest for imaging, not being biodegradable raised concerns for the long-term safety of these nano-agents [67]. In 2010, a platform was first developed to synthesize metal/polymer biodegradable nanoclusters smaller than 100 nm with strong NIR absorbance for multimodal application [67]. With the safety issue gaining greater attention in the development of nano-imaging contrast agents, imaging agents underwent further development. Soon after that, biodegradable polydisulfide dendrimer nanoclusters were developed as MRI contrast agents to overcome safety concerns related to nephrogenic systemic fibrosis [68].

Photoacoustic imaging has also emerged as a promising imaging platform with a high tissue penetration depth [69,70]. It applies both NIR laser and ultrasound for the imaging purpose. The use of polymers made it possible to develop tunable and biodegradable gold nanoparticles as contrast agents for both CT and photoacoustic imaging [71]. Recently, not only polymers, but also natural materials are being engineered as biodegradable imaging agents. Fathi et al. have recently developed a photoacoustic imaging nanoprobe from nanoprecipitation of biliverdin, a naturally occurring heme-based pigment, which can be completely biodegraded to biliverdin reductase, a ubiquitous enzyme found in the body [69]. Excitation at near-infrared wavelengths leads to a strong photoacoustic signal, while excitation with ultraviolet wavelengths results in fluorescence emission [69]. In vivo experiments demonstrated that these nanoparticles accumulate in lymph nodes, suggesting that they can be used as a means to detect metastasized cancer [69]. Similarly, MTP1, a tumor metastasis targeting peptide, has been employed to modify the indocyanine green (ICG)-loaded PEG-PLGA micelles for targeted imaging of cervical cancer and metastasis [72].

One major factor that limits the efficacy of particle-based agents is their rapid sequestration by the mononuclear phagocytic system [73]. Even though low-fouling polymers such as PEG can reduce the immune recognition and clearance, these nondegradable polymers can accumulate in the human body and may cause adverse effects after prolonged use [73]. To overcome this challenge, Bonnard et al. used a recombined protein with the amino acid repeat proline, alanine, and serine (PAS) cross-linked into particles with lysine (K) and polyglutamic acid (E) [73]. The obtained PASKE particles have a prolonged circulation time and can be rapidly degraded in the cell's lysosomal compartment [73]. When combined with near-infrared fluorescent molecules and an anti-glycoprotein IIb/IIIa single-chain antibody targeting activated platelets, the PASKE nanoparticles was able to image carotid artery thrombosis in a mouse model, demonstrating its potential as a promising biodegradable tool for molecular imaging of vascular diseases [73].

Not only in cancer field, nano-technology assisted imaging has also been applied in atherosclerosis and other cardiovascular disease [74,75]. With the help of a tumor homing peptide, micelles have been shown effective in targeting not only the tumor site, but also at the plaques of atherosclerosis. The targeting property is realized by adding a peptide that homes to plaques- a clot-binding peptide cysteine-arginine-glutamic acid-lysine-alanine (CREKA) [76]. When CREKA is directly bound to the MRI contrast agent, it has been shown to be effective in detecting breast tumor [77]. Depending on the loading agents, micelles can help diagnose atherosclerosis if loaded with dyes, and decrease the plaque size if loaded with drugs [76]. This peptide was identified as a tumor-homing peptide by in vivo phage library screening, and subsequently it was shown to bind to clotted plasma proteins in the blood vessels and stroma of tumors [76].

4.2. Theranostics

A new concept- theranostics- the ability of "see and treat"- has become a well sought-after model in developing new multifunctional nanomedicine. These smart nanoparticles combine imaging agents, payload drugs and targeting moieties to accomplish diagnosis together with therapy delivery [7]. They can be engineered to be triggered in response to environmental changes such as pH, temperature, light, and ultrasound [7]. Some smart PLGA-based nanoparticles have been developed, including PH-responsive, thermos-sensitive and light-responsive nanoparticles [78]. In cancer treatment, many biodegradable polyacrylamide nanocarriers are applied for theranostics [79–81]. Nanovesicles are being developed for photoacoustic imaging and photothermal therapy (PTT), a therapeutic method that induces cell death using the heat energy converted from absorbed light energy, to enable minimum invasive cancer therapy [82]. The disulfide bond at the terminus of PEG-b-PCL copolymer can allow dense packing of gold nanoparticles, therefore enabling simultaneous photoacoustic imaging as well as enhanced PTT [82]. The designs of the nanoparticles have also been continuously improved to enhance biodegradability and efficacy of PTT [83]. Using biodegradable photonic melanoidin nanoparticles, Lee et al. were able to image lymph nodes and GI track, and to perform tumor ablation and photothermal lipolysis [84]. Recently, biomimetic mineralization method has also been applied to develop biodegradable multifunctional anti-tumor nanoparticles. Using this concept, Fu et al. developed a biodegradable manganese-doped calcium phosphate nanoparticle that can be used both as an MRI contract agent and an anti-tumor drug [85].

Applying 3D printing technology, Ceylan et al. designed a gelatin hydrogel-based, magnetically powered and controlled microswimmer, responsive to the pathological markers in its microenvironment for theranostic cargo delivery in cancer diagnosis and treatment [86]. This microswimmer can be biodegraded by matrix metalloproteinase-2 (MMP-2) enzyme, an enzyme that is highly expressed in breast cancer [86]. At normal physiological concentrations, MMP-2 can degrade the microswimmer to soluble nontoxic molecules [86]. If the MMP-2 concentration reaches pathological level, the microswimmer rapidly responds by swelling and thereby boosting the release of the embedded cargo molecules [86]. Banik et al. recently reported a multifunctional dual-targeted HDL-mimicking PLGA nanocomplex with both mitochondria and macrophage-targeting surface functionalities loaded with MRI contrast agent to achieve target-specific MRI contrast enhancement as well as lipid removal property for the treatment of atherosclerosis [87].

4.3. Targeted Delivery System

Nanomedicine using biodegradable polymeric nanoparticles as drug delivery systems have been engineered to treat cancer via multiple approaches: to target cancer cells, or the blood vessels that supply the nutrients and oxygen that support tumor growth, or immune cells to promote anti-cancer immunotherapy [88]. The use of biodegradable nanoparticles for targeted anti-cancer therapies yielded some clinical trials [88]. The encapsulation approach using PLGA can help prolong the circulating time of drugs that are unstable under the physiological condition and to minimize the side effects of certain drugs [30]. For example, 9-Nitrocamptothecin (9-NC) is a family of anticancer agents with low stability at biological pH and low water solubility [89]. PLGA encapsulation improves the drug release profile of 9-NC up to 160 h [89].

Similarly, in the field of cardiovascular disease, targeted nanodrug delivery system is under active investigation. To minimize the adverse effects while maximizing the drug effects, nanoparticles could to be superior as drug delivery systems compared to conventional drugs. For the treatment of cardiovascular disease (CVDs), current goals are focused on restoring normal blood flow to the heart as well as the prevention of recurrent cardiovascular insults [20]. Antithrombotic therapy is the first-line treatments for the prevention of CVDs, but they also significantly increase the risk of bleeding [20]. It remains a great challenge to effectively balance the ischemic risk reduction and the risk of bleeding [90]. Situations like these call for the need of developing nanomedicine that can target the disease area for drug delivery yet minimizing the side effects. Several drugs are delivered via the liposome drug delivery system for the treatment of angina pectoris. Takahama et al. encapsulated amiodarone, an anti-arrhythmic drug, in conventional liposomes to treat rat models that had undergone cardiac ischemic/reperfusion procedure, and showed reduced morality rate in the treated group that was due to lethal arrhythmia and the negative hemodynamic changes- the common side effects of amiodarone [91]. PLA has also been applied to encapsulate the drug for restenosis [49]. Several anti-inflammatory nanomedicines have been developed for targeted treatment of atherosclerosis, ischemia/reperfusion and post myocardial infarction left ventricular remodeling [32]. However, so far the targeted therapy in cardiovascular diseases using nanomaterial-based drug delivery vehicles have only shown effectiveness in preclinical settings [92]. Limitations lie in the gap in the knowledge of clinical safety, the requirement of composition purity and long-term stability of payload, as well as challenges and cost in scaled up production [93]. Recently, by scaling up the animal models from murine to rabbit and porcine, Muldler's group has taken the imaging-assisted nanotherapy one step closer to be realized in clinical settings [94].

4.3.1. Antioxidant Delivery

Oxidative stress has been associated with cytotoxic effects of cellular exposure to engineered nanomaterials [95]. After entering human body, changes in structural and physicochemical properties of nanoparticles can lead to changes in biological activities including the generation of reactive oxygen species (ROS) [96]. In this respect, if the nano-drug delivery system can deliver agents that combat oxidative stress, it can alleviate cell injury induced by excessive ROS. Kang's group has developed a new type of nanoparticle, named PVAX, which was formulated from copolyoxalate containing vanillyl alcohol (VA), an antioxidant extracted from natural herbs [97,98]. VA and the H_2O_2-responsive peroxalate ester linkages are incorporated covalently in the backbone of PVAX [97,98]. When encountered high levels of H_2O_2 at the sites of ischemia/reperfusion (I/R), PVAX is degraded and releases VA, exerting anti-inflammatory and anti-apoptotic activities [97,98]. PVAX has shown effectiveness in different types of I/R injuries, including hind limb I/R, liver I/R as well as cardiac I/R [97,98]. Andrabi et al. used biodegradable nanoparticles (nano-SOD/CAT) to encapsulate the antioxidant enzymes superoxide dismutase (SOD) and catalase (CAT) to effectively deliver those enzymes at the lesion site to protect mitochondria from oxidative stress, therefore protecting the spinal cord from secondary injury [99]. Tapeinos et al. have developed biodegradable PLGA microspheres coated with collagen type I and MnO_2 nanoparticles to scavenge ROS and protect cells from apoptosis induced by oxidative stress [100]. Many other nanoparticles targeting oxidative stress as a theranostic strategy for CVD have also been actively developed and evaluated [101].

4.3.2. Gene Therapy

Gene therapy is a type a therapeutic approach that seeks to modify the expression of certain genes in order to alter certain biological properties, which has gained significant amount of interest in recent years [102]. Whether by replacing the disease-causing gene with a healthy gene, or inactivating the disease-causing gene, or introducing a new or modified gene to treat the existing disease, gene therapy requires precise targeting [102]. With the help of nanoparticles for targeted delivery, we are getting closer to the realization of gene therapy being used in the clinical setting.

Cationic PEG-PLA nanoparticles is used as a major delivery system to deliver small interference RNA (siRNA) [11]. PEG-PLA nanoparticles encapsulating siRNA can enter the cells to perform gene-specific knockdown [103]. However, challenges still remain for these biodegradable nanoparticle-assisted anti-cancer therapies to come to realization in the clinics. CALAA-01, an anti-solid tumor nanoparticle containing siRNA, showed great potential in its phase I clinical trial (NCT00689065), was terminated after phase Ib as two of the five patients enrolled had experienced dose-limiting toxicities [104,105]. CRLX101, another anti-tumor targeted nanoparticle for various cancers, continues to show promising results and its clinical trial (NCT02769962) is still actively recruiting patients [106]. In 2019, the first ever siRNA nanodrug for hereditary amyloidosis, Onpattro, was approved by the FDA [107]. Onpattro encapsulates the therapeutic siRNA moiety into a lipid nanoparticle, and delivers it directly to the liver to prevent the body from producing the disease-causing amyloid proteins [107].

Similarly, research on stem cell therapy utilizing nanoparticles is also on the rise. Because of the small size and target specificity of the nanoparticles, scientists are aiming to treat some neurological diseases using this strategy [108].

4.3.3. Oral Drug Delivery

Another active area of research using biodegradable nanomaterials is to make it possible for certain drugs that are normally either poorly or erratically absorbed in the digestive system to be administered orally. This approach can ease the administration process of many biologics, proteins and peptides. For example, insulin is a peptide that is digested in the stomach [109]. Almost a century after the discovery of insulin, it can still only be administered via subcutaneous injection, adding not only physical discomfort and infection risks, but also psychological burdens to the diabetic patients. Research has been underway to make oral administration of insulin possible in order improve the quality of life of the diabetic patients [109]. A specific formulation of 1.6% zinc insulin in PLGA was developed in 2010 in an effort to realize the oral administration of insulin [109]. Although this PLGA nanoparticle encapsulating insulin only showed 11.4% of the efficacy of zinc insulin via intraperitoneal delivery, it still shed light to a possible future of oral administration of insulin [109]. More recently, combinations of different biodegradable nanomaterials, including chitosan, have also been applied in the development of oral delivery system of insulin [110,111]. Some phase I/II clinical trials are also underway [110].

4.4. Implantable Device with Biodegradable Materials

Continuous efforts have been putting forward to improve the outcomes of implantable devices using biodegradable materials. Biodegradable nanoparticles not only are employed in the nanodrug delivery systems but can also be incorporated in the implantable devices such as orthopedic fixation devices (including fracture-fixation pins and plates, interference screws, suture anchors, craniomaxillofacial fixation devices and tacks for meniscal repair), and biodegradable stents for percutaneous coronary intervention [112,113]. A number of these devices have already been approved and are available in the market [112]. The biodegradable nanoparticles enable the implanted devices to gradually degrade while the host tissues undergo constructive remodeling, eventually replacing the implant [112].

Significant progress has been made especially in interventional cardiology [114]. New drug-eluting stents have been developed to not only minimize neointimal hyperplasia and reduce restenosis after revascularization, but also minimize stent thrombosis, a problem that was observed at higher frequency with the first generation stents [114]. Recently, Lih et al. developed a new approach to prevent acid-induced inflammatory responses associated with biodegradable PLGA, by neutralizing the acidic environment using oligo(lactide)-grafted magnesium hydroxide ($Mg(OH)_2$) nanoparticles [115]. They demonstrated in porcine models that incorporating the modified $Mg(OH)_2$ nanoparticles within degradable coatings on drug-eluting arterial stents could efficiently attenuate the inflammatory response and in-stent intimal thickening [115]. Their results suggested that modifications of biodegradable nanoparticles could be useful to broaden the applicability and improve clinical success of biodegradable devices used in various biomedical fields. Biodegradable stents were invented with the intention to replace bare metal stents due to the high risk of in-stent restenosis using the metal materials [116]. However, great challenge still persists to achieve the right balance of the polymer, drug and degradation rate in order to avoid acute or chronic recoil and maintain vessel patency after stent implantation [116].

Taken together, biodegradable nanomaterials have shown many advantages in various biomedical applications. Here in Table 2 we highlight some of the above mentioned advantages to illustrate the ability of these biodegradable nanomaterials in meeting different clinical needs.

Table 2. Biomedical applications of biodegradable nanomaterials.

Purpose	Application	Advantages of Biodegradable Nanomaterials	References
Imaging	MRI photoacoustic imaging	• Reduced contrast load • Stronger signals • Targeted imaging	[66–76]
Theranostics	photoacoustic imaging and photothermal therapy	• Safely facilitate real-time therapeutic efficacy towards individualized treatment strategies	[78–87]
Targeted Delivery (carried by liposomes, polymeric nanoparticles, dendrimers, or micelles)	Drug delivery	• Encapsulation of hydrophobic molecules • Reduced premature degradation • Improved drug uptake • Sustained drug concentrations within the therapeutic window • Reduced side effects	[88–101]
	Gene therapy	• Protect DNA from enzymatic degradation • Reduced rejection from host immune system	[102–107]
	Antigen delivery	• Controlled release of antigen/ligand • Protect the antigen load	[112]
Implants	Stents	• Reduced vessel occlusion, restenosis, or late stent thrombosis • Improved lesion imaging • No need for a secondary surgical removal	[112]
	Mesh	• No need for a secondary procedure to remove the mesh	[112]
	Suture	• Low immunogenicity and toxicity • Excellent biocompatibility Predictable biodegradation rates • Good mechanical properties	[112]

5. Current Status of Biodegradable Nanomaterials and Challenges Ahead

Nanotoxicity, defined as toxicity induced by nanomaterials, is still an important discipline of research as the human body is being increasingly exposed to foreign materials at a nanoscale with the development of nanomedicine either intentionally or unintentionally [95]. Even with biodegradable nanomaterials, safety assessment remains as one of the top priorities in the application of nanomedicine. The toxicity of nanomaterials has been largely decreased with the application of biodegradable materials, even when sometimes the toxicity of the payload is unavoidable for some treatments [117]. However, it needs to be noted that not all biodegradable materials are deemed safe for application in humans. Even with biodegradability, some nanoparticles may still have undesired effects on the blood coagulation system due to their physiochemical properties such as size, charge and hydrophobicity [118]. The PLGA and PLA for clinical applications are manufactured under current good manufacturing practice protocols regulation by the FDA to ensure efficacy, safety, and stability for pharmaceuticals [11]. However, poly-alkyl-cyanoacrylate (PAC), for instance, can be degraded by esterases in the body but the degradation process produces toxic components [8,119]. Given the relatively short history of nanomedicine, the long-term effects of newly developed nanoparticles still need to be carefully evaluated.

The bio-distribution and pharmacokinetics of nanoparticles are largely dependent on the size, shape and the surface charge of the nanoparticles applied [117,120]. The early challenge of premature denaturation and undesired biodistribution of nanoparticles due to non-specific protein adsorption forming a protein corona around the material when being exposed to the biological environment has been solved by PEGylation [121]. However, it is still crucial to control the degradation rate of the nanomaterials and the payload's releasing profile since these biodegradable nanomaterials will eventually be degraded. The choice of nanomaterial also influences the outcome of certain nanotherapeutics. For example, PCL has a much lower encapsulation efficiency for taxol, an anti-cancer drug, compared to PLGA (20% vs. 100%) [122]. However, PCL nanoparticles have better therapeutic efficiency and stability than PLGA nanoparticles [122].

There is continuous advancement of the nanotechnologies using biodegradable materials in nanomedicine. Literature search using the PubMed database revealed that about 60% of all the studies on biodegradable nanomaterials in nanomedicine were from the past 5 years. However, the translation of different novel biodegradable nanoparticle designs into clinical settings remains a huge challenge. Current nanoparticle production methods are still constrained by several limitations, including the relatively high cost of particle production with difficulty in synthesizing particles that are homogeneous in shape and size; the low drug encapsulation efficiency; the difficulty in large scale production and sterilization; and the lack of reliable method for releasing profile measurement with the potential problem of high initial burst release or incomplete drug release [11]. Furthermore, there is still huge unknown regarding the correlation between nanoparticles' properties and their in vivo behaviors, their long-term stability, and how would some residual materials used for nanoparticle modification affect the human body [11]. As many nanoparticles have their unique structures and compositions which lead to their unique properties, there is also unmet need to develop standardized test protocols as well as reference particles for validation [11].

All these challenges call for the need to collect comprehensive information of biodegradable nanomaterials, drugs, as well as human data for the optimal modification and application of the nanoparticles and drugs. Even with this systemic review, we are still at risk of falling in the underreporting bias category as the studies were screened and selected manually. Using machine learning and artificial intelligence (AI), the properties of different nanomaterials and different combinations can be screened and the behavior of combinatorial nano-bio interface can be predicted [123,124]. The screening and model development might also lead to new discoveries of potential biodegradable nanomaterials and nanoparticle designs. Therefore, as for the future of the development of biodegradable nanomaterials in nanomedicine, machine learning and AI will be a great asset to the realization of efficient bench to bedside translation as well as personalized nanomedicine. Given the fast development of this field in the past few years, it is likely that in the near future, more newly developed biodegradable nanoparticles, especially multimodal nanoparticles, will be evaluated in clinical trials for their potential translational use.

6. Conclusions

In this review, we discussed different types of biodegradable nanomaterials and their applications in the biomedical field. These materials have demonstrated superiority compared to non-degradable counterparts and hold great translational potential in various clinical settings. There is still great challenge in developing nanomedicine and more biodegradable nanomaterials remain to be explored and validated for their potential clinical use.

Author Contributions: Conceptualization, S.S. and P.M.K.; Writing, S.S. and P.M.K.; Funding Acquisition, P.M.K. All authors have read and agreed to the published version of the manuscript.

Funding: This study was supported in part by grants from the National Institutes of Health R44DK103389-01 (P.M.K.), and American Heart Association Grant in Aid 17GRNT33680110 (P.M.K.).

Conflicts of Interest: The authors declare no conflict of interest.

References

1. Rossi, M.; Cubadda, F.; Dini, L.; Terranova, M.L.; Aureli, F.; Sorbo, A.; Passeri, D. Scientific Basis of Nanotechnology, Implications for the Food Sector and Future Trends. *Trends Food Sci. Technol.* **2014**, *40*, 127–148. [CrossRef]
2. Salinas, F.M.; Smith, D.M.; Viswanathan, S. Nanotechnology: Ethical and Social Issues. *Nanotechnol. Ethical Soc. Implic.* **2012**, 125–153. [CrossRef]
3. Sahoo, S.K.; Parveen, S.; Panda, J.J. The Present and Future of Nanotechnology in Human Health Care. *Nanomed. Nanotechnol. Biol. Med.* **2007**, *3*, 20–31. [CrossRef] [PubMed]
4. Duhan, J.S.; Kumar, R.; Kumar, N.; Kaur, P.; Nehra, K.; Duhan, S. Nanotechnology: The New Perspective in Precision Agriculture. *Biotechnol. Rep.* **2017**, *15*, 11–23. [CrossRef] [PubMed]
5. Roco, M.C.; Mirkin, C.A.; Hersam, M.C. Nanotechnology Research Directions for Societal Needs in 2020: Summary of International Study. *J. Nanopart. Res.* **2011**, *13*, 897–919. [CrossRef]
6. Mehta, D.; Guvva, S.; Patil, M. Future Impact of Nanotechnology on Medicine and Dentistry. *J. Indian Soc. Periodontol.* **2008**, *12*, 34. [CrossRef] [PubMed]
7. Pelaz, B.; Alexiou, C.; Alvarez-Puebla, R.A.; Alves, F.; Andrews, A.M.; Ashraf, S.; Balogh, L.P.; Ballerini, L.; Bestetti, A.; Brendel, C.; et al. Diverse Applications of Nanomedicine. *ACS Nano* **2017**, *11*, 2313–2381. [CrossRef]
8. Kumari, A.; Yadav, S.K.; Yadav, S.C. Biodegradable Polymeric Nanoparticles Based Drug Delivery Systems. *Colloids Surf. B Biointerfaces* **2010**, *75*, 1–18. [CrossRef]
9. Wiwanitkit, V. Biodegradable Nanoparticles for Drug Delivery and Targeting. *Surf. Modif. Nanopart. Target. Drug Deliv.* **2019**, 167–181. [CrossRef]
10. Farah, S.; Anderson, D.G.; Langer, R. Physical and Mechanical Properties of PLA, and Their Functions in Widespread Applications—A Comprehensive Review. *Adv. Drug Deliv. Rev.* **2016**, *107*, 367–392. [CrossRef]
11. Lee, B.K.; Yun, Y.; Park, K. PLA Micro- and Nano-Particles. *Adv. Drug Deliv. Rev.* **2016**, *107*, 176–191. [CrossRef] [PubMed]
12. Ding, D.; Zhu, Q. Recent Advances of PLGA Micro/Nanoparticles for the Delivery of Biomacromolecular Therapeutics. *Mater. Sci. Eng. C* **2018**, *92*, 1041–1060. [CrossRef] [PubMed]
13. Espinoza, S.M.; Patil, H.I.; San Martin Martinez, E.; Casañas Pimentel, R.; Ige, P.P. Poly-ε-Caprolactone (PCL), a Promising Polymer for Pharmaceutical and Biomedical Applications: Focus on Nanomedicine in Cancer. *Int. J. Polym. Mater. Polym. Biomater.* **2020**, *69*, 85–126. [CrossRef]
14. Mohammed, M.A.; Syeda, J.T.M.; Wasan, K.M.; Wasan, E.K. An Overview of Chitosan Nanoparticles and Its Application in Non-Parenteral Drug Delivery. *Pharmaceutics* **2017**, *9*, 53. [CrossRef] [PubMed]
15. Han, J.; Zhao, D.; Li, D.; Wang, X.; Jin, Z.; Zhao, K. Polymer-Based Nanomaterials and Applications for Vaccines and Drugs. *Polymers (Basel)* **2018**, *10*, 31. [CrossRef] [PubMed]
16. Mahapatro, A.; Singh, D.K. Biodegradable Nanoparticles Are Excellent Vehicle for Site Directed In-Vivo Delivery of Drugs and Vaccines. *J. Nanobiotechnol.* **2011**, *9*, 55. [CrossRef]
17. Villemin, E.; Ong, Y.C.; Thomas, C.M.; Gasser, G. Polymer Encapsulation of Ruthenium Complexes for Biological and Medicinal Applications. *Nat. Rev. Chem.* **2019**, *3*, 261–282. [CrossRef]
18. Panyam, J.; Labhasetwar, V. Biodegradable Nanoparticles for Drug and Gene Delivery to Cells and Tissue. *Adv. Drug Deliv. Rev.* **2003**, *55*, 329–347. [CrossRef]
19. Ribeiro, A.M.; Amaral, C.; Veiga, F.; Figueiras, A. Polymeric Micelles as a Versatile Tool in Oral Chemotherapy. *Des. Dev. New Nanocarr.* **2018**, 293–329. [CrossRef]
20. Chandarana, M.; Curtis, A.; Hoskins, C. The Use of Nanotechnology in Cardiovascular Disease. *Appl. Nanosci.* **2018**, *8*, 1607–1619. [CrossRef]
21. Nabar, G.M.; Mahajan, K.D.; Calhoun, M.A.; Duong, A.D.; Souva, M.S.; Xu, J.; Czeisler, C.; Puduvalli, V.K.; Otero, J.J.; Wyslouzil, B.E.; et al. Micelle-Templated, Poly(Lactic-Co-Glycolic Acid) Nanoparticles for Hydrophobic Drug Delivery. *Int. J. Nanomed.* **2018**, *13*, 351–366. [CrossRef] [PubMed]

22. Singh, A.K.; Yadav, T.P.; Pandey, B.; Gupta, V.; Singh, S.P. Engineering Nanomaterials for Smart Drug Release. *Appl. Target. Nano Drugs Deliv. Syst.* **2019**, 411–449. [CrossRef]
23. Husseini, G.A.; Pitt, W.G. Micelles and Nanoparticles for Ultrasonic Drug and Gene Delivery. *Adv. Drug Deliv. Rev.* **2008**, *60*, 1137–1152. [CrossRef] [PubMed]
24. Tapeinos, C.; Battaglini, M.; Ciofani, G. Advances in the Design of Solid Lipid Nanoparticles and Nanostructured Lipid Carriers for Targeting Brain Diseases. *J. Control. Release* **2017**, *264*, 306–332. [CrossRef] [PubMed]
25. Sin, L.T.; Rahmat, A.R.; Rahman, W.A.W.A. Synthesis and Production of Poly(Lactic Acid). *Polylactic Acid* **2013**, 71–107. [CrossRef]
26. Larrañeta, E.; Lutton, R.E.M.; Woolfson, A.D.; Donnelly, R.F. Microneedle Arrays as Transdermal and Intradermal Drug Delivery Systems: Materials Science, Manufacture and Commercial Development. *Mater. Sci. Eng. R Rep.* **2016**, *104*, 1–32. [CrossRef]
27. Saini, P.; Arora, M.; Kumar, M.N.V.R. Poly(Lactic Acid) Blends in Biomedical Applications. *Adv. Drug Deliv. Rev.* **2016**, *107*, 47–59. [CrossRef]
28. Da Silva, J.; Jesus, S.; Bernardi, N.; Colaço, M.; Borges, O. Poly(D, L-Lactic Acid) Nanoparticle Size Reduction Increases Its Immunotoxicity. *Front. Bioeng. Biotechnol.* **2019**, *7*, 137. [CrossRef]
29. Batrakova, E.V.; Bronich, T.K.; Vetro, J.A.; Kabanov, A.V. Polymer Micelles as Drug Carriers. *Nanopart. Drug Carr.* **2006**, 57–93. [CrossRef]
30. Rezvantalab, S.; Drude, N.I.; Moraveji, M.K.; Güvener, N.; Koons, E.K.; Shi, Y.; Lammers, T.; Kiessling, F. PLGA-Based Nanoparticles in Cancer Treatment. *Front. Pharmacol.* **2018**, *9*. [CrossRef]
31. Makadia, H.K.; Siegel, S.J. Poly Lactic-Co-Glycolic Acid (PLGA) as Biodegradable Controlled Drug Delivery Carrier. *Polymers (Basel)* **2011**, *3*, 1377–1397. [CrossRef] [PubMed]
32. Katsuki, S.; Matoba, T.; Koga, J.; Nakano, K.; Egashira, K. Anti-Inflammatory Nanomedicine for Cardiovascular Disease. *Front. Cardiovasc. Med.* **2017**, *4*, 87. [CrossRef] [PubMed]
33. Gunatillake, P.A.; Adhikari, R.; Gadegaard, N. Biodegradable Synthetic Polymers for Tissue Engineering. *Eur. Cells Mater.* **2003**, *5*, 1–16. [CrossRef] [PubMed]
34. Danhier, F.; Ansorena, E.; Silva, J.M.; Coco, R.; Le Breton, A.; Préat, V. PLGA-Based Nanoparticles: An Overview of Biomedical Applications. *J. Control. Release* **2012**, *161*, 505–522. [CrossRef] [PubMed]
35. Zhong, H.; Chan, G.; Hu, Y.; Hu, H.; Ouyang, D. A Comprehensive Map of FDA-Approved Pharmaceutical Products. *Pharmaceutics* **2018**, *10*, 263. [CrossRef] [PubMed]
36. Park, K.; Skidmore, S.; Hadar, J.; Garner, J.; Park, H.; Otte, A.; Soh, B.K.; Yoon, G.; Yu, D.; Yun, Y.; et al. Injectable, Long-Acting PLGA Formulations: Analyzing PLGA and Understanding Microparticle Formation. *J. Control. Release* **2019**, *304*, 125–134. [CrossRef] [PubMed]
37. Cho, H.; Gao, J.; Kwon, G.S. PEG-b-PLA Micelles and PLGA-b-PEG-b-PLGA Sol–Gels for Drug Delivery. *J. Control. Release* **2016**, *240*, 191–201. [CrossRef]
38. Zhang, K.; Tang, X.; Zhang, J.; Lu, W.; Lin, X.; Zhang, Y.; Tian, B.; Yang, H.; He, H. PEG-PLGA Copolymers: Their Structure and Structure-Influenced Drug Delivery Applications. *J. Control. Release* **2014**, *183*, 77–86. [CrossRef]
39. Otsuka, H.; Nagasaki, Y.; Kataoka, K. PEGylated Nanoparticles for Biological and Pharmaceutical Applications. *Adv. Drug Deliv. Rev.* **2003**, *55*, 403–419. [CrossRef]
40. Trivedi, R.; Kompella, U.B. Nanomicellar Formulations for Sustained Drug Delivery: Strategies and Underlying Principles. *Nanomedicine* **2010**, *5*, 485–505. [CrossRef]
41. Tang, H.; Zhao, W.; Yu, J.; Li, Y.; Zhao, C. Recent Development of PH-Responsive Polymers for Cancer Nanomedicine. *Molecules* **2019**, *24*, 4. [CrossRef] [PubMed]
42. Sánchez-González, S.; Diban, N.; Urtiaga, A. Hydrolytic Degradation and Mechanical Stability of Poly(ε-Caprolactone)/Reduced Graphene Oxide Membranes as Scaffolds for in Vitro Neural Tissue Regeneration. *Membranes (Basel)* **2018**, *8*, 12. [CrossRef] [PubMed]
43. Heimowska, A.; Morawska, M.; Bocho-Janiszewska, A. Biodegradation of Poly(ε-Caprolactone) in Natural Water Environments. *Polish J. Chem. Technol.* **2017**, *19*, 120–126. [CrossRef]
44. Zhang, H.; Tong, S.Y.; Zhang, X.Z.; Cheng, S.X.; Zhuo, R.X.; Li, H. Novel Solvent-Free Methods for Fabrication of Nano- And Microsphere Drug Delivery Systems from Functional Biodegradable Polymers. *J. Phys. Chem. C* **2007**, *111*, 12681–12685. [CrossRef]

45. Grossen, P.; Witzigmann, D.; Sieber, S.; Huwyler, J. PEG-PCL-Based Nanomedicines: A Biodegradable Drug Delivery System and Its Application. *J. Control. Release* **2017**, *260*, 46–60. [CrossRef] [PubMed]
46. Islam, N.; Dmour, I.; Taha, M.O. Degradability of Chitosan Micro/Nanoparticles for Pulmonary Drug Delivery. *Heliyon* **2019**, *5*, e01684. [CrossRef]
47. Szymańska, E.; Winnicka, K. Stability of Chitosan—A Challenge for Pharmaceutical and Biomedical Applications. *Mar. Drugs* **2015**, *13*, 1819–1846. [CrossRef]
48. Mandal, K.; Jana, D.; Ghorai, B.K.; Jana, N.R. Functionalized Chitosan with Self-Assembly Induced and Subcellular Localization-Dependent Fluorescence "switch on" Property. *New J. Chem.* **2018**, *42*, 5774–5784. [CrossRef]
49. Singh, B.; Garg, T.; Goyal, A.K.; Rath, G. Recent Advancements in the Cardiovascular Drug Carriers. *Artif. Cells Nanomed. Biotechnol.* **2016**, *44*, 216–225. [CrossRef]
50. Santos, A.; Veiga, F.; Figueiras, A. Dendrimers as Pharmaceutical Excipients: Synthesis, Properties, Toxicity and Biomedical Applications. *Materials (Basel)* **2019**, *13*, 65. [CrossRef]
51. Abbasi, E.; Aval, S.F.; Akbarzadeh, A.; Milani, M.; Nasrabadi, H.T.; Joo, S.W.; Hanifehpour, Y.; Nejati-Koshki, K.; Pashaei-Asl, R. Dendrimers: Synthesis, Applications, and Properties. *Nanoscale Res. Lett.* **2014**, *9*, 1–10. [CrossRef] [PubMed]
52. Duncan, R.; Izzo, L. Dendrimer Biocompatibility and Toxicity. *Adv. Drug Deliv. Rev.* **2005**, *57*, 2215–2237. [CrossRef] [PubMed]
53. Luo, K.; Li, C.; Li, L.; She, W.; Wang, G.; Gu, Z. Arginine Functionalized Peptide Dendrimers as Potential Gene Delivery Vehicles. *Biomaterials* **2012**, *33*, 4917–4927. [CrossRef] [PubMed]
54. Pattni, B.S.; Chupin, V.V.; Torchilin, V.P. New Developments in Liposomal Drug Delivery. *Chem. Rev.* **2015**, *115*, 10938–10966. [CrossRef]
55. Li, Z.; Tan, S.; Li, S.; Shen, Q.; Wang, K. Cancer Drug Delivery in the Nano Era: An Overview and Perspectives (Review). *Oncol. Rep.* **2017**, *38*, 611–624. [CrossRef]
56. Sercombe, L.; Veerati, T.; Moheimani, F.; Wu, S.Y.; Sood, A.K.; Hua, S. Advances and Challenges of Liposome Assisted Drug Delivery. *Front. Pharmacol.* **2015**, *6*, 286. [CrossRef]
57. Dong, C.; Ma, A.; Shang, L. Nanoparticles for Postinfarct Ventricular Remodeling. *Nanomedicine* **2018**, *13*, 3037–3050. [CrossRef]
58. Jesorka, A.; Orwar, O. Liposomes: Technologies and Analytical Applications. *Annu. Rev. Anal. Chem.* **2008**, *1*, 801–832. [CrossRef]
59. Mazur, F.; Bally, M.; Städler, B.; Chandrawati, R. Liposomes and Lipid Bilayers in Biosensors. *Adv. Colloid Interface Sci.* **2017**, *249*, 88–99. [CrossRef]
60. Xiao, Y.; Liu, Q.; Clulow, A.J.; Li, T.; Manohar, M.; Gilbert, E.P.; de Campo, L.; Hawley, A.; Boyd, B.J. PEGylation and Surface Functionalization of Liposomes Containing Drug Nanocrystals for Cell-Targeted Delivery. *Colloids Surf. B Biointerfaces* **2019**, *182*, 110362. [CrossRef]
61. Feng, L.; Mumper, R.J. A Critical Review of Lipid-Based Nanoparticles for Taxane Delivery. *Cancer Lett.* **2013**, *334*, 157–175. [CrossRef]
62. Dave, V.; Tak, K.; Sohgaura, A.; Gupta, A.; Sadhu, V.; Reddy, K.R. Lipid-Polymer Hybrid Nanoparticles: Synthesis Strategies and Biomedical Applications. *J. Microbiol. Methods* **2019**, *160*, 130–142. [CrossRef]
63. Chiang, Y.T.; Lyu, S.Y.; Wen, Y.H.; Lo, C.L. Preparation and Characterization of Electrostatically Crosslinked Polymer–Liposomes in Anticancer Therapy. *Int. J. Mol. Sci.* **2018**, *19*, 1615. [CrossRef]
64. Sahoo, N.; Sahoo, R.K.; Biswas, N.; Guha, A.; Kuotsu, K. Recent Advancement of Gelatin Nanoparticles in Drug and Vaccine Delivery. *Int. J. Biol. Macromol.* **2015**, *81*, 317–331. [CrossRef]
65. Kommareddy, S.; Shenoy, D.B.; Amiji, M.M. Gelatin Nanoparticles and Their Biofunctionalization. *Nanotechnol. Life Sci.* **2007**. [CrossRef]
66. Seeliger, E.; Sendeski, M.; Rihal, C.S.; Persson, P.B. Contrast-Induced Kidney Injury: Mechanisms, Risk Factors, and Prevention. *Eur. Heart J.* **2012**, *33*, 2007–2015. [CrossRef]
67. Tam, J.M.; Tam, J.O.; Murthy, A.; Ingram, D.R.; Ma, L.L.; Travis, K.; Johnston, K.P.; Sokolov, K.V. Controlled Assembly of Biodegradable Plasmonic Nanoclusters for Near-Infrared Imaging and Therapeutic Applications. *ACS Nano* **2010**, *4*, 2178–2184. [CrossRef]
68. Huang, C.H.; Nwe, K.; Al Zaki, A.; Brechbiel, M.W.; Tsourkas, A. Biodegradable Polydisulfide Dendrimer Nanoclusters as MRI Contrast Agents. *ACS Nano* **2012**, *6*, 9416–9424. [CrossRef]

69. Fathi, P.; Knox, H.J.; Sar, D.; Tripathi, I.; Ostadhossein, F.; Misra, S.K.; Esch, M.B.; Chan, J.; Pan, D. Biodegradable Biliverdin Nanoparticles for Efficient Photoacoustic Imaging. *ACS Nano* **2019**, *13*, 7690–7704. [CrossRef]
70. Beard, P. Biomedical Photoacoustic Imaging. *Interface Focus* **2011**, *1*, 602–631. [CrossRef]
71. Cheheltani, R.; Ezzibdeh, R.M.; Chhour, P.; Pulaparthi, K.; Kim, J.; Jurcova, M.; Hsu, J.C.; Blundell, C.; Litt, H.I.; Ferrari, V.A.; et al. Tunable, Biodegradable Gold Nanoparticles as Contrast Agents for Computed Tomography and Photoacoustic Imaging. *Biomaterials* **2016**, *102*, 87–97. [CrossRef]
72. Wei, R.; Jiang, G.; Lv, M.Q.; Tan, S.; Wang, X.; Zhou, Y.; Cheng, T.; Gao, X.; Chen, X.; Wang, W.; et al. TMTP1-Modified Indocyanine Green-Loaded Polymeric Micelles for Targeted Imaging of Cervical Cancer and Metastasis Sentinel Lymph Node in Vivo. *Theranostics* **2019**, *9*, 7325–7344. [CrossRef]
73. Bonnard, T.; Jayapadman, A.; Putri, J.A.; Cui, J.; Ju, Y.; Carmichael, C.; Angelovich, T.A.; Cody, S.H.; French, S.; Pascaud, K.; et al. Low-Fouling and Biodegradable Protein-Based Particles for Thrombus Imaging. *ACS Nano* **2018**, *12*, 6988–6996. [CrossRef]
74. Mulder, W.J.M.; Jaffer, F.A.; Fayad, Z.A.; Nahrendorf, M. Imaging and Nanomedicine in Inflammatory Atherosclerosis. *Sci. Transl. Med.* **2014**, *6*, 239sr1. [CrossRef]
75. Lobatto, M.E.; Fuster, V.; Fayad, Z.A.; Mulder, W.J.M. Perspectives and Opportunities for Nanomedicine in the Management of Atherosclerosis. *Nat. Rev. Drug Discov.* **2011**, *10*, 835–852. [CrossRef]
76. Peters, D.; Kastantin, M.; Kotamraju, V.R.; Karmali, P.P.; Gujraty, K.; Tirrell, M.; Ruoslahti, E. Targeting Atherosclerosis by Using Modular, Multifunctional Micelles. *Proc. Natl. Acad. Sci. USA* **2009**, *106*, 9815–9819. [CrossRef]
77. Zhou, Z.; Qutaish, M.; Han, Z.; Schur, R.M.; Liu, Y.; Wilson, D.L.; Lu, Z.R. MRI Detection of Breast Cancer Micrometastases with a Fibronectin-Targeting Contrast Agent. *Nat. Commun.* **2015**, *6*. [CrossRef]
78. Swider, E.; Koshkina, O.; Tel, J.; Cruz, L.J.; de Vries, I.J.M.; Srinivas, M. Customizing Poly(Lactic-Co-Glycolic Acid) Particles for Biomedical Applications. *Acta Biomater.* **2018**, *73*, 38–51. [CrossRef]
79. Wang, S.; Kim, G.; Lee, Y.E.K.; Hah, H.J.; Ethirajan, M.; Pandey, R.K.; Kopelman, R. Multifunctional Biodegradable Polyacrylamide Nanocarriers for Cancer Theranostics-A "see and Treat" Strategy. *ACS Nano* **2012**, *6*, 6843–6851. [CrossRef]
80. Mir, M.; Ahmed, N.; ur Rehman, A. Recent Applications of PLGA Based Nanostructures in Drug Delivery. *Colloids Surfaces B Biointerfaces* **2017**, *159*, 217–231. [CrossRef]
81. Chan, J.M.W.; Wojtecki, R.J.; Sardon, H.; Lee, A.L.Z.; Smith, C.E.; Shkumatov, A.; Gao, S.; Kong, H.; Yang, Y.Y.; Hedrick, J.L. Self-Assembled, Biodegradable Magnetic Resonance Imaging Agents: Organic Radical-Functionalized Diblock Copolymers. *ACS Macro Lett.* **2017**, *6*, 176–180. [CrossRef]
82. Huang, P.; Lin, J.; Li, W.; Rong, P.; Wang, Z.; Wang, S.; Wang, X.; Sun, X.; Aronova, M.; Niu, G.; et al. Biodegradable Gold Nanovesicles with an Ultrastrong Plasmonic Coupling Effect for Photoacoustic Imaging and Photothermal Therapy. *Angew. Chem.-Int. Ed.* **2013**, *52*, 13958–13964. [CrossRef]
83. Lyu, Y.; Zeng, J.; Jiang, Y.; Zhen, X.; Wang, T.; Qiu, S.; Lou, X.; Gao, M.; Pu, K. Enhancing Both Biodegradability and Efficacy of Semiconducting Polymer Nanoparticles for Photoacoustic Imaging and Photothermal Therapy. *ACS Nano* **2018**, *12*, 1801–1810. [CrossRef]
84. Lee, M.Y.; Lee, C.; Jung, H.S.; Jeon, M.; Kim, K.S.; Yun, S.H.; Kim, C.; Hahn, S.K. Biodegradable Photonic Melanoidin for Theranostic Applications. *ACS Nano* **2016**, *10*, 822–831. [CrossRef]
85. Fu, L.H.; Hu, Y.R.; Qi, C.; He, T.; Jiang, S.; Jiang, C.; He, J.; Qu, J.; Lin, J.; Huang, P. Biodegradable Manganese-Doped Calcium Phosphate Nanotheranostics for Traceable Cascade Reaction-Enhanced Anti-Tumor Therapy. *ACS Nano* **2019**, *13*, 13985–13994. [CrossRef]
86. Ceylan, H.; Yasa, I.C.; Yasa, O.; Tabak, A.F.; Giltinan, J.; Sitti, M. 3D-Printed Biodegradable Microswimmer for Theranostic Cargo Delivery and Release. *ACS Nano* **2019**, *13*, 3353–3362. [CrossRef]
87. Banik, B.; Surnar, B.; Askins, B.W.; Banerjee, M.; Dhar, S. Dual-Targeted Synthetic Nanoparticles for Cardiovascular Diseases. *ACS Appl. Mater. Interfaces* **2020**, *12*, 6852–6862. [CrossRef]
88. Karlsson, J.; Vaughan, H.J.; Green, J.J. Biodegradable Polymeric Nanoparticles for Therapeutic Cancer Treatments. *Annu. Rev. Chem. Biomol. Eng.* **2018**, *9*, 105–127. [CrossRef]
89. Derakhshandeh, K.; Erfan, M.; Dadashzadeh, S. Encapsulation of 9-Nitrocamptothecin, a Novel Anticancer Drug, in Biodegradable Nanoparticles: Factorial Design, Characterization and Release Kinetics. *Eur. J. Pharm. Biopharm.* **2007**, *66*, 34–41. [CrossRef]

90. Onwordi, E.N.C.; Gamal, A.; Zaman, A. Anticoagulant Therapy for Acute Coronary Syndromes. *Interv. Cardiol. Rev.* **2018**, *13*, 87–92. [CrossRef]
91. Takahama, H.; Shigematsu, H.; Asai, T.; Matsuzaki, T.; Sanada, S.; Fu, H.Y.; Okuda, K.; Yamato, M.; Asanuma, H.; Asano, Y.; et al. Liposomal Amiodarone Augments Anti-Arrhythmic Effects and Reduces Hemodynamic Adverse Effects in an Ischemia/ Reperfusion Rat Model. *Cardiovasc. Drugs Ther.* **2013**, *27*, 125–132. [CrossRef]
92. Singh, A.P.; Biswas, A.; Shukla, A.; Maiti, P. Targeted Therapy in Chronic Diseases Using Nanomaterial-Based Drug Delivery Vehicles. *Signal Transduct. Target. Ther.* **2019**, *4*. [CrossRef]
93. Flores, A.M.; Ye, J.; Jarr, K.U.; Hosseini-Nassab, N.; Smith, B.R.; Leeper, N.J. Nanoparticle Therapy for Vascular Diseases. *Arterioscler. Thromb. Vasc. Biol.* **2019**, *39*, 635–646. [CrossRef]
94. Binderup, T.; Duivenvoorden, R.; Fay, F.; Van Leent, M.M.T.; Malkus, J.; Baxter, S.; Ishino, S.; Zhao, Y.; Sanchez-Gaytan, B.; Teunissen, A.J.P.; et al. Imaging-Assisted Nanoimmunotherapy for Atherosclerosis in Multiple Species. *Sci. Transl. Med.* **2019**, *11*. [CrossRef]
95. Shvedova, A.A.; Kagan, V.E.; Fadeel, B. Close Encounters of the Small Kind: Adverse Effects of Man-Made Materials Interfacing with the Nano-Cosmos of Biological Systems. *Annu. Rev. Pharmacol. Toxicol.* **2010**, *50*, 63–88. [CrossRef]
96. Manke, A.; Wang, L.; Rojanasakul, Y. Mechanisms of Nanoparticle-Induced Oxidative Stress and Toxicity. *BioMed Res. Int.* **2013**, *2013*. [CrossRef]
97. Lee, D.; Bae, S.; Hong, D.; Lim, H.; Yoon, J.H.; Hwang, O.; Park, S.; Ke, Q.; Khang, G.; Kang, P.M. H_2O_2-Responsive Molecularly Engineered Polymer Nanoparticles as Ischemia/Reperfusion-Targeted Nanotherapeutic Agents. *Sci. Rep.* **2013**, *3*, 2233. [CrossRef]
98. Bae, S.; Park, M.; Kang, C.; Dilmen, S.; Kang, T.H.; Kang, D.G.; Ke, Q.; Lee, S.U.; Lee, D.; Kang, P.M. Hydrogen Peroxide-Responsive Nanoparticle Reduces Myocardial Ischemia/Reperfusion Injury. *J. Am. Heart Assoc.* **2016**, *5*, e003697. [CrossRef]
99. Andrabi, S.S.; Yang, J.; Gao, Y.; Kuang, Y.; Labhasetwar, V. Nanoparticles with Antioxidant Enzymes Protect Injured Spinal Cord from Neuronal Cell Apoptosis by Attenuating Mitochondrial Dysfunction. *J. Control. Release* **2020**, *317*, 300–311. [CrossRef]
100. Tapeinos, C.; Larrañaga, A.; Sarasua, J.R.; Pandit, A. Functionalised Collagen Spheres Reduce H_2O_2 Mediated Apoptosis by Scavenging Overexpressed ROS. *Nanomed. Nanotechnol. Biol. Med.* **2018**, *14*, 2397–2405. [CrossRef]
101. Kim, K.S.; Song, C.G.; Kang, P.M. Targeting Oxidative Stress Using Nanoparticles as a Theranostic Strategy for Cardiovascular Diseases. *Antioxid. Redox Signal.* **2019**, *30*, 733–746. [CrossRef] [PubMed]
102. Chen, J.; Guo, Z.; Tian, H.; Chen, X. Production and Clinical Development of Nanoparticles for Gene Delivery. *Mol. Ther.-Methods Clin. Dev.* **2016**, *3*, 16023. [CrossRef] [PubMed]
103. Gu, G.; Hu, Q.; Feng, X.; Gao, X.; Menglin, J.; Kang, T.; Jiang, D.; Song, Q.; Chen, H.; Chen, J. PEG-PLA Nanoparticles Modified with APTEDB Peptide for Enhanced Anti-Angiogenic and Anti-Glioma Therapy. *Biomaterials* **2014**, *35*, 8215–8226. [CrossRef] [PubMed]
104. Davis, M.E.; Zuckerman, J.E.; Choi, C.H.J.; Seligson, D.; Tolcher, A.; Alabi, C.A.; Yen, Y.; Heidel, J.D.; Ribas, A. Evidence of RNAi in Humans from Systemically Administered SiRNA via Targeted Nanoparticles. *Nature* **2010**, *464*, 1067–1070. [CrossRef]
105. Zuckerman, J.E.; Gritli, I.; Tolcher, A.; Heidel, J.D.; Lim, D.; Morgan, R.; Chmielowski, B.; Ribas, A.; Davis, M.E.; Yen, Y. Correlating Animal and Human Phase Ia/Ib Clinical Data with CALAA-01, a Targeted, Polymer-Based Nanoparticle Containing SiRNA. *Proc. Natl. Acad. Sci. USA* **2014**, *111*, 11449–11454. [CrossRef]
106. Voss, M.H.; Hussain, A.; Vogelzang, N.; Lee, J.L.; Keam, B.; Rha, S.Y.; Vaishampayan, U.; Harris, W.B.; Richey, S.; Randall, J.M.; et al. A Randomized Phase II Trial of CRLX101 in Combination with Bevacizumab versus Standard of Care in Patients with Advanced Renal Cell Carcinoma. *Ann. Oncol.* **2017**, *28*, 2754–2760. [CrossRef]
107. Rai, R.; Alwani, S.; Badea, I. Polymeric Nanoparticles in Gene Therapy: New Avenues of Design and Optimization for Delivery Applications. *Polymers (Basel)* **2019**, *11*, 745. [CrossRef]
108. Zhang, G.; Khan, A.A.; Wu, H.; Chen, L.; Gu, Y.; Gu, N. The Application of Nanomaterials in Stem Cell Therapy for Some Neurological Diseases. *Curr. Drug Targets* **2018**, *19*, 279–298. [CrossRef]

109. Díaz, A.; David, A.; Pérez, R.; González, M.L.; Báez, A.; Wark, S.E.; Zhang, P.; Clearfield, A.; Colón, J.L. Nanoencapsulation of Insulin into Zirconium Phosphate for Oral Delivery Applications. *Biomacromolecules* **2010**, *11*, 2465–2470. [CrossRef]
110. Han, Y.; Gao, Z.; Chen, L.; Kang, L.; Huang, W.; Jin, M.; Wang, Q.; Bae, Y.H. Multifunctional Oral Delivery Systems for Enhanced Bioavailability of Therapeutic Peptides/Proteins. *Acta Pharm. Sin. B* **2019**, *9*, 902–922. [CrossRef]
111. Safari, M.; Kamari, Y.; Ghiaci, M.; Sadeghi-aliabadi, H.; Mirian, M. Synthesis and Characterization of Insulin/Zirconium Phosphate@TiO$_2$ Hybrid Composites for Enhanced Oral Insulin Delivery Applications. *Drug Dev. Ind. Pharm.* **2017**, *43*, 862–870. [CrossRef] [PubMed]
112. Li, C.; Guo, C.; Fitzpatrick, V.; Ibrahim, A.; Zwierstra, M.J.; Hanna, P.; Lechtig, A.; Nazarian, A.; Lin, S.J.; Kaplan, D.L. Design of Biodegradable, Implantable Devices towards Clinical Translation. *Nat. Rev. Mater.* **2020**, *5*, 61–81. [CrossRef]
113. Tyler, B.; Gullotti, D.; Mangraviti, A.; Utsuki, T.; Brem, H. Polylactic Acid (PLA) Controlled Delivery Carriers for Biomedical Applications. *Adv. Drug Deliv. Rev.* **2016**, *107*, 163–175. [CrossRef]
114. Lee, D.H.; de la Torre Hernandez, J.M. The Newest Generation of Drug-Eluting Stents and Beyond. *Eur. Cardiol. Rev.* **2018**, *13*, 54–59. [CrossRef]
115. Lih, E.; Kum, C.H.; Park, W.; Chun, S.Y.; Cho, Y.; Joung, Y.K.; Park, K.S.; Hong, Y.J.; Ahn, D.J.; Kim, B.S.; et al. Modified Magnesium Hydroxide Nanoparticles Inhibit the Inflammatory Response to Biodegradable Poly(Lactide- Co-Glycolide) Implants. *ACS Nano* **2018**, *12*, 6917–6925. [CrossRef]
116. Di Mario, C.; Ferrante, G. Biodegradable Drug-Eluting Stents: Promises and Pitfalls. *Lancet* **2008**, *371*, 873–874. [CrossRef]
117. Cassano, D.; Pocoví-Martínez, S.; Voliani, V. Ultrasmall-in-Nano Approach: Enabling the Translation of Metal Nanomaterials to Clinics. *Bioconjug. Chem.* **2018**, *29*, 4–16. [CrossRef]
118. Ilinskaya, A.N.; Dobrovolskaia, M.A. Nanoparticles and the Blood Coagulation System. Part II: Safety Concerns. *Nanomedicine* **2013**, *8*, 969–981. [CrossRef]
119. Sulheim, E.; Iversen, T.G.; Nakstad, V.T.; Klinkenberg, G.; Sletta, H.; Schmid, R.; Hatletveit, A.R.; Wågbø, A.M.; Sundan, A.; Skotland, T.; et al. Cytotoxicity of Poly(Alkyl Cyanoacrylate) Nanoparticles. *Int. J. Mol. Sci.* **2017**, *18*, 2454. [CrossRef]
120. Blanco, E.; Shen, H.; Ferrari, M. Principles of Nanoparticle Design for Overcoming Biological Barriers to Drug Delivery. *Nat. Biotechnol.* **2015**, *33*, 941–951. [CrossRef]
121. Bobo, D.; Robinson, K.J.; Islam, J.; Thurecht, K.J.; Corrie, S.R. Nanoparticle-Based Medicines: A Review of FDA-Approved Materials and Clinical Trials to Date. *Pharm. Res.* **2016**, *33*, 2373–2387. [CrossRef]
122. Kim, S.Y.; Lee, Y.M. Taxol-Loaded Block Copolymer Nanospheres Composed of Methoxy Poly(Ethylene Glycol) and Poly(ε-Caprolactone) as Novel Anticancer Drug Carriers. *Biomaterials* **2001**, *22*, 1697–1704. [CrossRef]
123. Ho, D.; Wang, P.; Kee, T. Artificial Intelligence in Nanomedicine. *Nanoscale Horiz.* **2019**, *4*, 365–377. [CrossRef]
124. Cai, P.; Zhang, X.; Wang, M.; Wu, Y.L.; Chen, X. Combinatorial Nano-Bio Interfaces. *ACS Nano* **2018**, *12*, 5078–5084. [CrossRef]

© 2020 by the authors. Licensee MDPI, Basel, Switzerland. This article is an open access article distributed under the terms and conditions of the Creative Commons Attribution (CC BY) license (http://creativecommons.org/licenses/by/4.0/).

Review

Application of Biodegradable and Biocompatible Nanocomposites in Electronics: Current Status and Future Directions

Haichao Liu [1,†], Ranran Jian [2,†], Hongbo Chen [3], Xiaolong Tian [3], Changlong Sun [4], Jing Zhu [5], Zhaogang Yang [6,*], Jingyao Sun [1,2,*] and Chuansheng Wang [1,3,*]

1. Academic Division of Engineering, Qingdao University of Science & Technology, Qingdao 266061, China
2. College of Mechanical and Electrical Engineering, Beijing University of Chemical Technology, Beijing 100029, China
3. College of Electromechanical Engineering, Qingdao University of Science & Technology, Qingdao 266061, China
4. College of Sino-German Science and Technology, Qingdao University of Science & Technology, Qingdao 266061, China
5. College of Pharmacy, The Ohio State University, Columbus, OH 43210, USA
6. Department of Radiation Oncology, The University of Texas Southwestern Medical Center, Dallas, TX 75390, USA

* Correspondence: Zhaogang.Yang@UTSouthwestern.edu (Z.Y.); sunjingyao@mail.buct.edu.cn (J.S.); wcsmta@qust.edu.cn (C.W.); Tel.: +1-214-645-6873 (Z.Y.); +86-10-6443-5015 (J.S.); +86-136-0896-6169 (C.W.)

† These authors contributed equally to this work.

Received: 15 May 2019; Accepted: 24 June 2019; Published: 29 June 2019

Abstract: With the continuous increase in the production of electronic devices, large amounts of electronic waste (E-waste) are routinely being discarded into the environment. This causes serious environmental and ecological problems because of the non-degradable polymers, released hazardous chemicals, and toxic heavy metals. The appearance of biodegradable polymers, which can be degraded or dissolved into the surrounding environment with no pollution, is promising for effectively relieving the environmental burden. Additionally, biodegradable polymers are usually biocompatible, which enables electronics to be used in implantable biomedical applications. However, for some specific application requirements, such as flexibility, electric conductivity, dielectric property, gas and water vapor barrier, most biodegradable polymers are inadequate. Recent research has focused on the preparation of nanocomposites by incorporating nanofillers into biopolymers, so as to endow them with functional characteristics, while simultaneously maintaining effective biodegradability and biocompatibility. As such, bionanocomposites have broad application prospects in electronic devices. In this paper, emergent biodegradable and biocompatible polymers used as insulators or (semi)conductors are first reviewed, followed by biodegradable and biocompatible nanocomposites applied in electronics as substrates, (semi)conductors and dielectrics, as well as electronic packaging, which is highlighted with specific examples. To finish, future directions of the biodegradable and biocompatible nanocomposites, as well as the challenges, that must be overcome are discussed.

Keywords: biodegradable; biocompatible; electronics; nanocomposites

1. Introduction

Electronic products have enhanced our lives and brought about changes in almost all areas, including communications, manufacturing, entertainment, and health care [1]. With the rapid renewal of electronic products, such as smartphones and tablets, the life of electronic products is becoming shorter. As a result, an increasing amount of electronic waste (E-waste) is routinely discarded [2,3].

The fastest growing type of E-waste is solid waste. Not only is solid E-waste comprised of a large amount of non-degradable polymers, but it also releases hazardous chemicals and toxic heavy metals, both of which are damaging to the environment and ecology [4,5]. For certain electronic products, this damage would start with raw material procurement and continue throughout the whole life cycle [2].

Biodegradable electronics may be an effective solution for E-waste management, since they can be degraded or dissolved into the surrounding environment with no pollution. This endows the electronics with environmental safety and disposability [6–8], by simultaneously decreasing the cost for recycling operations and the health risks associated with harmful emissions [9–12].

Additionally, biodegradable materials are usually biocompatible, which enables electronics to be used in implantable biomedical applications. Biocompatibility allows the materials to directly contact tissues or skin without generating adverse effects [13–17]. Furthermore, electronics which are both biodegradable and biocompatible can be dissolved or resorbed safely by human body at controlled rates after treatment or diagnosis is completed. Eliminating the need for a second surgery to retrieve the device simultaneously decreases the associated infection risks [18].

Besides biodegradability and biocompatibility, some other characteristics, including flexibility, mechanical properties, electric conductivity, and gas and vapor barrier properties, are also essential for specific applications in electronics. However, many polymers cannot completely meet these performance requirements. Therefore, recent research has focused on incorporating nanofillers with excellent properties into polymers so as to improve their performance capabilities [19–22].

This paper aims to carefully demonstrate the development and potential of the biodegradable and biocompatible nanocomposites in electronic applications. It will first review emergent biodegradable and biocompatible polymers used as insulators or (semi)conductors, and then highlight specific examples of nanocomposites used in electronics as substrates, conductors, semiconductors, and dielectrics, as well as electronic packaging [23].

2. Biodegradable and Biocompatible Polymers

Biopolymers are the basis of biodegradable and biocompatible nanocomposites. They can be classified as natural-based polymers and synthetic polymers [13]. Natural-based polymers refer to those which come from nature. Table 1 shows an overview of biodegradable and biocompatible polymers used to fabricate electronics. In this section, biodegradable and biocompatible polymers will be introduced according to their conductivity, since the electrical property directly determines their application directions.

Table 1. Summary of biopolymers mentioned in this review.

Category	Polymer Material	Electrical Property	Biodegradable/ Biocompatible	Applications
Natural Polymers	Cellulose	Insulator	Both	Substrate [24,25]; Dielectric [26]
	Silk	Insulator	Both	Substrate [27,28]; Dielectric [29]
	Shellac	Insulator	Both	Substrate [30]; Dielectric [30,31]
	Gelatin	Insulator	Both	Substrate [32,33]; Dielectric [34–36]
Synthetic Polymer	Poly(vinyl alcohol) (PVA)	Insulator	Biocompatible	Substrate [37,38]; Dielectric [39,40]
	Polydimethylsiloxane (PDMS)	Insulator	Biocompatible	Substrate [41]; Dielectric [42–45]
	Polylactide (PLA)	Insulator	Both	Substrate [46–48]; Dielectric [49]
	Polycaprolactone (PCL)	Insulator	Both	Dielectric [49]
	Poly(glycerol-co-sebacate) (PGS)	Insulator	Both	Dielectric [50]
	Poly(lactic-co-glycolic acid) (PLGA)	Insulator	Both	Substrate [51]; Dielectric [52]
	Polyaniline (PANI)	Conductor (doped)	Biocompatible	Conductor [53]
	Polypyrrole (PPy)	Conductor (doped)	Biocompatible	Conductor [54]
	Poly(3,4-ethylenedioxythiophene) (PEDOT)	Conductor (doped)	Biocompatible	Conductor [55]

2.1. Insulated Polymers

Cellulose, as a macromolecule polysaccharide composed of glucose, is the oldest and cheapest biodegradable natural source polymer. It is inexpensive, biodegradable, abundant, easily available, and lightweight, and thus is considered to be a potential substitute for the substrate materials of various electronic devices, including organic field-effect transistors (OFETs), organic light-emitting diodes (OLEDs), and solar cells [56–60]. For example, Zhang et al. [61] introduced a MoS_2 phototransistor with a flexible and transparent paper substrate (fabricated from cellulose), as shown in Figure 1. The phototransistor has a high transparency with an average transmittance of 82%. Aside from its use as a substrate, cellulose can also be used to fabricate dielectrics [26,62,63]. Dai et al. [64] fabricated a class of all solid-state ionic dielectrics using cellulose nanopaper. These dielectrics show high transparency, low surface roughness, good thermal durability, and excellent mechanical properties. The successful applications of cellulose as substrates and dielectric materials demonstrate its potential for use in flexible, environmentally friendly and biodegradable electronic devices.

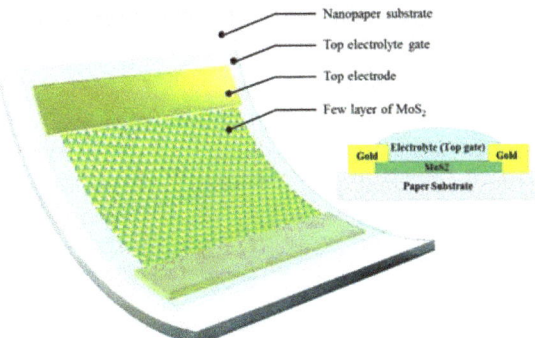

Figure 1. Three-dimensional schematic and cross-sectional view of the MoS_2 phototransistor, fabricated on flexible and transparent cellulose nanopaper. Reproduced with permission from [61], RSC, 2016.

Silk is a polypeptide polymer mainly composed of fibroin and sericin [65]. Because of its outstanding mechanical properties, flexibility, processability, and chemical stability, silk is an ideal backbone for flexible and stretchable electronics [66]. Moreover, silk is non-toxic, completely biodegradable and bioresorbable. It can also be safely implanted into the human body with no immune response, which allows it to be used for implantable electronic therapeutic devices. Kim et al. [67] successfully fabricated an ultrathin electronic sensor array on silk, and tested its performance in vivo by placing it onto exposed brain tissue. The silk was safely dissolved and resorbed, forming a conformal coating on folded brain tissue with the sensor array. Other studies have also demonstrated the successful application of silk as a substrate in implantable electronics [28,68] and food sensors [69]. Applications of silk in dielectrics were also reported [29,70–77]. Liang et al. [70] fabricated organic thin-film transistors (OTFTs) with silk as their dielectric layer. The silk dielectric layer annealed at 40 °C, and had the smallest particles and least aggregation. The mobility of the OTFTs was 2.06×10^{-3} $cm^2\ V^{-1}\ s^{-1}$, and the highest on/off ratio was 10^3.

Shellac is a natural resin collected from the secretion of the female lac bug after they ingest the sap of their host trees. Shellac can not only be extracted through a variety of polar organic solutions, but it can also be synthesized from a multitude of compositional grades and shades [78,79]. Similar to the aforementioned biopolymers, shellac is biodegradable and can be used as an electronic substrate and a dielectric [30]. Irimia-Vladu et al. [80] reported an organic thin-film transistor (OTFT), which is built on a smooth and uniform shellac film substrate prepared by drop-casting. The OTFT exhibits a mobility of $10^{-2}\ cm^2\ V^{-1}\ s^{-1}$, partially attributed to the outstanding barrier and insulation properties of the shellac film substrate. In addition, shellac also shows excellent dielectric properties. Baek et al. [31]

fabricated semiconducting copolymer-based OFETs with shellac and poly(4-vinylphenol) (PVP) as the dielectric materials. The shellac dielectric layer facilitated electron transport at the interface with copolymer channels, endowing the OFETs with superior performances.

Gelatin is another protein-based material, derived from the degradation of collagen in connective tissues, such as animal skin, bone, sarcolemma, and muscle. It is fully biocompatible and biodegradable and most commonly used for oral drug capsules [81]. Nowadays, gelatin is also the basis of many substrates and dielectrics of high-performance electronics [32–35]. Electronics mounted on hard gelatin substrates can be easily ingested for specific biomedical applications. When used as the gate dielectric in oxide FETs, gelatin could yield a specific capacitance over 0.93 µF cm^{-2} as a result of the formation of electric-double-layers [35].

In addition to natural-based polymers, some synthetic polymers also possess excellent biodegradability and biocompatibility. Poly(vinyl alcohol) (PVA) is one such synthetic polymer and has been widely used in the substrates and dielectrics of electronics [37,38,82,83]. Kim et al. [84] reported an integrated device on the surface of a thin polydimethylsiloxane (PDMS) foil with a water-soluble PVA substrate designed to measure the electrical signals produced by human body. The integrated device includes a set of multifunctional sensors, transistors, capacitors, photo-detectors, oscillators, light-emitting diodes, radio-frequency inductors, and wireless power transmitter coils [85–89]. After the integrated device is mounted on skin, the PVA substrate can easily be washed away. The device also can be peeled off. Furthermore, Afsharimani and Nysten [39] prepared PVA thin films by spin-coating and utilized them as polymer gate dielectrics to fabricate transistors. The transistors show ambipolar behavior with hole and electron mobilities in a low voltage range, indicating a promising potential future in dielectrics.

PDMS is a transparent elastic polymer with excellent biocompatibility [90–92]. It has been approved by the US National Heart, Lung, and Blood Institute to be a discriminatory tool for validating the evaluation of biomaterials [38]. Because of its elasticity and biocompatibility, PDMS has been widely used in flexible electronics, and it shows great potential in implantable electronics [93–97]. Delivopoulos et al. [41] developed an implantable monitoring device to record and distinguish two types of bladder afferent activity. The device could survive under immersion in warm saline for three months, exhibiting excellent stability. Disappointingly, PDMS cannot biodegrade easily, greatly limiting its applications.

In addition to the abovementioned polymers, there are some other insulated biodegradable or biocompatible polymers that can be used in electronics, such as starch [98,99], chitosan [100,101], albumen [102], and poly (glycerol-co-sebacate) (PGS) [103], which will not be introduced in detail.

Generally, almost all the insulated biopolymers can be used to fabricate both substrates and dielectrics. When used as substrates, biopolymers should be flexible, lightweight, and processable. However, incorporating nanofillers into substrate materials would seriously decrease the flexibility of the substrate, which is the developing direction for stretchable electronics. Thus, nanofillers are not usually added.

When used as dielectrics, biopolymers must exhibit a significant dielectric property. However, biopolymers cannot always satisfy the requirements of a standard dielectric layer in electronics. Adding certain nanofillers into the biopolymer matrix would significantly improve the dielectric properties. This application will be carefully reviewed in the following section.

2.2. Conductive and Semiconductive Polymers

The active materials in electronic devices are usually semiconducting to achieve a certain degree of controllable conductivity, which is the basic principle of most electronics. A conductive or semiconductive polymer is a kind of polymer material with a conjugated π-bond, which can change the polymer from an insulator to a conductor by chemical or electrochemical doping. The basis of the electrical conductivity in conjugated polymers is the delocalization of electrons along the polymer backbone, through the overlap of π-orbitals as well as π-π stacking between polymer chains. Compared with inorganic (semi)conductors, the main advantages of conjugated polymers are mechanical flexibility and lower cost in processing, which allows for inexpensive manufacturing [104]. Many highly conjugated polymers have been developed and applied for (semi)conductive components in various electronics [105].

In addition to electronic conductivity, ionic conductivity also exists in certain polymers, such as melanin and chitosan. Ionic-conducting materials, extensively researched for fuel cell applications, have recently been recognized as having great potential in biocompatible electronics. Additionally, many conducting polymers which can conduct both ionic and electronic currents are extremely well suited to be bioelectronic interface materials. A demonstration of a proton-conducting chitosan thin-film transistor device controlled by the electronic field effect of a gate is a functional realization of the electronic/protonic interface [106].

Melanin is a bio-pigment in animals, plants, and protozoa, formed by a series of chemical reactions of tyrosine or 3,4-dihydroxyphenylalanine. It is a biodegradable and biocompatible natural polymer exhibiting charge transport properties [107–110], which are believed to possess a mixed protonic/electronic property. The protonic/electronic property is influenced by redox reactions, which can be manipulated by changing the hydration state of the material [111]. When prepared into films, the conductivity of melanin reaches the order of 10^{-8} S cm^{-1} in a dehydrated state, and up to 10^{-3} S cm^{-1} in a fully hydrated state. Bettinger et al. [107] demonstrated a tissue engineering application with melanin as the biodegradable semiconducting material. The melanin film in its fully hydrated state possesses a conductivity of 7×10^{-5} S cm^{-1}. The fabricated melanin implant exhibits a similar inflammatory response compared with the silicone implant, and it can be completely degraded in vivo after eight weeks, which makes it more promising.

At present, while the application of (semi)conductive natural polymers is still limited, that of synthetic (semi)conductive polymers is relatively mature, such as polyaniline (PANI), polypyrrole (PPy), poly(3,4-ethylenedioxythiophene) (PEDOT). These conjugated polymers exhibit good biocompatibility in biological applications, but their biodegradability is relatively poor. One strategy to combat this is to blend conjugated polymers with biodegradable, insulating polymers to fabricate partially biodegradable composites. The relative composition can be varied to maximize electric conductivity and minimize the proportion of the non-degradable conjugated component.

PANI has attracted great attention because of its high electric conductivity. Beyond that it also has other beneficial characteristics, including facile synthesis, excellent thermal and environmental stability, controllable electric conductivity, appealing electrochemical properties, and reversible doping/dedoping characteristics [112]. PANI has promising future applications in flexible electronics, such as elastic electrodes and strain-sensors [113–115]. For example, Huang et al. [116] developed a smart pH self-adjusting switching system using a layer-structured silver nanowire/PANI nanocomposite film, which was fabricated via an easy vertical spinning method. The as-prepared nanocomposite film shows a high electric conductivity of 1.03×10^4 S cm^{-1} at the silver nanowire areal density of 0.84 mg cm^{-2}. In addition to electric conductivity, PANI also shows good biocompatibility to cells and tissues, which has been demonstrated in vitro [53] and in vivo [117].

PPy is among the first-studied conductive polymers and has been used widely in bioelectronics and biosensors. It is usually prepared by the oxidation of pyrrole, which can be achieved using ferric chloride or electrochemical polymerization. In the oxidation process, the conductivity of PPy can be greatly affected by the conditions and reagents because dopants could offer additional properties. For example, introducing poly(glutamic acid) as a dopant into PPy would provide pendant carboxylic acid groups, which would further improve electrical conductivity [118]. Similar to PANI, PPy also shows good biocompatibility both in vitro and in vivo [119], but suffers from poor biodegradability.

PEDOT is a conductive polymer based on 3,4-ethylenedioxythiophene (EDOT) monomer. It is produced by oxidation, starting with the preparation of the radical cation of EDOT monomer, which attacks a neutral EDOT, followed by deprotonation. It has optical transparency in its conducting state, high stability, and moderate band gap and low redox potential [120,121]. PEDOT nanotubes with interfacial conducting properties were successfully utilized for neural recording [122]. Richardson-Burns et al. [123] demonstrated the electrochemical polymerization of PEDOT around living neuronal cells with no toxic effects. Furthermore, PEDOT combined with poly(styrene-sulfonate) (PSS) (PEDOT:PSS) has proven to be an excellent system with good conductivity, good stability, high optical transparency, and low toxicity. Thus, it is widely used in electronic circuits, electrostatic packaging, OLEDs, sensing, and photovoltaic devices [124–126]. Yang et al. [127] prepared silver nanowire (AgNW)-PEDOT:PSS composite transparent flexible electrodes (FTEs) through a Mayer rod coating method. The AgNW-PEDOT:PSS composite FTEs exhibited high optoelectrical properties, with a sheet resistance of 12 Omega sq^{-1} and a transmittance of 96% at 550 nm. Unfortunately, the biodegradability of PEDOT is also low.

In general, melanin is the only conductive polymer that exhibits both biodegradability and biocompatibility, but its availability and mechanical properties are not sufficient, which limits its broad application. Synthesized polymers with electric properties, including PANI, PPy, and PEDOT, are biocompatible but not biodegradable. Thus, the application of bare (semi)conductive biopolymers in electronic devices is greatly limited. To obtain (semi)conductive biomaterials with excellent biodegradability, incorporating conductive nanofillers into biodegradable biopolymers is an efficient solution. This field of research will be carefully reviewed in the following section.

3. Applications of Nanocomposites for Electronics

3.1. Substrates

Electronic devices usually consist of a solid substrate and several functional components, such as semiconducting layers, dielectric layers, electrodes, and capsulations. All of these components can be fabricated with biopolymers, replacing the traditional polymers, which are not environmentally friendly. The substrate's role is to support other layers. As such, it is thicker and larger, and generates more E-waste. It also commonly electrically isolates the electronics to prevent undesirable crosstalk, and thus the substrate material is usually insulated. Figure 2 shows some electronic devices whose substrates were fabricated with different insulated biodegradable or biocompatible polymers.

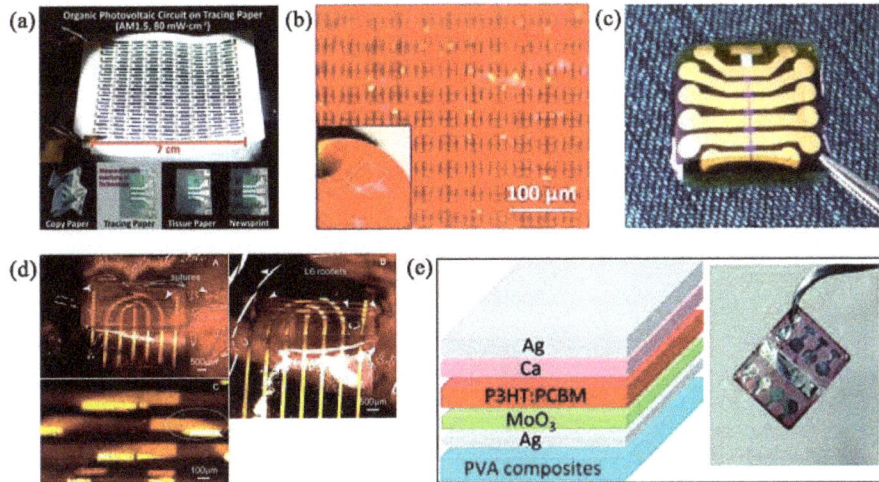

Figure 2. (**a**) Organic photovoltaic circuits fabricated on various paper substrates. Reproduced with permission from [60], Wiley-VCH, 2011. (**b**) split ring resonators fabricated on the silk substrate, wrapped on an apple. Reproduced with permission from [69], Wiley-VCH, 2012. (**c**) biodegradable transistors fabricated on shellac substrate. Reproduced with permission from [80], Wiley-VCH, 2012. (**d**) pressure sensor with polydimethylsiloxane (PDMS) substrate, mounted on rat spinal cord. Reproduced with permission from [41], RSC, 2012. (**e**) schematic device structure and optical image of the transient organic solar cells with poly(vinyl alcohol) (PVA) substrates. Reproduced with permission from [82], RSC, 2017.

Flexibility is of paramount importance for the substrates of stretchable electronics. Generally, adding nanofillers into substrate materials would seriously decrease the flexibility of the substrate, thus, substrate material is usually pure polymer without nanofillers. The most common biodegradable nanomaterial applied in substrates is nanocellulose (NC). Depending on the preparation methods, NCs can be classified as cellulose nanocrystals (CNCs) or cellulose nanofibrils (CNFs) [128–130]. NCs possess a large variety of superior characteristics, such as biodegradability, environmental sustainability, inherent renewability, simplified disposal, distinctive morphology, outstanding chemical-modification capabilities, and extraordinary mechanical strength [128,131–135]. They have attracted great attention in recent years [131,136–139]. Because of their diverse properties, morphologies, and forms, NCs have great potential in a variety of applications, including biomaterial engineering, batteries and solar cells, textiles and clothing, food, packaging industries, and electronic devices [140–144].

In the field of electronics, many devices with NCs as substrates have been reported. Park et al. [145] displayed a flexible, transparent, and nontoxic phototransistor for detecting visible light, which was fabricated on biodegradable CNF substrates. They carried out mechanical bending tests with radii ranging from 100 to 5 mm and cyclic bending tests of up to 2000 cycles at a fixed radius of 5 mm. The bending test proved excellent operational stability. Combined with the phototransistors' flexibility, transparency, and biodegradability, this report indicates the significant potential of NCs as low-cost and environmentally friendly sensors. Cheng et al. [146] synthesized O-(2,3-Dihydroxypropyl) cellulose (DHPC) by the homogeneous etherification of cellulose in 7 wt.% NaOH/12 wt.% urea aqueous solution, and then introduced stiff tunicate cellulose nanocrystals (TCNCs) into the DHPC, in order to construct tough nanocomposite papers. Owing to the excellent interfacial compatibility between TCNCs and DHPC, the nanocomposite papers had smooth surfaces, high transparency, and excellent mechanical properties, enabling them to be used as the substrates of biodegradable and wearable electronics. A fabrication schematic of the cellulose-based nanocomposite papers and their properties is shown in Figure 3.

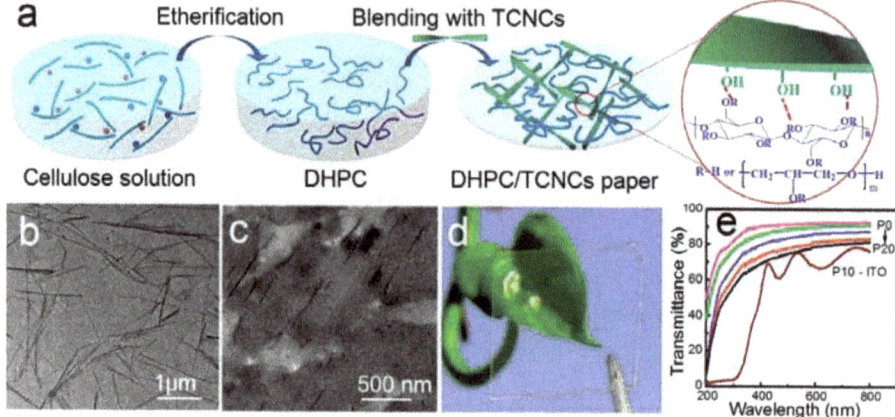

Figure 3. (**a**) The fabrication process of cellulose-based nanocomposite papers; (**b**) transmission electron microscope (TEM) image of tunicate cellulose nanocrystals (TCNCs); (**c**) TEM image of P10; (**d**) photograph of P10; (**e**) optical transmittance of neat O-(2,3-Dihydroxypropyl) cellulose (DHPC) and nanocomposite papers under UV-vis light. Reproduced with permission from [146], ACS, 2018.

Jung et al. [147] utilized biodegradable and flexible CNF papers as substrates and constructed many electronic devices, including flexible microwave and digital electronics, gallium arsenide microwave devices, and consumer wireless workhorses. Figure 4 shows the fabrication process of GaAs devices built on CNF papers.

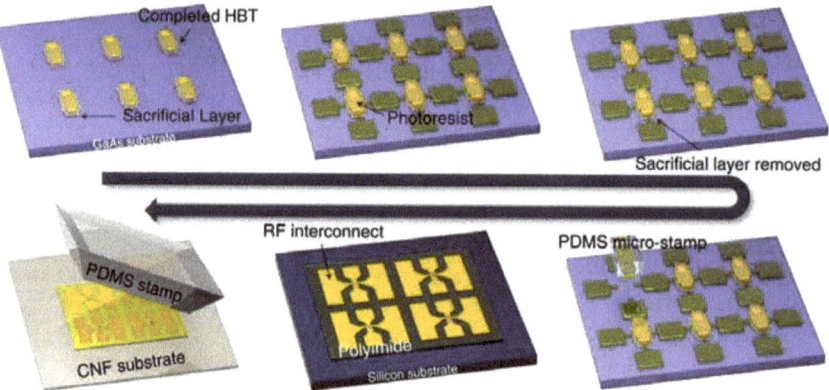

Figure 4. The fabrication process of GaAs devices built on cellulose nanofibril (CNF) paper. Reproduced with permission from [147], NPG, 2015.

3.2. Conductors and Semiconductors

Many biodegradable or biocompatible polymers, including melanin, PANI, PPy, and PPDOT can be used as conductors or semiconductors in electronic devices. Nevertheless, the conductivities of these polymers are usually not sufficiently efficient. To address this problem, some functional nanofillers have been incorporated into conductive polymers to improve their conductivities. Nanofillers have even been added into insulated polymers to endow them with conductivities [148,149]. Among these nanofillers, graphene and carbon nanotubes (CNTs) are most widely used [150–152].

Graphene is a single layer of hybridized carbon atoms arranged in a two-dimensional lattice, which can be manufactured by peeling graphite nanosheets. It has outstanding thermal, optical, mechanical, and electrical properties, attributed to its special structure [153–156]. The carrier mobility of graphene at room temperature is about 15,000 cm^2 V^{-1} s^{-1}, 10 times higher than that of silicon. The excellent conductivity makes it an efficient nanofiller to improve the electrical conductivity of polymers.

Wang et al. [157] reported a healable and multifunctional E-tattoo based on a graphene/silk fibroin (SF)/Ca^{2+} combination. The flexible E-tattoos are fabricated by printing or writing with a graphene/SF/Ca^{2+} suspension. The graphene sheets are uniformly distributed in the matrix, forming an electrically conductive path, which can sensitively respond to the changes of the surrounding environment, including strain, temperature, and humidity. This property enables the E-tattoo to be used as a sensor, monitoring these variables with high sensitivity, fast response, and excellent stability. In addition, the E-tattoo exhibits excellent healable properties. After being damaged by water, the E-tattoo can remarkably and completely heal in only 0.3 s, as a result of the effective reformation of hydrogen and coordination bonds at the fractured interface. The fabrication process and applications of E-tattoos as humidity and temperature sensors are shown in Figure 5.

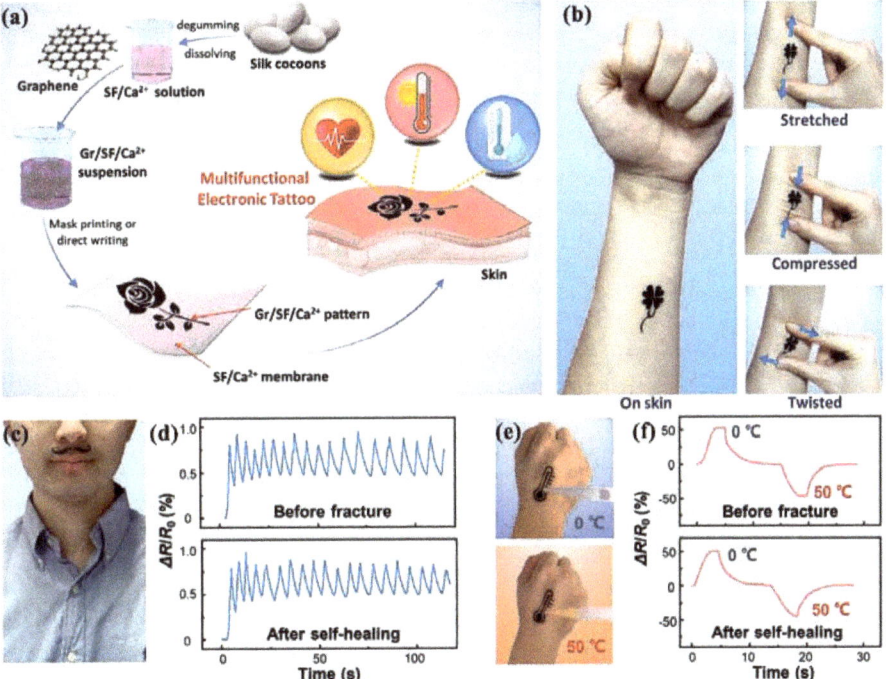

Figure 5. (a) The fabrication illustration of a Gr/silk fibroin (SF)/Ca^{2+} E-tattoo; (b) E-tattoo mounted on the forearm and its variations in stretched, compressed, and twisted states; (c) E-tattoo mounted on the upper lip for monitoring respiration; (d) comparison of the relative resistance between unbroken and healed humidity sensors; (e) E-tattoo mounted on the hand for monitoring temperature; (f) ccomparison of relative resistance between unbroken and healed temperature sensors. Reproduced with permission from [157], Wiley-VCH, 2019.

Ling et al. [158] also utilized graphene to develop novel conductive nanocomposites and to fabricate sensors. They used SF as a matrix and prepared graphene/SF nanocomposites through a uniformly dispersed and highly stable graphene/SF suspension system. The prepared graphene/SF nanocomposites maintain not only the electronic advantages of graphene but also the mechanical properties of SF. Their electrical resistances are sensitive to deformation, body movement, humidity, and changes in the chemical environment, showing a promising future for effective applications as wearable sensors, intelligent skins, and human–machine interfaces.

Similarly, Scaffaro et al. [159] prepared a piezoresistive sensor by exploiting amphiphilic graphene oxide (GO) to endow the polylactide (PLA)-poly (ethylene-glycol) (PEG) blends with electrical properties sensitive to changes in pressure and strain. The responsivity of the biodegradable pressure sensor is 35 µA MPa^{-1} from 0.6 to 8.5 MPa, and 19 µA MPa^{-1} from 8.5 to 25 MPa, while in lower pressure ranges (around 0.16–0.45 MPa) the responsivity reaches 220 µA MPa^{-1}. Additionally, the presence of GO acts as a compatibilizer, providing stiffness and strength without any negative impact on toughness. It provides the stability of mechanical properties for up to 40 days.

CNTs, as one-dimensional nanomaterials, have abnormal mechanical, electrical, and chemical properties. Recently, in an in-depth study, CNTs revealed broad prospective applications. CNTs are formed by crimping graphene sheets, and can be classified into single-walled CNTs (SWCNTs) and multi-walled CNTs (MWCNTs), according to the number of layers of the graphene sheets [160]. The P-electrons of the carbon atoms on CNTs form a wide system of delocalized π bonds, endowing CNTs with special electrical properties due to the significant conjugation effect. According to theoretical prediction, the conductivity of CNTs depends on their diameter and the helical angle of the wall. When the diameter of CNTs is greater than 6 nm, their conductivity is relatively lower; when less than 6 nm, CNTs can be regarded as one-dimensional quantum wires with good conductivity. Their excellent properties make CNTs desirable for high-strength, conductive nanocomposites based on sustainable resources and polymer materials [161,162]. Many efforts have been made to utilize CNTs by preparing conductive nanocomposites and then using them in biodegradable electronic devices [163–168].

Dionigi et al. [169] prepared a conductive nanocomposite with SF and SWCNTs using a novel wet templating method, which combines the excellent mechanical properties and biocompatibility of SF with the electric conductivity and stiffness of SWCNTs. The prepared SF-SWCNT nanocomposites exhibit a periodic structure in which SWCNTs are regularly and uniformly distributed in the SF matrix. The film based on the SF-SWCNT nanocomposites possesses a conductivity only one order of magnitude lower than the bare SWCNTs. Remarkably, the SF-SWCNT nanocomposite enables the growth of primary rat Dorsal Root Ganglion neurons. Figure 6 displays the fabrication process of the nSF-SWCNT film and its electrical properties.

Sivanjineyulu et al. [170] prepared poly(butylene succinate) (PBS)/PLA blend-based nanocomposites with CNTs as reinforcing nanofillers. The electrical resistivity values of PBS, PLA, or their blends are all higher than 1013 Ω square^{-1}, illustrating that they are electrically insulated. When CNT is added into the PBS/PLA blend, the electrical resistivity is greatly decreased. Even with only a 3 phr CNT loading, the electrical resistivity of the blend decreased by up to 11 orders of magnitude as a result of the formation of a semi-conductive network structure in the nanocomposite system.

Valentini et al. [168] reported a photo-responsive device with a semiconducting polymer film built on semitransparent and conductive biodegradable poly(3-hydroxybutyrate) (PHB)/CNT substrates. The biodegradable PHB/CNT nanocomposite can be prepared using SWCNTs or MWCNTs through a simple solvent-casting approach. The electrical resistance value measured on the PHB-SWCNT sample is around 3×10^7 Ohm, while that on the PHB-MWCNT sample is around 2×10^8 Ohm.

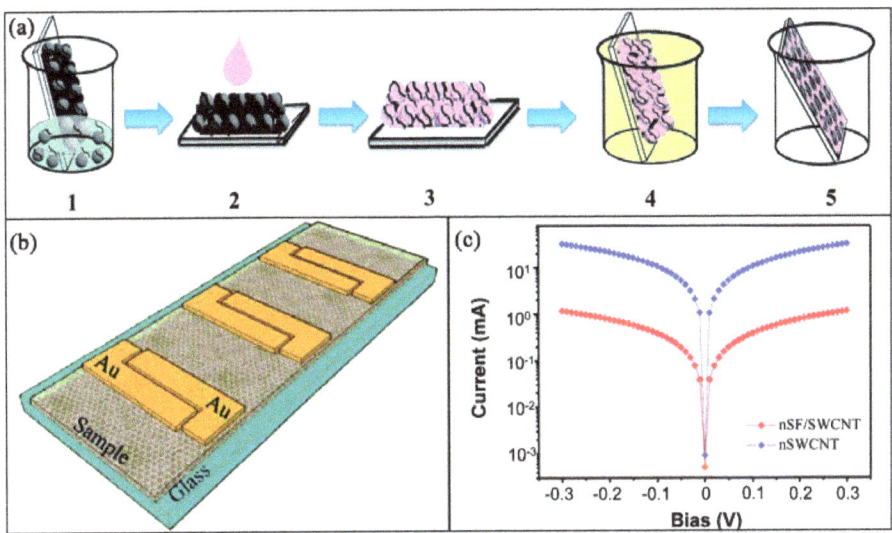

Figure 6. (a) Schematic illustration of the fabrication process of the porous SF-SWCNT films; (b) device, including two gold pads which contact the film; (c) comparison of the I-V characteristics at 0.3 V for the SF-SWCNT and SWCNT films. Reproduced with permission from [169], RSC, 2014.

In addition to adding graphene or CNTs as nanofillers into polymer matrices independently, some researchers also added both graphene and CNTs simultaneously into a polymer's matrix. For example, Miao et al. [55] reported a biodegradable and flexible transparent electrode, in which the prepared conductive nanocomposites had a 3D interconnected SWCNT-pristine graphene (PG)-PEDOT network architecture and was structured using Nacre-inspired interface designs. The one-dimensional SWCNT and the two-dimensional PG sheets were tightly cross-linked at the junction interface by PEDOT chains. The fabrication process of the transparent electrode is shown in Figure 7. The formation of the SWCNT-PG-PEDOT continuously conductive network results in a low electrical resistance, as well as excellent flexibility. Even after hundreds of bending cycles, the electrical resistance of the electrode only increases by less than 3%. Moreover, the fabricated electrode exhibits an outstanding optoelectronic property: typically, a sheet resistance of 46 Ω square^{-1} with a transmittance of 83.5% at a typical wavelength of 550 nm. More importantly, the conductive nanocomposites are incorporated with an edible starch-chitosan substrate, which leads to perfect biodegradability: it could be rapidly degraded in a lysozyme solution at room temperature, with no toxic residues produced.

Chen et al. [171] prepared Polycaprolactone (PCL)/MWCNT nanocomposites by blending GO sheets and MWCNTs into PCL, where GO acts as an adjuvant for regulating the dispersion state of MWCNTs and thus balances the electrical and mechanical properties of the nanocomposites. Strong π-π interactions between MWCNTs and GO nanosheets make it easy for MWCNTs to be adsorbed onto the surfaces of GO nanosheets, thereby forming GO/MWCNT hybrids, which hinder the aggregation of MWCNTs in PCL. Based on this mechanism, the dispersion of GO/MWCNT hybrids in PCL is greatly affected by the GO/MWCNT ratio. The dispersion states of MWCNTs in PCL were divided into PCL/MWCNT, PCL/GO/MWCNT (1/4), and PCL/GO/MWCNT (2/1) in their research, representing severe, low, and almost no aggregation of MWCNTs, respectively. Among the three dispersion states, the PCL/GO/MWCNT nanocomposites with a GO/MWCNT ratio of 2/1 showed the best MWCNT dispersion in PCL matrix, and thus the highest tensile strength and elongation at break. However, the PCL/GO/MWCNT (1/4) nanocomposites achieved the best electrical conductivity. This is attributed to the relatively low MWCNT aggregation.

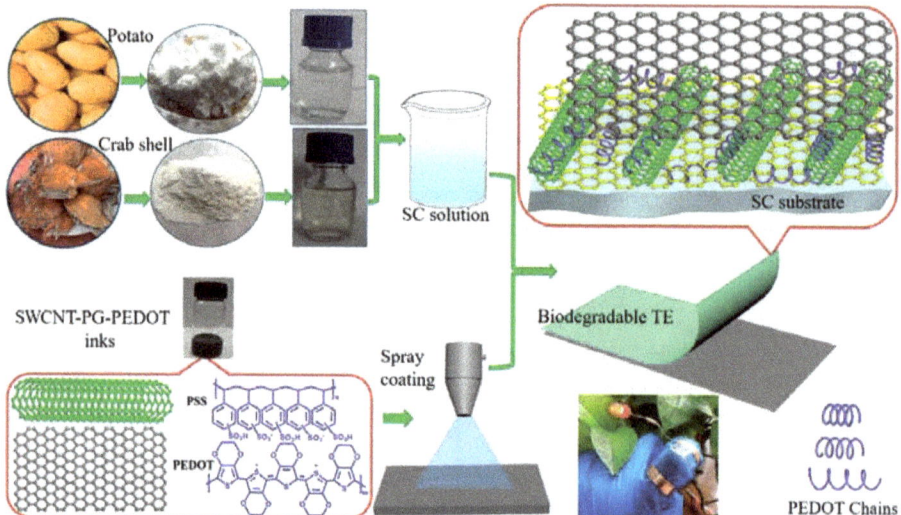

Figure 7. Fabrication process of the biodegradable and flexible 3D interconnected single-walled carbon nanotubes-pristine graphene-poly(3,4-ethylenedioxythiophene) (SWCNT-PG-PEDOT) based transparent electrode. Reproduced with permission from [55], ACS, 2018.

Other nanofillers. Aside from carbon nanotubes and graphene, metal nanowires can also improve the conductivity of the nanocomposites. Li et al. [172] reported biodegradable poly(citrates-siloxane) (PCS) elastomers reinforced by ultralong copper sulfide nanowires (CSNWs). The CSNWs were uniformly distributed throughout the PCS matrix because of the hydrophobic-hydrophobic interaction between them. The content of the CSNWs directly influences the electric conductivity of the nanocomposites, which could reach a high value of 5×10^{-4} S cm^{-1} when the addition content of CSNWs is 30%. PCS-CSNW also exhibits a high degree of biocompatibility, which decreases the inflammatory reaction of cells. Additionally, it possesses a unique photo-luminescent property and strong near-infrared (NIR) photo-thermal capacity, which allows in vivo thermal imaging and biodegradation tracking with high resolution. PCS-CSNW could assist in the effective killing of cancer cells via a selective NIR-induced photo-thermal therapy [173]. Therefore, PCS-CSNW is a promising material in the area of next-generation implanted electronics, tissue engineering or regenerative medicine for biomedical applications. A processing illustration and physicochemical structure characterizations of the PCS-CSNW nanocomposites are shown in Figure 8.

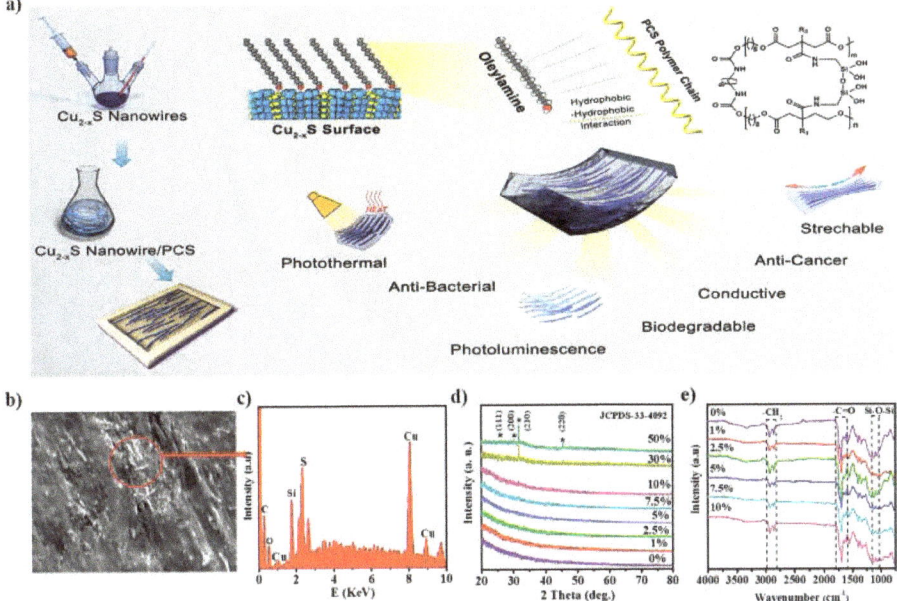

Figure 8. A processing illustration and physicochemical structure characterizations of the poly(citrates-siloxane)-copper sulfide nanowires (PCS-CSNW) nanocomposites: (**a**) processing illustration and potential applications for biomedicine; (**b**) scanning electron microscope (SEM) images; (**c**) energy dispersion spectrum (EDS) spectrum; (**d**) X-ray diffraction (XRD) patterns; (**e**) Fourier transform infrared (FTIR) spectra between 4000 and 650 cm^{-1}. Reproduced with permission from [172], Elsevier, 2019.

Not all nanofillers in conductive nanocomposites are designed to improve electrical properties. They may also be used only to strengthen materials, and electrical conductivity is instead achieved by conductive polymers. Han et al. [112] reported conductive hybrid elastomers fabricated with a natural rubber (NR) matrix and nanostructured CNF-PANI complexes, in which PANI provides the conductivity, while CNFs strengthen the material. The CNF-PANI complexes were prepared via oxidative polymerization of aniline monomers on CNF surface, and then evenly dispersed into NR latex to fabricate CNF-PANI/NR elastomers using latex co-coagulation. The presence of CNFs in the nanocomposites constructs a reinforcing network and simultaneously supports the 3D conductive network in NR matrix. The fabricated bio-based elastomers with homogeneous structures showed inherent flexibility, improved mechanical properties, decent stretchability, low density, and desired electric conductivity (up to 8.95×10^{-1} S m^{-1}). Then, the elastomer was used to fabricate a strain sensor with high sensitivity and repeatability, which could monitor the motion of the human body in real time. The elastomer-based electrode with 20 phr of PANI presented superior electrochemical properties. Its specific capacitance could reach a maximum of 110 F g^{-1} with a relatively low capacitance degradation of 22% after 1200 cycles at a current density of 0.3 A g^{-1}. The processing illustration of the conductive CNF-PANI/NR elastomers and their properties are shown in Figure 9.

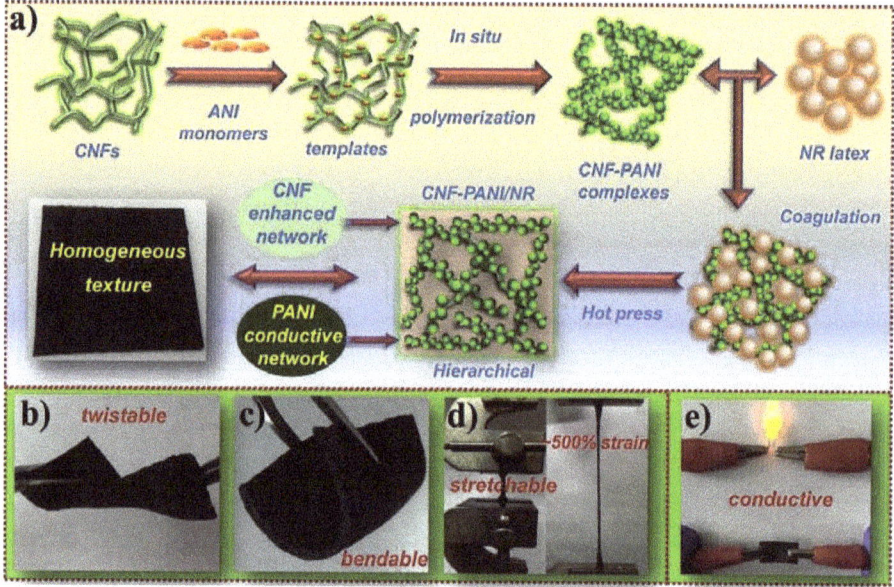

Figure 9. The processing illustration of the conductive cellulose nanofibril-polyaniline/natural rubber (CNF-PANI/NR) elastomers and their properties: (**a**) processing illustration of the conductive CNF-PANI/NR hybrid elastomers; (**b**) flexibility of the CNF-PANI/NR elastomers; (**c**) bendability of the CNF-PANI/NR elastomers; (**d**) stretchability of the CNF-PANI/NR elastomers; (**e**) conductivity of the CNF-PANI/NR elastomers. Reproduced with permission from [112], Elsevier, 2019.

Aside from the nanocomposites previously discussed, there are many other biodegradable and biocompatible nanocomposites used to enhance electronic conductors, which will not be carefully discussed here [174–180].

3.3. Dielectrics

Dielectric materials are usually electrically insulated and can be polarized when an electric field is applied. Under the action of an electric field, the electric charges in dielectric materials will slightly deviate from their equilibrium positions, rather than flowing and forming current as in conductive materials. The slight movement of positive and negative charges produces an internal electric field opposite to the direction of the applied eternal electric field, thereby reducing the total electric field in the dielectric material. For example, the dielectric layer in OFETs produces induced electric charges in the semiconducting channel when the gate voltage is applied. The dielectric constant and breakdown voltage of the dielectric layer are two crucial parameters that must be carefully considered for low-voltage operation and long-time stability [1].

Many naturally-based and synthesized substrate materials can be used to make dielectrics, such as cellulose, silk, shellac, gelatin, PVA, PDMS, PGS, Poly(lactic-co-glycolic acid) (PLGA), PCL, and PLA. Regarding dielectric polymers, adding a small amount of nanofillers into them to prepare nanocomposites would further enhance their dielectric performance. The chemical structure, surface morphology, and preparation method, as well as the additives, all have an effect on the dielectric properties of the nanocomposites [39,181–183]. Conductive and high dielectric constant nanofillers, such as CNTs, GO, Al_2O_3, and SiO_2, all have the potential to improve dielectric performance. Kashi et al. [184] investigated the effect of graphene nanosheets on the dielectric performance of biodegradable nanocomposites and found that the presence of graphene nanosheets could heighten the dielectric constant of polymers to a large extent.

Deshmukh et al. [185] prepared bio-based nanocomposites by blending cellulose acetate (CA) with Al_2O_3 nanoparticles (Al_2O_3 NPs) and investigated the microstructure, morphology, thermal, and dielectric properties of the CA/Al_2O_3 nanocomposites. In the solution blending process, the Al_2O_3 NPs were uniformly dispersed in the CA matrix and showed good intermolecular interaction. The incorporation of Al_2O_3 NPs significantly enhanced the dielectric properties of CA. For instance, in the condition of 50 Hz and 30 °C, when loaded with 25 wt.% Al_2O_3, the dielectric constant increased from 8.63 to 27.57 and the dielectric loss increased from 0.26 to 0.64. However, the values of tan δ for all the samples were all very low (below 1).

Zeng et al. [186] reported flexible dielectric papers based on biodegradable CNFs and CNTs for dielectric energy storage. They successfully prepared highly ordered, homogeneous CNF/CNT papers through a simple vacuum-assisted self-assembly technique. When the CNT loading was 4.5 wt.%, the dielectric constant of the CNF/CNT paper was 3198 at a frequency of 1.0 kHz, which is far higher than 15 for the neat CNF paper. The significant enhancement resulted from the formation of microcapacitor networks in the papers by neighboring conductive CNTs and insulating CNFs. The excellent dielectric constant also improved the dielectric energy storage capability (0.81 ± 0.1 J cm^{-3}). In addition, the CNF/CNT papers showed a high degree of flexibility and enhanced mechanical strength. The preparation process and dielectric properties of the CNF/CNT papers are shown in Figure 10.

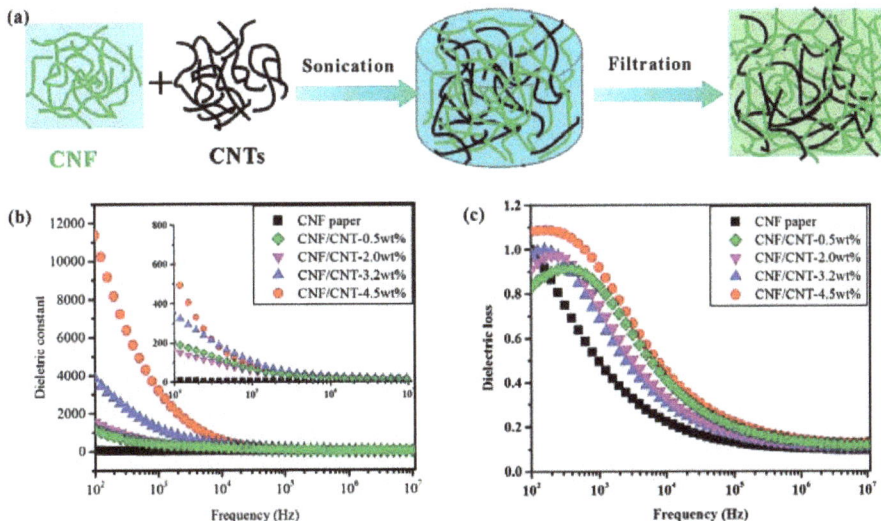

Figure 10. (a) Preparation process of the CNF/CNT papers; (b,c) frequency dependence of dielectric constant and loss of the CNF/CNT papers with different CNT loadings. Reproduced with permission from [186], RSC, 2016.

Choudhary [187] prepared polymer nanocomposite films with a biodegradable polymer blend matrix of PVA and poly(vinyl pyrrolidone) (PVP) and dispersed amorphous silica (SiO_2) nanoparticles using the aqueous solution-cast method. It was found that the presence of the dispersed SiO_2 nanoparticles in the PVA–PVP blend matrix decreased the size of PVA crystallites, and forced the surface morphology of the nanocomposite films to turn from smooth to relatively rough. The dielectric constant of the nanocomposite films decreased as the SiO_2 content increased to 3 wt.%. However, when the SiO_2 content was 5 wt.%, the dielectric constant was close to that of the pure polymer blend matrix. Additionally, temperature had an effect on the dielectric constant. The dielectric constant of the nanocomposite film increased non-linearly with the increase of temperature.

Deshmukh et al. [188] also prepared SiO_2 nanoparticle-reinforced PVA and PVP blend nanocomposite films. SiO_2 nanoparticles were homogeneously dispersed in the PVA/PVP blend matrix in a solution-blending process. The dielectric constant and dielectric loss of the PVA/PVP/SiO$_2$ nanocomposite films were tested under a frequency range of 10^{-2} Hz to 20 MHz and temperature range of 40–150 °C. In the testing conditions, the dielectric constants of the prepared nanocomposites were higher than those of PVA/PVP blends. With 25 wt.% SiO$_2$ content, the dielectric constant reached a maximum of 125 (10^{-2} Hz, 150 °C) and the dielectric loss was 1.1 (10^{-2} Hz, 70 °C). Deshmukh et al. obtained better dielectric properties compared to Choudhary [187], with the same PVA/PVP/SiO$_2$ system. The results reported by Deshmukh et al. show that SiO$_2$ could significantly improve the dielectric properties of polymers, and the solution-casting method they utilized has great potential for flexible organic electronics.

Deshmukh et al. [183] fabricated flexible dielectric nanocomposites, which are composed of water soluble PPy (WPPy), PVA, and GO, and then characterized them at different GO contents (0.5–3 wt.%). Because of the presence of GO and its uniform dispersion in the polymer matrix, the nanocomposites show a significant improvement in the dielectric constant with low dielectric loss. With a GO loading of 3 wt.%, frequency of 50 Hz, and temperature of 150 °C, the dielectric constant increased from 27.93 for WPPy/PVA blend to 155.18 for nanocomposites, and the dielectric loss increased from 2.01 for WPPy/PVA blend to 4.71 for nanocomposites.

As can be seen from the above references in this section, adding nanofillers, such as CNTs, GO, Al$_2$O$_3$, and SiO$_2$, into polymers can enhance the dielectric properties of those polymers. With biocompatible and biodegradable polymer matricies, the newly developed nanocomposites would enable the feasible fabrication of dielectrics with high-performance capabilities, flexibility, and environmental friendliness.

4. Electronics Packaging

Unlike substrates, (semi)conductors, or dielectrics, materials for electronics packaging require different functions for various operational environments, such as cyclical mechanical bending, aqueous solutions, elevated temperatures, electromagnetic shielding, electrostatic prevention. Most polymeric substrate materials, such as PLGA, PCL, parylene-c, and poly(vinyl acetate) PVAc, can be used to form strain-resistant packaging layers to prevent the rapid degradation of devices [52,189–192].

One particular concern is to ensure that the packaging materials are able to repel gas and water vapor over a period of several months, because many conjugated organic compounds used as (semi)conductors are easy to oxidize and lose their function in ambient and aqueous environments. Therefore, the gas and water vapor barrier property is the most important consideration in the packaging of electronics. Biodegradable polymers are likely be used to meet this demand due to their high crystallinity, high hydrophobicity, and facile processing [193]. An adequate packaging polymer is poly(L-lactide) (PLLA), which satisfies the aforementioned demand and can be easily prepared by melt casting or using ordinary organic solvents. Adding nanofillers into polymers is an efficient method to enhance gas and vapor barrier properties, and many nanofillers, including organophilic layered double hydroxides (OLDH) nanosheets [194–196], montmorillonite [197], and GO [198], have been used for this purpose.

Xie et al. [194] incorporated OLDH nanosheets into a biodegradable PVA matrix via a solution casting method and prepared PVA/OLDH films. The OLDH nanosheets, which were intercalated with aliphatic long-chain molecules as reinforcing agents, were homogeneously dispersed in PVA matrix and formed strong interfacial interactions with the PVA chains, resulting in significant enhancements of optical property, mechanical performance, thermal stability, and water vapor barrier property. Even when only 0.5 wt.% OLDH was loaded in PVA, water vapor permeability could decrease by 24.22%. The significant improvement of the water vapor barrier property results from the homogeneous dispersion of OLDH nanosheets, which causes the paths for water vapor diffusion to be tortuous and thus decreases the water vapor permeability. The experiments demonstrated that the PVA/OLDH nanocomposite films have a wide variety of potential applications in the field of electronics packaging. A schematic illustration for the preparation of PVA/OLDH films is shown in Figure 11.

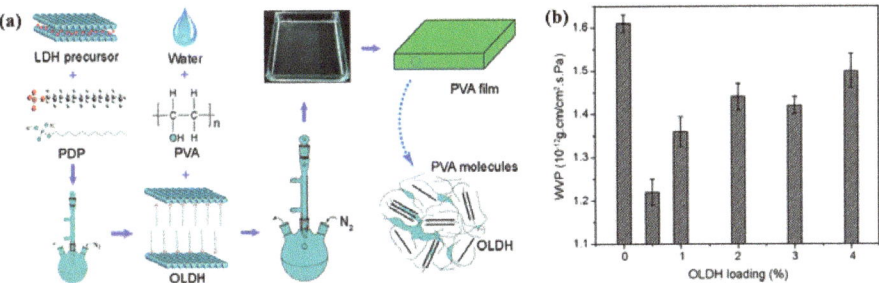

Figure 11. (a) Processing of the PVA/organophilic layered double hydroxides (OLDH) films; (b) water vapor permeability variation as a function of OLDH loading. Reproduced with permission from [194], Springer, 2017.

Xie et al. [195] synthesized a series of biodegradable nanocomposite films based on poly(butylene adipate-co-terephthalate) (PBAT), and reinforced them with OLDH nanosheets. The OLDH nanosheets were pre-synthesized by solvent-free high-energy ball milling and dispersed uniformly in the PBAT matrix. Compared with pure PBAT films, PBAT/OLDH films with 1 wt.% OLDH loading exhibited improved thermal, optical, mechanical, and water vapor barrier properties, including a 37% reduction in haze and a 41.9% increase in nominal tensile strain at break. The feasibility of scale-up production, outstanding processability, manufacturing scalability, mechanical property, optical transparency, and water vapor barrier properties indicate a promising future application of the PBAT/OLDH nanocomposite films as biodegradable packaging films. Figure 12 shows the schematic illustration of the fabrication process for the PBAT/OLDH nanocomposite films.

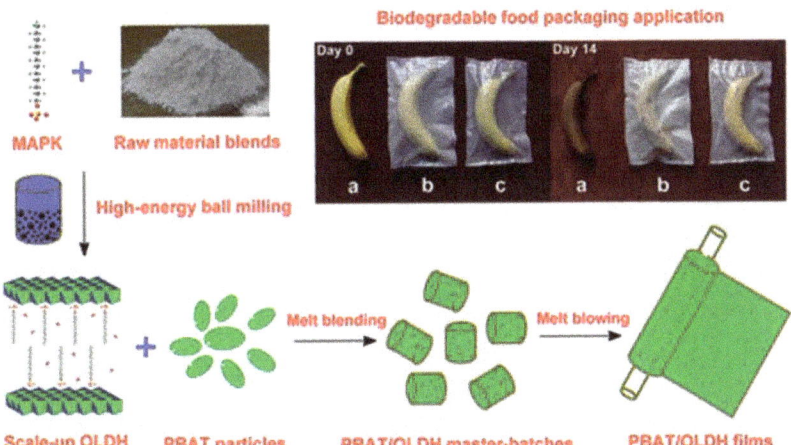

Figure 12. Fabrication process for the poly(butylene adipate-co-terephthalate) (PBAT)/OLDH nanocomposite films. Reproduced with permission from [195], ACS, 2018.

Aside from the OLDH nanosheets, montmorillonite is another commonly used nanofiller for the improvement of barrier performance. Wang and Jing [197] prepared biodegradable montmorillonite/chitosan nanocomposites and coated them onto the traditional package paper so as to expand the potential application of the paper. They found that montmorillonite/chitosan nanocomposite showed superior water vapor barrier properties, especially with a high montmorillonite and dispersant content, dispersion rate, and coating weight. In addition, the montmorillonite/chitosan nanocomposite coated with a lower content of montmorillonite or with a higher dispersion speed and dispersant content had better smoothness and elongation. However, the addition of OLDH nanosheets had a bad impact on the formation process.

GOs can also be used as nanofillers for enhancing the gas and water vapor barrier performance of polymer systems. Ren et al. [198] introduced an extremely low amount of GO nanosheets into biodegradable poly(butylene adipate-co-terephthalate) (PBAT) and the barrier performance was significantly improved. The permeability coefficients of oxygen and water vapor decreased, exceeding 70% and 36% with a GO nanosheet loading of 0.35 vol.%. The enhanced barrier performance was attributed to the excellent impermeability and homogeneous dispersion of GO nanosheets, as well as the strong interfacial adhesion between the GO nanosheets and PBAT matrix.

In addition to having a gas and water barrier, some special electronic devices need be packaged with materials having electromagnetic shielding and antistatic functions. In the electromagnetic shielding field of electronics packaging, nanocomposites have great advantages compared with pure polymers because of the addition of conductive nanofillers [199–202]. For example, Kuang et al. [201] developed lightweight high-strength PLLA/MWCNT nanocomposites foams with an efficient, environmentally friendly, and inexpensive method, using a pressure-induced flow technique and solid-state supercritical CO_2 foaming. The nanocomposite foams have a density as low as 0.3 g cm^{-3}, possess an electric conductivity of 3.4 S m^{-1}, and an electromagnetic interference (EMI) shielding efficiency (SE) of around 23 dB in the range of 8.00–12.48 GHz. The corresponding average specific EMI SE reaches 77 dB g^{-1} cm^3, far exceeding those of metals and many carbon-based composites with similar densities and thickness. Absorption was proven to be the major mechanism of EMI shielding for the PLLA/MWCNT nanocomposite foams, which is shown in Figure 13. In addition, the nanocomposite foams also show superior compressive stress. The prepared biodegradable PLLA/MWCNT nanocomposites foams are suitable for EMI shielding in electronics packaging.

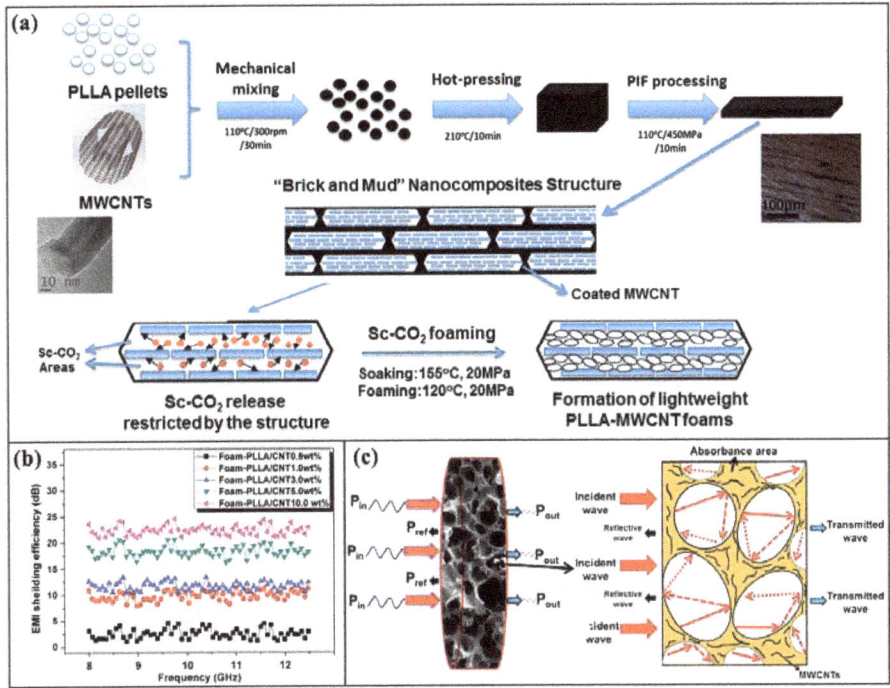

Figure 13. (a) Fabrication process of the lightweight poly(L-lactide)/single-walled carbon nanotubes (PLLA/MWCNT) nanocomposite foams using a combinatorial technology of pressure induced flow processing and Sc-CO$_2$ foaming; (b) electromagnetic interference (EMI), shielding efficiency (SE) of PLLA/MWCNT nanocomposite foams in the frequency ranges of 8.00–12.48 GHz; (c) schematic illustration of electromagnetic microwave dissipation in the PLLA/MWCNT nanocomposite foams. Reproduced with permission from [201], Elsevier, 2016.

In antistatic aspects of packaging, nanocomposites also have great advantages compared with pure polymers. Shih et al. [203] prepared PBS/MWCNT nanocomposites through a melt–blending method. The MWCNTs were firstly modified with N,N'-dicyclohexylcarbodiimide (DCC) dehydrating agents, and then uniformly dispersed in organic solvents. The PBS/MWCNT nanocomposites were subsequently prepared via the facile melt–blending process. The prepared PBS/MWCNT nanocomposites exhibit a surface resistivity of 7.30 × 10^6 Ω, 10^9 folds lower in value compared with the neat PBS sample. At an MWCNTs loading of 3 wt.%, the PBS/MWCNT nanocomposites showed an excellent anti-static capacity, indicating a promising potential in electronic packaging materials for anti-static function. Figure 14 shows the results of anti-static test.

Figure 14. Anti-static test by Shih et al. Reproduced with permission from [203], Elsevier, 2008.

5. Summary and Outlook

Biodegradable nanocomposites have been widely investigated and used for fabricating components of green electronics, and provide an efficient solution for E-waste management and environment protection. Furthermore, green electronics also exhibit promising potential for biomedical applications of transient electronic devices. Although functional nanomaterials have significantly enhanced the overall performances of biodegradable polymers, some particular characteristics, including electric conductivity and flexibility, as well as biodegradability and biocompatibility, still need be further improved.

Flexibility mainly depends on the properties of the polymer matrix. Synthetic polymers display superiority in flexibility, as well as better conformal contact between the implantable electronics and dynamic tissue surface, but they generally suffer from worse biodegradability or biocompatibility. To achieve an excellent performance for all the requirements still remains an arduous challenge. Developing novel polymers derived from natural materials with enhanced mechanical properties or blending the natural-based polymers and synthetic polymers together may be feasible methods for future improvement.

Electric conductivity is affected by both the polymer matrix and nanofillers. Synthetic conjugated polymers are usually not biodegradable, and thus a conjugation-breaking degradation strategy through mechanisms such as oxidation, ultraviolet (UV) exposure, or enzymes, without affecting conductivities is crucial. However, the trade-off between material degradability and device stability is difficult to balance. Natural-based conductive polymers can be easily biodegradable, but they suffer unsatisfied conductivity and bad mechanical properties, which seriously limits their applications. Adding functional nanofillers, such as graphene and CNTs, into a polymer matrix can significantly enhance the electric conductivity. The electric conductivity is deeply affected by the dispersion state of nanofillers in the polymer matrix as well as the interfacial morphology, which has been studied by many researchers and will still be an important research direction.

Author Contributions: Conceptualization, Z.Y., J.S. and C.W., Writing, H.L., R.J., H.C., X.T., C.S., J.Z., Z.Y., J.S. and C.W.

Funding: This research was funded by Shandong Provincial Natural Science Foundation, China (No. ZR2016XJ003), National Natural Science Foundation of China (No. 51575287), the Fundamental Research Funds for Central Universities (No. JD1910) and the Talents Introduction Project in Beijing University of Chemical Technology (No. buctrc201909).

Acknowledgments: We would like to thank Emma Parks (Carnegie Mellon University, Departments of Chemical and Biomedical Engineering) for her careful and critical reading of this manuscript.

Conflicts of Interest: The authors declare no conflict of interest.

References

1. Cao, Y.; Uhrich, K.E. Biodegradable and biocompatible polymers for electronic applications: A review. *J. Bioact. Compat. Polym.* **2019**, *34*, 3–15. [CrossRef]
2. Ogunseitan, O.A.; Schoenung, J.M.; Saphores, J.-D.M.; Shapiro, A.A. The Electronics Revolution: From E-Wonderland to E-Wasteland. *Science* **2009**, *326*, 670–671. [CrossRef] [PubMed]
3. Zou, Z.N.; Zhu, C.P.; Li, Y.; Lei, X.F.; Zhang, W.; Xiao, J.L. Rehealable, fully recyclable, and malleable electronic skin enabled by dynamic covalent thermoset nanocomposite. *Sci. Adv.* **2018**, *4*, 1–8. [CrossRef] [PubMed]
4. Lei, T.; Guan, M.; Liu, J.; Lin, H.-C.; Pfattner, R.; Shaw, L.; McGuire, A.F.; Huang, T.-C.; Shao, L.; Cheng, K.-T.; et al. Biocompatible and totally disintegrable semiconducting polymer for ultrathin and ultralightweight transient electronics. *Proc. Natl. Acad. Sci. USA* **2017**, *114*, 5107–5112. [CrossRef] [PubMed]
5. Fu, K.K.; Wang, Z.; Dai, J.; Carter, M.; Hu, L. Transient Electronics: Materials and Devices. *Chem. Mater.* **2016**, *28*, 3527–3539. [CrossRef]
6. Tafazoli, S.; Rafiemanzelat, F.; Hassanzadeh, F.; Rostami, M. Synthesis and characterization of novel biodegradable water dispersed poly(ether-urethane)s and their MWCNT-AS nanocomposites functionalized with aspartic acid as dispersing agent. *Iran. Polym. J.* **2018**, *27*, 755–774. [CrossRef]

7. Salehpour, S.; Jonoobi, M.; Ahmadzadeh, M.; Siracusa, V.; Rafieian, F.; Oksman, K. Biodegradation and ecotoxicological impact of cellulose nanocomposites in municipal solid waste composting. *Int. J. Biol. Macromol.* **2018**, *111*, 264–270. [CrossRef] [PubMed]
8. Rao, K.M.; Kumar, A.; Rao, K.S.V.K.; Haider, A.; Han, S.S. Biodegradable Tragacanth Gum Based Silver Nanocomposite Hydrogels and Their Antibacterial Evaluation. *J. Polym. Environ.* **2018**, *26*, 778–788. [CrossRef]
9. Gao, X.; Huang, L.; Wang, B.; Xu, D.; Zhong, J.; Hu, Z.; Zhang, L.; Zhou, J. Natural Materials Assembled, Biodegradable, and Transparent Paper-Based Electret Nanogenerator. *ACS Appl. Mater. Interfaces* **2016**, *8*, 35587–35592. [CrossRef] [PubMed]
10. Ko, J.; Nguyen, L.T.H.; Surendran, A.; Tan, B.Y.; Ng, K.W.; Leong, W.L. Human Hair Keratin for Biocompatible Flexible and Transient Electronic Devices. *ACS Appl. Mater. Interfaces* **2017**, *9*, 43004–43012. [CrossRef] [PubMed]
11. Lee, D.; Lim, Y.W.; Im, H.G.; Jeong, S.; Ji, S.; Kim, Y.H.; Choi, G.M.; Park, J.U.; Lee, J.Y.; Jin, J.; et al. Bioinspired Transparent Laminated Composite Film for Flexible Green Optoelectronics. *ACS Appl. Mater. Interfaces* **2017**, *9*, 24161–24168. [CrossRef] [PubMed]
12. Song, Y.; Kim, S.; Heller, M.J. An Implantable Transparent Conductive Film with Water Resistance and Ultrabendability for Electronic Devices. *ACS Appl. Mater. Interfaces* **2017**, *9*, 42302–42312. [CrossRef] [PubMed]
13. Armentano, I.; Dottori, M.; Fortunati, E.; Mattioli, S.; Kenny, J.M. Biodegradable polymer matrix nanocomposites for tissue engineering: A review. *Polym. Degrad. Stab.* **2010**, *95*, 2126–2146. [CrossRef]
14. Pinto, V.C.; Costa-Almeida, R.; Rodrigues, I.; Guardao, L.; Soares, R.; Guedes, R.M. Exploring the in vitro and in vivo compatibility of PLA, PLA/GNP and PLA/CNT-COOH biodegradable nanocomposites: Prospects for tendon and ligament applications. *J. Biomed. Mater. Res. Part A* **2017**, *105*, 2182–2190. [CrossRef] [PubMed]
15. Abd El-Kareem, S.A.; Abd Elsamie, G.H.; Abd-Alkareem, A.S. Sono-photodynamic modality for cancer treatment using biodegradable bio-conjugated sonnelux nanocomposite in tumor-bearing mice: Activated cancer therapy using light and ultrasound. *Biochem. Biophys. Res. Commun.* **2018**, *503*, 1075–1086. [CrossRef] [PubMed]
16. Li, Y.N.; Guo, Y.; Niu, W.; Chen, M.; Xue, Y.M.; Ge, J.; Ma, P.X.; Lei, B. Biodegradable Multifunctional Bioactive Glass-Based Nanocomposite Elastomers with Controlled Biomineralization Activity, Real-Time Bioimaging Tracking, and Decreased Inflammatory Response. *ACS Appl. Mater. Interfaces* **2018**, *10*, 17722–17731. [CrossRef] [PubMed]
17. Sha, L.; Chen, Z.; Chen, Z.; Zhang, A.; Yang, Z. Polylactic Acid Based Nanocomposites: Promising Safe and Biodegradable Materials in Biomedical Field. *Int. J. Polym. Sci.* **2016**, *2016*, 1–11. [CrossRef]
18. Feig, V.R.; Tran, H.; Bao, Z. Biodegradable polymeric materials in degradable electronic devices. *ACS Cent. Sci.* **2018**, *4*, 337–348. [CrossRef] [PubMed]
19. Chivrac, F.; Pollet, E.; Averous, L. Progress in nano-biocomposites based on polysaccharides and nanoclays. *Mater. Sci. Eng. R Rep.* **2009**, *67*, 1–17. [CrossRef]
20. Lim, L.-T.; Auras, R.; Rubino, M. Processing technologies for poly (lactic acid). *Prog. Polym. Sci.* **2008**, *33*, 820–852. [CrossRef]
21. Reddy, M.M.; Vivekanandhan, S.; Misra, M.; Bhatia, S.K.; Mohanty, A.K. Biobased plastics and bionanocomposites: Current status and future opportunities. *Prog. Polym. Sci.* **2013**, *38*, 1653–1689. [CrossRef]
22. Sun, J.Y.; Shen, J.J.; Chen, S.K.; Cooper, M.A.; Fu, H.B.; Wu, D.M.; Yang, Z.G. Nanofiller Reinforced Biodegradable PLA/PHA Composites: Current Status and Future Trends. *Polymers* **2018**, *10*, 505. [CrossRef] [PubMed]
23. Bettinger, C.J.; Bao, Z. Biomaterials-based organic electronic devices. *Polym. Int.* **2010**, *59*, 563–567. [CrossRef] [PubMed]
24. Eder, F.; Klauk, H.; Halik, M.; Zschieschang, U.; Schmid, G.; Dehm, C. Organic electronics on paper. *Appl. Phys. Lett.* **2004**, *84*, 2673–2675. [CrossRef]
25. Bollström, R.; Määttänen, A.; Tobjörk, D.; Ihalainen, P.; Kaihovirta, N.; Österbacka, R.; Peltonen, J.; Toivakka, M. A multilayer coated fiber-based substrate suitable for printed functionality. *Org. Electron.* **2009**, *10*, 1020–1023. [CrossRef]

26. Petritz, A.; Wolfberger, A.; Fian, A.; Griesser, T.; Irimia-Vladu, M.; Stadlober, B. Cellulose-Derivative-Based Gate Dielectric for High-Performance Organic Complementary Inverters. *Adv. Mater.* **2015**, *27*, 7645–7656. [CrossRef]
27. Kim, D.-H.; Kim, Y.-S.; Amsden, J.; Panilaitis, B.; Kaplan, D.L.; Omenetto, F.G.; Zakin, M.R.; Rogers, J.A. Silicon electronics on silk as a path to bioresorbable, implantable devices. *Appl. Phys. Lett.* **2009**, *95*, 1–3. [CrossRef]
28. You, X.; Pak, J.J. Graphene-based field effect transistor enzymatic glucose biosensor using silk protein for enzyme immobilization and device substrate. *Sens. Actuators B Chem.* **2014**, *202*, 1357–1365. [CrossRef]
29. Wang, C.H.; Hsieh, C.Y.; Hwang, J.C. Flexible Organic Thin-Film Transistors with Silk Fibroin as the Gate Dielectric. *Adv. Mater.* **2011**, *23*, 1630–1634. [CrossRef]
30. Irimia-Vladu, M.; Glowacki, E.D.; Schwabegger, G.; Leonat, L.; Akpinar, H.Z.; Sitter, H.; Bauer, S.; Sariciftci, N.S. Natural resin shellac as a substrate and a dielectric layer for organic field-effect transistors. *Green Chem.* **2013**, *15*, 1473–1476. [CrossRef]
31. Baek, S.W.; Ha, J.W.; Yoon, M.; Hwang, D.H.; Lee, J. Shellac Films as a Natural Dielectric Layer for Enhanced Electron Transport in Polymer Field-Effect Transistors. *ACS Appl. Mater. Interfaces* **2018**, *10*, 18948–18955. [CrossRef] [PubMed]
32. Xiao, H.; Cebe, P.; Weiss, A.S.; Omenetto, F.; Kaplan, D.L. Protein-based composite materials. *Mater. Today* **2012**, *15*, 208–215.
33. Irimia-Vladu, M.; Troshin, P.A.; Reisinger, M.; Shmygleva, L.; Kanbur, Y.; Schwabegger, G.; Bodea, M.; Schwodiauer, R.; Mumyatov, A.; Fergus, J.W.; et al. Biocompatible and Biodegradable Materials for Organic Field-Effect Transistors. *Adv. Funct. Mater.* **2010**, *20*, 4069–4076. [CrossRef]
34. Ning, N.; Wang, Z.; Yao, Y.; Zhang, L.; Tian, M. Enhanced electromechanical performance of bio-based gelatin/glycerin dielectric elastomer by cellulose nanocrystals. *Carbohyd. Polym.* **2015**, *130*, 262–267. [CrossRef] [PubMed]
35. He, Y.; Sun, J.; Qian, C.; Kong, L.-a.; Jiang, J.; Yang, J.; Li, H.; Gao, Y. Solution-processed natural gelatin was used as a gate dielectric for the fabrication of oxide field-effect transistors. *Org. Electron.* **2016**, *38*, 357–361. [CrossRef]
36. Zhuang, X.M.; Zhang, D.Y.; Wang, X.L.; Yu, X.G.; Yu, J.S. Biocompatible and degradable gelatin dielectric based low-operating voltage organic transistors for ultra-high sensitivity NH_3 detection. *Appl. Phys. Lett.* **2018**, *113*, 1–4. [CrossRef]
37. Li, Y.; Neoh, K.G.; Kang, E.T. Controlled Release of Heparin from Polypyrrole-Poly(vinyl alcohol) Assembly by Electrical Stimulation. *J. Biomed. Mater. Res. Part A* **2005**, *73A*, 171–181. [CrossRef] [PubMed]
38. Belanger, M.C.; Marois, Y. Hemocompatibility, biocompatibility, inflammatory and in vivo studies of primary reference materials low-density polyethylene and polydimethylsiloxane: a review. *J. Biomed. Mater. Res.* **2001**, *58*, 467–477. [CrossRef]
39. Afsharimani, N.; Nysten, B. Hybrid gate dielectrics: A comparative study between polyvinyl alcohol/SiO_2 nanocomposite and pure polyvinyl alcohol thin-film transistors. *Bull. Mater. Sci.* **2019**, *42*, 26. [CrossRef]
40. Nawaz, A.; Hummelgen, I.A. Poly(vinyl alcohol) gate dielectric in organic field-effect transistors. *J. Mater. Sci. Mater. Electron.* **2019**, *30*, 5299–5326. [CrossRef]
41. Delivopoulos, E.; Chew, D.J.; Minev, I.R.; Fawcett, J.W.; Lacour, S.P. Concurrent recordings of bladder afferents from multiple nerves using a microfabricated PDMS microchannel electrode array. *Lab Chip* **2012**, *12*, 2540–2551. [CrossRef] [PubMed]
42. Shi, J.; Chan-Park, M.B.; Li, C.M. Adhesive-Free Transfer of Gold Patterns to PDMS-Based Nanocomposite Dielectric for Printed High-Performance Organic Thin-Film Transistors. *ACS Appl. Mater. Interfaces* **2011**, *3*, 1880–1886. [CrossRef] [PubMed]
43. Chen, J.; Guo, H.; He, X.; Liu, G.; Xi, Y.; Shi, H.; Hu, C. Enhancing Performance of Triboelectric Nanogenerator by Filling High Dielectric Nanoparticles into Sponge PDMS Film. *ACS Appl. Mater. Interfaces* **2016**, *8*, 736–744. [CrossRef] [PubMed]
44. He, X.; Mu, X.; Wen, Q.; Wen, Z.; Yang, J.; Hu, C.; Shi, H. Flexible and transparent triboelectric nanogenerator based on high performance well-ordered porous PDMS dielectric film. *Nano Res.* **2016**, *9*, 3714–3724. [CrossRef]

45. Rajitha, G.; Dash, R.K. Optically transparent and high dielectric constant reduced graphene oxide (RGO)-PDMS based flexible composite for wearable and flexible sensors. *Sens. Actuator A Phys.* **2018**, *277*, 26–34. [CrossRef]
46. Shi, X.W.; Dai, X.; Cao, Y.; Li, J.W.; Huo, C.G.; Wang, X.L. Degradable Poly(lactic acid)/Metal-Organic Framework Nanocomposites Exhibiting Good Mechanical, Flame Retardant, and Dielectric Properties for the Fabrication of Disposable Electronics. *Ind. Eng. Chem. Res.* **2017**, *56*, 3887–3894. [CrossRef]
47. Li, H.; Zhao, C.C.; Wang, X.X.; Meng, J.P.; Zou, Y.; Noreen, S.; Zhao, L.M.; Liu, Z.; Ouyang, H.; Tan, P.C.; et al. Fully Bioabsorbable Capacitor as an Energy Storage Unit for Implantable Medical Electronics. *Adv. Sci.* **2019**, *6*, 1–9.
48. Wu, X.H.; Ma, Y.; Zhang, G.Q.; Chu, Y.L.; Du, J.; Zhang, Y.; Li, Z.; Duan, Y.R.; Fan, Z.Y.; Huang, J. Thermally Stable, Biocompatible, and Flexible Organic Field-Effect Transistors and Their Application in Temperature Sensing Arrays for Artificial Skin. *Adv. Funct. Mater.* **2015**, *25*, 2138–2146. [CrossRef]
49. Meriakri, V.V.; Parkhomenko, M.P.; Kalenov, D.S.; Fedoseev, N.A.; Sh, Z. Dielectric properties of biocompatible and biodegradated poly-caprolactone, polylactide and its nanocomposites in the millimeter wave range. *Elektromagn. Volny Elektron. Sist.* **2012**, *17*, 30–33.
50. Boutry, C.M.; Nguyen, A.; Lawal, Q.O.; Chortos, A.; Rondeau-Gagné, S.; Bao, Z. A sensitive and biodegradable pressure sensor array for cardiovascular monitoring. *Adv. Mater.* **2015**, *27*, 6954–6961. [CrossRef]
51. Bettinger, C.J.; Bao, Z. Organic Thin-Film Transistors Fabricated on Resorbable Biomaterial Substrates. *Adv. Mater.* **2010**, *22*, 651–655. [CrossRef] [PubMed]
52. Chang, J.-K.; Chang, H.-P.; Guo, Q.; Koo, J.; Wu, C.-I.; Rogers, J.A. Biodegradable electronic systems in 3D, heterogeneously integrated formats. *Adv. Mater.* **2018**, *30*, 1–10. [CrossRef] [PubMed]
53. Bidez, P.R.; Li, S.; MacDiarmid, A.G.; Venancio, E.C.; Wei, Y.; Lelkes, P.I. Polyaniline, an electroactive polymer, supports adhesion and proliferation of cardiac myoblasts. *J. Biomater. Sci. Polym. Ed.* **2006**, *17*, 199–212. [CrossRef]
54. George, P.M.; Lyckman, A.W.; Lavan, D.A.; Anita, H.; Yuika, L.; Rupali, A.; Chris, T.; Alexander, P.M.; Robert, L.; Mriganka, S. Fabrication and biocompatibility of polypyrrole implants suitable for neural prosthetics. *Biomaterials* **2005**, *26*, 3511–3519. [CrossRef] [PubMed]
55. Miao, J.; Liu, H.; Li, Y.; Zhang, X. Biodegradable Transparent Substrate Based on Edible Starch-Chitosan Embedded with Nature-Inspired Three-Dimensionally Interconnected Conductive Nanocomposites for Wearable Green Electronics. *ACS Appl. Mater. Interfaces* **2018**, *10*, 23037–23047. [CrossRef]
56. Fortunato, E.; Correia, N.; Barquinha, P.; Pereira, L.; Goncalves, G.; Martins, R. High-Performance Flexible Hybrid Field-Effect Transistors Based on Cellulose Fiber Paper. *IEEE Electron Device Lett.* **2008**, *29*, 988–990. [CrossRef]
57. Zschieschang, U.; Yamamoto, T.; Takimiya, K.; Kuwabara, H.; Ikeda, M.; Sekitani, T.; Someya, T.; Klauk, H. Organic electronics on banknotes. *Adv. Mater.* **2011**, *23*, 654–658. [CrossRef] [PubMed]
58. Peng, B.; Ren, X.; Wang, Z.; Wang, X.; Roberts, R.C.; Chan, P.K. High performance organic transistor active-matrix driver developed on paper substrate. *Sci. Rep.* **2014**, *4*, 1–7. [CrossRef]
59. Leonat, L.; White, M.S.; Głowacki, E.D.; Scharber, M.C.; Zillger, T.; Rühling, J.; Hübler, A.; Sariciftci, N.S. 4% efficient polymer solar cells on paper substrates. *J. Phys. Chem. C* **2014**, *118*, 16813–16817. [CrossRef]
60. Barr, M.C.; Rowehl, J.A.; Lunt, R.R.; Xu, J.; Wang, A.; Boyce, C.M.; Im, S.G.; Bulović, V.; Gleason, K.K. Direct monolithic integration of organic photovoltaic circuits on unmodified paper. *Adv. Mater.* **2011**, *23*, 3500–3505. [CrossRef]
61. Zhang, Q.; Bao, W.; Gong, A.; Gong, T.; Ma, D.; Wan, J.; Dai, J.; Munday, J.N.; He, J.-H.; Hu, L.; et al. A highly sensitive, highly transparent, gel-gated MoS2 phototransistor on biodegradable nanopaper. *Nanoscale* **2016**, *8*, 14237–14242. [CrossRef] [PubMed]
62. Petritz, A.; Wolfberger, A.; Fian, A.; Irimia-Vladu, M.; Haase, A.; Gold, H.; Rothlander, T.; Griesser, T.; Stadlober, B. Cellulose as biodegradable high-k dielectric layer in organic complementary inverters. *Appl. Phys. Lett.* **2013**, *103*, 1–5. [CrossRef]
63. Cunha, I.; Barras, R.; Grey, P.; Gaspar, D.; Fortunato, E.; Martins, R.; Pereira, L. Reusable Cellulose-Based Hydrogel Sticker Film Applied as Gate Dielectric in Paper Electrolyte-Gated Transistors. *Adv. Funct. Mater.* **2017**, *27*, 1–11. [CrossRef]

64. Dai, S.L.; Chu, Y.L.; Liu, D.P.; Cao, F.; Wu, X.H.; Zhou, J.C.; Zhou, B.L.; Chen, Y.T.; Huang, J. Intrinsically ionic conductive cellulose nanopapers applied as all solid dielectrics for low voltage organic transistors. *Nat. Commun.* **2018**, *9*, 1–10. [CrossRef] [PubMed]
65. Acharya, C.; Ghosh, S.K.; Kundu, S. Silk fibroin film from non-mulberry tropical tasar silkworms: A novel substrate for in vitro fibroblast culture. *Acta Biomater.* **2009**, *5*, 429–437. [CrossRef] [PubMed]
66. Zhu, B.; Wang, H.; Leow, W.R.; Cai, Y.; Loh, X.J.; Han, M.-Y.; Chen, X. Silk fibroin for flexible electronic devices. *Adv. Mater.* **2016**, *28*, 4250–4265. [CrossRef] [PubMed]
67. Kim, D.-H.; Viventi, J.; Amsden, J.J.; Xiao, J.; Vigeland, L.; Kim, Y.-S.; Blanco, J.A.; Panilaitis, B.; Frechette, E.S.; Contreras, D.; et al. Dissolvable films of silk fibroin for ultrathin conformal bio-integrated electronics. *Nat. Mater.* **2010**, *9*, 511–517. [CrossRef] [PubMed]
68. Tao, H.; Hwang, S.-W.; Marelli, B.; An, B.; Moreau, J.E.; Yang, M.; Brenckle, M.A.; Kim, S.; Kaplan, D.L.; Rogers, J.A. Silk-based resorbable electronic devices for remotely controlled therapy and in vivo infection abatement. *Proc. Natl. Acad. Sci. USA* **2014**, *111*, 17385–17389. [CrossRef] [PubMed]
69. Tao, H.; Brenckle, M.A.; Yang, M.; Zhang, J.; Liu, M.; Siebert, S.M.; Averitt, R.D.; Mannoor, M.S.; McAlpine, M.C.; Rogers, J.A. Silk-based conformal, adhesive, edible food sensors. *Adv. Mater.* **2012**, *24*, 1067–1072. [CrossRef] [PubMed]
70. Liang, F.C.; Huang, Y.H.; Kuo, C.C.; Cho, C.J.; Rwei, S.P.; Jia, Q.; Ding, Y.S.; Chen, Y.G.; Borsali, R. Thermally deposited silk fibroin as the gate dielectric layer in organic thin-film transistors based on conjugated polymer. *React. Funct. Polym.* **2018**, *131*, 368–377. [CrossRef]
71. Tsai, L.-S.; Hwang, J.-C.; Lee, C.-Y.; Lin, Y.-T.; Tsai, C.-L.; Chang, T.-H.; Chueh, Y.-L.; Meng, H.-F. Solution-based silk fibroin dielectric in n-type C-60 organic field-effect transistors: Mobility enhancement by the pentacene interlayer. *Appl. Phys. Lett.* **2013**, *103*, 1–4. [CrossRef]
72. Li, H.-Q.; Yu, J.-S.; Huang, W.; Shi, W.; Huang, J. High performance pentacene organic field-effect transistors consisting of biocompatible PMMA/silk fibroin bilayer dielectric. *Chin. Phys. B* **2014**, *23*, 1–4. [CrossRef]
73. Shi, L.; Xu, X.; Ma, M.; Li, L. High-performance, low-operating voltage, and solution-processable organic field-effect transistor with silk fibroin as the gate dielectric. *Appl. Phys. Lett.* **2014**, *104*, 1–4. [CrossRef]
74. Li, X.; Shi, W.; Yu, X.; Yu, J. Performance improvement of organic field-effect transistor based nitrogen dioxide gas sensor using biocompatible PMMA/silk fibroin bilayer dielectric. *J. Mater. Sci.-Mater. Electron.* **2015**, *26*, 7948–7954. [CrossRef]
75. Park, M.H.; Kim, J.; Lee, S.C.; Cho, S.Y.; Kim, N.R.; Kang, B.; Song, E.; Cho, K.; Jin, H.-J.; Lee, W.H. Critical role of silk fibroin secondary structure on the dielectric performances of organic thin-film transistors. *RSC Adv.* **2016**, *6*, 5907–5914. [CrossRef]
76. Zhuang, X.; Huang, W.; Yang, X.; Han, S.; Li, L.; Yu, J. Biocompatible/Degradable Silk Fibroin:Poly(Vinyl Alcohol)-Blended Dielectric Layer Towards High-Performance Organic Field-Effect Transistor. *Nanoscale Res. Lett.* **2016**, *11*, 1–8. [CrossRef]
77. Lee, J.H.; Kwak, H.W.; Park, M.H.; Hwang, J.; Kim, J.W.; Jang, H.W.; Jin, H.-J.; Lee, W.H. Understanding hydroscopic properties of silk fibroin and its use as a gate-dielectric in organic field-effect transistors. *Org. Electron.* **2018**, *59*, 213–219. [CrossRef]
78. Weinberger, H.; Gardner, W.H. Chemical composition of shellac. *Ind. Eng. Chem.* **1938**, *30*, 454–458. [CrossRef]
79. Altman, G.H.; Diaz, F.; Jakuba, C.; Calabro, T.; Horan, R.L.; Chen, J.; Lu, H.; Richmond, J.; Kaplan, D.L. Silk-based biomaterials. *Biomaterials* **2003**, *24*, 401–416. [CrossRef]
80. Irimia-Vladu, M.; Głowacki, E.D.; Troshin, P.A.; Schwabegger, G.; Leonat, L.; Susarova, D.K.; Krystal, O.; Ullah, M.; Kanbur, Y.; Bodea, M.A. Indigo-a natural pigment for high performance ambipolar organic field effect transistors and circuits. *Adv. Mater.* **2012**, *24*, 375–380. [CrossRef]
81. Zhuang, J.; Wu, D.-M.; Xu, H.; Huang, Y.; Liu, Y.; Sun, J.-Y. Edge Effect in Hot Embossing and its Influence on Global Pattern Replication of Polymer-Based Microneedles. *Int. Polym. Proc.* **2019**, *34*, 231–238. [CrossRef]
82. Xi, H.; Chen, D.; Lv, L.; Zhong, P.; Lin, Z.; Chang, J.; Wang, H.; Wang, B.; Ma, X.; Zhang, C. High performance transient organic solar cells on biodegradable polyvinyl alcohol composite substrates. *RSC Adv.* **2017**, *7*, 52930–52937. [CrossRef]
83. Yoon, J.; Han, J.; Choi, B.; Lee, Y.; Kim, Y.; Park, J.; Lim, M.; Kang, M.-H.; Kim, D.H.; Kim, D.M.; et al. Three-Dimensional Printed Poly(vinyl alcohol) Substrate with Controlled On-Demand Degradation for Transient Electronics. *ACS Nano* **2018**, *12*, 6006–6012. [CrossRef] [PubMed]

84. Kim, D.-H.; Lu, N.; Ma, R.; Kim, Y.-S.; Kim, R.-H.; Wang, S.; Wu, J.; Won, S.M.; Tao, H.; Islam, A.; et al. Epidermal Electronics. *Science* **2011**, *333*, 838–843. [CrossRef] [PubMed]
85. Zhuang, J.; Hu, W.; Fan, Y.; Sun, J.; He, X.; Xu, H.; Huang, Y.; Wu, D. Fabrication and testing of metal/polymer microstructure heat exchangers based on micro embossed molding method. *Microsyst. Technol.* **2019**, *25*, 381–388. [CrossRef]
86. Jingyao, S.; Daming, W.; Ying, L.; Zhenzhou, Y.; Pengsheng, G. Rapid fabrication of micro structure on polypropylene by plate to plate isothermal hot embossing method. *Polym. Eng. Sci.* **2018**, *58*, 952–960. [CrossRef]
87. Sun, J.; Zhuang, J.; Jiang, H.; Huang, Y.; Zheng, X.; Liu, Y.; Wu, D. Thermal dissipation performance of metal-polymer composite heat exchanger with V-shape microgrooves: A numerical and experimental study. *Appl. Therm. Eng.* **2017**, *121*, 492–500. [CrossRef]
88. Sun, J.; Wu, D.; Liu, Y.; Dai, L.; Jiang, C. Numerical simulation and experimental study of filling process of micro prism by isothermal hot embossing in solid-like state. *Adv. Polym. Tech.* **2018**, *37*, 1581–1591. [CrossRef]
89. Wu, D.; Sun, J.; Liu, Y.; Yang, Z.; Xu, H.; Zheng, X.; Gou, P. Rapid fabrication of microstructure on PMMA substrate by the plate to plate Transition-Spanning isothermal hot embossing method nearby glass transition temperature. *Polym. Eng. Sci.* **2017**, *57*, 268–274. [CrossRef]
90. He, X.; Huang, Y.; Wan, C.; Zheng, X.; Kormakov, S.; Gao, X.; Sun, J.; Zheng, X.; Wu, D. Enhancing thermal conductivity of polydimethylsiloxane composites through spatially confined network of hybrid fillers. *Compos. Sci. Technol.* **2019**, *172*, 163–171. [CrossRef]
91. Sun, J.Y.; Wang, X.B.; Wu, J.H.; Jiang, C.; Shen, J.J.; Cooper, M.A.; Zheng, X.T.; Liu, Y.; Yang, Z.G.; Wu, D.M. Biomimetic Moth-eye Nanofabrication: Enhanced Antireflection with Superior Self-cleaning Characteristic. *Sci. Rep.* **2018**, *8*, 1–10. [CrossRef]
92. Wu, H.W.; Zhu, J.; Huang, Y.; Wu, D.M.; Sun, J.Y. Microfluidic-Based Single-Cell Study: Current Status and Future Perspective. *Molecules* **2018**, *23*, 2347. [CrossRef] [PubMed]
93. Hassan, G.; Bae, J.; Hassan, A.; Ali, S.; Lee, C.H.; Choi, Y. Ink-jet printed stretchable strain sensor based on graphene/ZnO composite on micro-random ridged PDMS substrate. *Compos. Part A Appl. Sci. Manuf.* **2018**, *107*, 519–528. [CrossRef]
94. Wang, D.; Ba, D.; Hao, Z.; Li, Y.; Sun, F.; Liu, K.; Du, G.; Mei, Q. A novel approach for PDMS thin films production towards application as substrate for flexible biosensors. *Mater. Lett.* **2018**, *221*, 228–231. [CrossRef]
95. Wu, D.; Gao, X.; Sun, J.; Wu, D.; Liu, Y.; Kormakov, S.; Zheng, X.; Wu, L.; Huang, Y.; Guo, Z. Spatial Confining Forced Network Assembly for preparation of high-performance conductive polymeric composites. *Compos. Part A Appl. Sci. Manuf.* **2017**, *102*, 88–95. [CrossRef]
96. Cui, J.L.; Zhang, B.Z.; Duan, J.P.; Guo, H.; Tang, J. Flexible Pressure Sensor with Ag Wrinkled Electrodes Based on PDMS Substrate. *Sensors* **2016**, *16*, 2131. [CrossRef] [PubMed]
97. Ko, E.H.; Kim, H.J.; Lee, S.M.; Kim, T.W.; Kim, H.K. Stretchable Ag electrodes with mechanically tunable optical transmittance on wavy-patterned PDMS substrates. *Sci. Rep.* **2017**, *7*, 1–12. [CrossRef]
98. Jeong, H.; Baek, S.; Han, S.; Jang, H.; Kim, S.H.; Lee, H.S. Novel Eco-Friendly Starch Paper for Use in Flexible, Transparent, and Disposable Organic Electronics. *Adv. Funct. Mater.* **2018**, *28*, 1–9. [CrossRef]
99. Misman, M.A.; Azura, A.R.; Sidek, O. Validation of an Electronic Sensor Network (ESN) Control Chamber for Monitoring the Soil Decomposition Process of Sago Starch-filled Natural Rubber Latex Films. *J. Test. Eval.* **2015**, *43*, 1–10. [CrossRef]
100. Lin, Y.H.; Kang, P.L.; Xin, W.; Yen, C.S.; Hwang, L.C.; Chen, C.J.; Liu, J.T.; Chang, S.J. Preparation and evaluation of chitosan biocompatible electronic skin. *Comput. Ind.* **2018**, *100*, 1–6. [CrossRef]
101. Chao, J.Y.; Zhu, L.Q.; Xiao, H.; Yuan, Z.G. Protonic/electronic hybrid oxide transistor gated by chitosan and its full-swing low voltage inverter applications. *J. Appl. Phys.* **2015**, *118*, 1–5. [CrossRef]
102. Chang, J.-W.; Wang, C.-G.; Huang, C.-Y.; Tsai, T.-D.; Guo, T.-F.; Wen, T.-C. Chicken Albumen Dielectrics in Organic Field-Effect Transistors. *Adv. Mater.* **2011**, *23*, 4077–4081. [CrossRef] [PubMed]
103. Najafabadi, A.H.; Tamayol, A.; Annabi, N.; Ochoa, M.; Mostafalu, P.; Akbari, M.; Nikkhah, M.; Rahimi, R.; Dokmeci, M.R.; Sonkusale, S.; et al. Biodegradable Nanofibrous Polymeric Substrates for Generating Elastic and Flexible Electronics. *Adv. Mater.* **2014**, *26*, 5823–5830. [CrossRef] [PubMed]
104. Yang, Z.W.; Xu, H.; Huang, Y.; Sun, J.Y.; Wu, D.M.; Gao, X.L.; Zhang, Y.J. Measuring Mechanism and Applications of Polymer-Based Flexible Sensors. *Sensors* **2019**, *19*, 1403. [CrossRef] [PubMed]

105. Thomas, S.W.; Joly, G.D.; Swager, T.M.J.C.R. Chemical sensors based on amplifying fluorescent conjugated polymers. *Chem. Rev.* **2007**, *107*, 1339–1386. [CrossRef] [PubMed]
106. Zhong, C.; Deng, Y.; Roudsari, A.F.; Kapetanovic, A.; Anantram, M.; Rolandi, M. A polysaccharide bioprotonic field-effect transistor. *Nat. Commun.* **2011**, *2*, 1–5. [CrossRef] [PubMed]
107. Bettinger, C.J.; Bruggeman, J.P.; Misra, A.; Borenstein, J.T.; Langer, R. Biocompatibility of biodegradable semiconducting melanin films for nerve tissue engineering. *Biomaterials* **2009**, *30*, 3050–3057. [CrossRef]
108. Mostert, A.B.; Powell, B.J.; Pratt, F.L.; Hanson, G.R.; Sarna, T.; Gentle, I.R.; Meredith, P. Role of semiconductivity and ion transport in the electrical conduction of melanin. *Proc. Natl. Acad. Sci. USA* **2012**, *109*, 8943–8947. [CrossRef]
109. Abbas, M.; D'Amico, F.; Morresi, L.; Pinto, N.; Ficcadenti, M.; Natali, R.; Ottaviano, L.; Passacantando, M.; Cuccioloni, M.; Angeletti, M.; et al. Structural, electrical, electronic and optical properties of melanin films. *Eur. Phys. J. E* **2009**, *28*, 285–291. [CrossRef]
110. Vahidzadeh, E.; Kalra, A.P.; Shankar, K. Melanin-based electronics: From proton conductors to photovoltaics and beyond. *Biosens. Bioelectron.* **2018**, *122*, 127–139. [CrossRef]
111. Sheliakina, M.; Mostert, A.B.; Meredith, P. Decoupling Ionic and Electronic Currents in Melanin. *Adv. Funct. Mater.* **2018**, *28*, 1–7. [CrossRef]
112. Han, J.Q.; Lu, K.Y.; Yue, Y.Y.; Mei, C.T.; Huang, C.B.; Wu, Q.L.; Xu, X.W. Nanocellulose-templated assembly of polyaniline in natural rubber-based hybrid elastomers toward flexible electronic conductors. *Ind. Crop. Prod.* **2019**, *128*, 94–107. [CrossRef]
113. Hashemi, M.; Rahmanifar, M.S.; El-Kady, M.F.; Noori, A.; Mousavi, M.F.; Kaner, R.B. The use of an electrocatalytic redox electrolyte for pushing the energy density boundary of a flexible polyaniline electrode to a new limit. *Nano Energy* **2018**, *44*, 489–498. [CrossRef]
114. Liu, T.Y.; Finn, L.; Yu, M.H.; Wang, H.Y.; Zhai, T.; Lu, X.H.; Tong, Y.X.; Li, Y. Polyaniline and Polypyrrole Pseudocapacitor Electrodes with Excellent Cycling Stability. *Nano Lett.* **2014**, *14*, 2522–2527. [CrossRef] [PubMed]
115. Reddy, K.R.; Karthik, K.V.; Prasad, S.B.B.; Soni, S.K.; Jeong, H.M.; Raghu, A.V. Enhanced photocatalytic activity of nanostructured titanium dioxide/polyaniline hybrid photocatalysts. *Polyhedron* **2016**, *120*, 169–174. [CrossRef]
116. Huang, G.W.; Xiao, H.M.; Fu, S.Y. Electrical Switch for Smart pH Self-Adjusting System Based on Silver Nanowire/Polyaniline Nanocomposite Film. *ACS Nano* **2015**, *9*, 3234–3242. [CrossRef] [PubMed]
117. Kamalesh, S.; Tan, P.; Wang, J.; Lee, T.; Kang, E.-T.; Wang, C.-H. Biocompatibility of electroactive polymers in tissues. *J. Biomed. Mater. Res.* **2000**, *52*, 467–478. [CrossRef]
118. Song, H.-K.; Toste, B.; Ahmann, K.; Hoffman-Kim, D.; Palmore, G. Micropatterns of positive guidance cues anchored to polypyrrole doped with polyglutamic acid: A new platform for characterizing neurite extension in complex environments. *Biomaterials* **2006**, *27*, 473–484. [CrossRef] [PubMed]
119. Wang, X.; Gu, X.; Yuan, C.; Chen, S.; Zhang, P.; Zhang, T.; Yao, J.; Chen, F.; Chen, G. Evaluation of biocompatibility of polypyrrole in vitro and in vivo. *J. Biomed. Mater. Res. Part A* **2004**, *68*, 411–422. [CrossRef] [PubMed]
120. Meng, A.L.; Yuan, X.C.; Li, Z.J.; Zhao, K.; Sheng, L.Y.; Li, Q.D. Direct growth of 3D porous (Ni-Co)(3)S(4)nanosheets arrays on rGO-PEDOT hybrid film for high performance non-enzymatic glucose sensing. *Sensor Actuat. B-Chem.* **2019**, *291*, 9–16. [CrossRef]
121. Nitta, K.; Tsumaki, M.; Kawano, T.; Terashima, K.; Ito, T. Printing PEDOT from EDOT via plasma-assisted inkjet printing. *J. Phys. D. Appl. Phys.* **2019**, *52*, 1–6. [CrossRef]
122. Abidian, M.R.; Ludwig, K.A.; Marzullo, T.C.; Martin, D.C.; Kipke, D.R. Interfacing Conducting Polymer Nanotubes with the Central Nervous System: Chronic Neural Recording using Poly (3,4-ethylenedioxythiophene) Nanotubes. *Adv. Mater.* **2009**, *21*, 3764–3770. [CrossRef] [PubMed]
123. Richardson-Burns, S.M.; Hendricks, J.L.; Foster, B.; Povlich, L.K.; Kim, D.-H.; Martin, D.C. Polymerization of the conducting polymer poly (3, 4-ethylenedioxythiophene)(PEDOT) around living neural cells. *Biomaterials* **2007**, *28*, 1539–1552. [CrossRef] [PubMed]
124. Mahato, S.; Gerling, L.G.; Voz, C.; Alcubilla, R.; Puigdollers, J. High efficiency ITO-free hybrid solar cell using highly conductive PEDOT:PSS with co-solvent and surfactant treatments. *Mater. Lett.* **2017**, *186*, 165–167. [CrossRef]

125. Kayser, L.V.; Lipomi, D.J. Stretchable Conductive Polymers and Composites Based on PEDOT and PEDOT:PSS. *Adv. Mater.* **2019**, *31*, 1–13. [CrossRef]
126. Khodagholy, D.; Doublet, T.; Gurfinkel, M.; Quilichini, P.; Ismailova, E.; Leleux, P.; Herve, T.; Sanaur, S.; Bernard, C.; Malliaras, G.G. Highly Conformable Conducting Polymer Electrodes for In Vivo Recordings. *Adv. Mater.* **2011**, *23*, H268–H272. [CrossRef]
127. Yang, H.; Bai, S.C.; Chen, T.R.; Zhang, Y.; Wang, H.F.; Guo, X.Z. Facile fabrication of large-scale silver nanowire-PEDOT:PSS composite flexible transparent electrodes for flexible touch panels. *Mater. Res. Express* **2019**, *6*, 1–8. [CrossRef]
128. Moon, R.J.; Martini, A.; Nairn, J.; Simonsen, J.; Youngblood, J. Cellulose nanomaterials review: structure, properties and nanocomposites. *Chem. Soc. Rev.* **2011**, *40*, 3941–3994. [CrossRef]
129. Habibi, Y. Key advances in the chemical modification of nanocelluloses. *Chem. Soc. Rev.* **2014**, *43*, 1519–1542. [CrossRef]
130. Kargarzadeh, H.; Huang, J.; Lin, N.; Ahmad, I.; Mariano, M.; Dufresne, A.; Thomas, S.; Gałęski, A. Recent developments in nanocellulose-based biodegradable polymers, thermoplastic polymers, and porous nanocomposites. *Prog. Polym. Sci.* **2018**, *87*, 197–227. [CrossRef]
131. Habibi, Y.; Lucia, L.A.; Rojas, O.J. Cellulose nanocrystals: chemistry, self-assembly, and applications. *Chem. Rev.* **2010**, *110*, 3479–3500. [CrossRef] [PubMed]
132. Klemm, D.; Kramer, F.; Moritz, S.; Lindström, T.; Ankerfors, M.; Gray, D.; Dorris, A. Nanocelluloses: A new family of nature-based materials. *Angew. Chem. Int. Ed.* **2011**, *50*, 5438–5466. [CrossRef] [PubMed]
133. Lin, N.; Dufresne, A. Nanocellulose in biomedicine: Current status and future prospect. *Eur. Polym. J.* **2014**, *59*, 302–325. [CrossRef]
134. Duran, N.; Paula Lemes, A.; B Seabra, A. Review of cellulose nanocrystals patents: preparation, composites and general applications. *Recent Pat. Nanotech.* **2012**, *6*, 16–28. [CrossRef]
135. Zhu, H.; Fang, Z.; Preston, C.; Li, Y.; Hu, L. Transparent paper: fabrications, properties, and device applications. *Energy Environ. Sci.* **2014**, *7*, 269–287. [CrossRef]
136. Lin, N.; Huang, J.; Dufresne, A. Preparation, properties and applications of polysaccharide nanocrystals in advanced functional nanomaterials: a review. *Nanoscale* **2012**, *4*, 3274–3294. [CrossRef] [PubMed]
137. Lavoine, N.; Desloges, I.; Dufresne, A.; Bras, J. Microfibrillated cellulose-Its barrier properties and applications in cellulosic materials: A review. *Carbohyd. Polym.* **2012**, *90*, 735–764. [CrossRef]
138. Khalil, H.A.; Bhat, A.; Yusra, A.I. Green composites from sustainable cellulose nanofibrils: A review. *Carbohyd. Polym.* **2012**, *87*, 963–979. [CrossRef]
139. Khalil, H.A.; Davoudpour, Y.; Islam, M.N.; Mustapha, A.; Sudesh, K.; Dungani, R.; Jawaid, M. Production and modification of nanofibrillated cellulose using various mechanical processes: a review. *Carbohyd. Polym.* **2014**, *99*, 649–665. [CrossRef]
140. Eichhorn, S.J.; Dufresne, A.; Aranguren, M.; Marcovich, N.; Capadona, J.; Rowan, S.; Weder, C.; Thielemans, W.; Roman, M.; Renneckar, S. Current international research into cellulose nanofibres and nanocomposites. *J. Mater. Sci.* **2010**, *45*, 1–33. [CrossRef]
141. Svagan, A.J.; Busko, D.; Avlasevich, Y.; Glasser, G.; Baluschev, S.; Landfester, K. Photon energy upconverting nanopaper: a bioinspired oxygen protection strategy. *ACS Nano* **2014**, *8*, 8198–8207. [CrossRef] [PubMed]
142. Orsolini, P.; Michen, B.; Huch, A.; Tingaut, P.; Caseri, W.R.; Zimmermann, T. Characterization of pores in dense nanopapers and nanofibrillated cellulose membranes: a critical assessment of established methods. *ACS Appl. Mater. Interfaces* **2015**, *7*, 25884–25897. [CrossRef] [PubMed]
143. Zhong, J.; Zhu, H.; Zhong, Q.; Dai, J.; Li, W.; Jang, S.-H.; Yao, Y.; Henderson, D.; Hu, Q.; Hu, L. Self-powered human-interactive transparent nanopaper systems. *ACS Nano* **2015**, *9*, 7399–7406. [CrossRef] [PubMed]
144. Nakayama, Y.; Mori, T.; Tsuruta, R.; Yamanaka, S.; Yoshida, K.; Imai, K.; Koganezawa, T.; Hosokai, T. Surface crystallographic structures of cellulose nanofiber films and overlayers of pentacene. *Jpn. J. Appl. Phys.* **2018**, *57*, 1–4. [CrossRef]
145. Park, J.; Seo, J.-H.; Yeom, S.-W.; Yao, C.; Yang, V.W.; Cai, Z.; Jhon, Y.M.; Ju, B.-K. Flexible and Transparent Organic Phototransistors on Biodegradable Cellulose Nanofibrillated Fiber Substrates. *Adv. Opt. Mater.* **2018**, *6*, 1–10. [CrossRef]
146. Cheng, Q.Y.; Ye, D.D.; Yang, W.T.; Zhang, S.H.; Chen, H.Z.; Chang, C.Y.; Zhang, L.N. Construction of Transparent Cellulose-Based Nanocomposite Papers and Potential Application in Flexible Solar Cells. *ACS Sustain. Chem. Eng.* **2018**, *6*, 8040–8047. [CrossRef]

147. Jung, Y.H.; Chang, T.H.; Zhang, H.L.; Yao, C.H.; Zheng, Q.F.; Yang, V.W.; Mi, H.Y.; Kim, M.; Cho, S.J.; Park, D.W.; et al. High-performance green flexible electronics based on biodegradable cellulose nanofibril paper. *Nat. Commun.* **2015**, *6*, 1–11. [CrossRef]
148. Huang, Y.; Kormakov, S.; He, X.; Gao, X.; Zheng, X.; Liu, Y.; Sun, J.; Wu, D. Conductive Polymer Composites from Renewable Resources: An Overview of Preparation, Properties, and Applications. *Polymers* **2019**, *11*, 187. [CrossRef]
149. He, X.; Huang, Y.; Liu, Y.; Zheng, X.; Kormakov, S.; Sun, J.; Zhuang, J.; Gao, X.; Wu, D. Improved thermal conductivity of polydimethylsiloxane/short carbon fiber composites prepared by spatial confining forced network assembly. *J. Mater. Sci.* **2018**, *53*, 14299–14310. [CrossRef]
150. Sun, J.; Zhuang, J.; Shi, J.; Kormakov, S.; Liu, Y.; Yang, Z.; Wu, D. Highly elastic and ultrathin nanopaper-based nanocomposites with superior electric and thermal characteristics. *J. Mater. Sci.* **2019**, *54*, 8436–8449. [CrossRef]
151. Sun, J.; Zhao, Y.; Yang, Z.; Shen, J.; Cabrera, E.; Lertola, M.J.; Yang, W.; Zhang, D.; Benatar, A.; Castro, J.M.; et al. Highly stretchable and ultrathin nanopaper composites for epidermal strain sensors. *Nanotechnology* **2018**, *29*, 355304. [CrossRef] [PubMed]
152. Zhao, Y.; Cabrera, E.D.; Zhang, D.; Sun, J.; Kuang, T.; Yang, W.; Lertola, M.J.; Benatar, A.; Castro, J.M.; Lee, L.J. Ultrasonic processing of MWCNT nanopaper reinforced polymeric nanocomposites. *Polymer* **2018**, *156*, 85–94. [CrossRef]
153. Ramadas, M.; Bharath, G.; Ponpandian, N.; Ballamurugan, A. Investigation on biophysical properties of Hydroxyapatite/Graphene oxide (HAp/GO) based binary nanocomposite for biomedical applications. *Mater. Chem. Phys.* **2017**, *199*, 179–184. [CrossRef]
154. Tan, B.; Thomas, N.L.J.J.O.M.S. A review of the water barrier properties of polymer/clay and polymer/graphene nanocomposites. *J. Membr. Sci.* **2016**, *514*, 595–612. [CrossRef]
155. Young, R.J.; Kinloch, I.A.; Gong, L.; Novoselov, K.S. The mechanics of graphene nanocomposites: a review. *Compos. Sci. Technol.* **2012**, *72*, 1459–1476. [CrossRef]
156. Sherlala, A.; Raman, A.; Bello, M.; Asghar, A. A review of the applications of organo-functionalized magnetic graphene oxide nanocomposites for heavy metal adsorption. *Chemosphere* **2018**, *193*, 1004–1017. [CrossRef] [PubMed]
157. Wang, Q.; Ling, S.; Liang, X.; Wang, H.; Lu, H.; Zhang, Y. Self-Healable Multifunctional Electronic Tattoos Based on Silk and Graphene. *Adv. Funct. Mater.* **2019**, *29*, 1808695. [CrossRef]
158. Ling, S.; Wang, Q.; Zhang, D.; Zhang, Y.; Mu, X.; Kaplan, D.L.; Buehler, M.J. Integration of stiff graphene and tough silk for the design and fabrication of versatile electronic materials. *Adv. Funct. Mater.* **2018**, *28*, 1705291. [CrossRef]
159. Scaffaro, R.; Maio, A.; Lo Re, G.; Parisi, A.; Busacca, A. Advanced piezoresistive sensor achieved by amphiphilic nanointerfaces of graphene oxide and biodegradable polymer blends. *Compos. Sci. Technol.* **2018**, *156*, 166–176. [CrossRef]
160. Chen, Z.; Zhang, A.; Wang, X.; Zhu, J.; Fan, Y.; Yu, H.; Yang, Z. The Advances of Carbon Nanotubes in Cancer Diagnostics and Therapeutics. *J. Nanomater.* **2017**, *2017*, 1–13. [CrossRef]
161. Daenen, M.; Zhang, L.; Erni, R.; Williams, O.A.; Hardy, A.; Van Bael, M.K.; Wagner, P.; Haenen, K.; Nesládek, M.; Van Tendeloo, G. Diamond Nucleation by Carbon Transport from Buried Nanodiamond TiO2 Sol-Gel Composites. *Adv. Mater.* **2009**, *21*, 670–673. [CrossRef]
162. Hapuarachchi, T.D.; Peijs, T. Multiwalled carbon nanotubes and sepiolite nanoclays as flame retardants for polylactide and its natural fibre reinforced composites. *Compos. Part A: Appl. Sci. Manuf.* **2010**, *41*, 954–963. [CrossRef]
163. Ma, P.; Jiang, L.; Ye, T.; Dong, W.; Chen, M. Melt free-radical grafting of maleic anhydride onto biodegradable poly (lactic acid) by using styrene as a comonomer. *Polymers* **2014**, *6*, 1528–1543. [CrossRef]
164. Hwang, J.Y.; Kim, H.S.; Kim, J.H.; Shin, U.S.; Lee, S.H. Carbon Nanotube Nanocomposites with Highly Enhanced Strength and Conductivity for Flexible Electric Circuits. *Langmuir* **2015**, *31*, 7844–7851. [CrossRef] [PubMed]
165. Sadykov, N.R.; Peshkov, D.A.; D'Yachkov, P.N. Combined Effect of External Periodic and Constant Electric Fields on Electron Transport in Carbon Nanotubes and Nanoribbons with Metallic Conductivity. *J. Phys. Soc. Jpn.* **2017**, *86*, 1–7. [CrossRef]

166. Badard, M.; Combessis, A.; Allais, A.; Flandin, L. Electric field as a tuning key to process carbon nanotube suspensions with controlled conductivity. *Polymer* **2016**, *82*, 198–205. [CrossRef]
167. Osazuwa, O.; Vasileiou, A.A.; Kontopoulou, M.; Docoslis, A. Electric-field induced filler association dynamics and resulting improvements in the electrical conductivity of polyester/multiwall carbon nanotube composites. *Polym. Compos.* **2017**, *38*, 1571–1578. [CrossRef]
168. Valentini, L.; Fabbri, P.; Messori, M.; Degli Esposti, M.; Bon, S.B. Multilayer Films Composed of Conductive Poly(3-hydroxybutyrate)/Carbon Nanotubes Bionanocomposites and a Photoresponsive Conducting Polymer. *J. Polym. Sci. Part B Polym. Phys.* **2014**, *52*, 596–602. [CrossRef]
169. Dionigi, C.; Posati, T.; Benfenati, V.; Sagnella, A.; Pistone, A.; Bonetti, S.; Ruani, G.; Dinelli, F.; Padeletti, G.; Zamboni, R.; et al. A nanostructured conductive bio-composite of silk fibroin-single walled carbon nanotubes. *J. Mater. Chem. B* **2014**, *2*, 1424–1431. [CrossRef]
170. Sivanjineyulu, V.; Behera, K.; Chang, Y.-H.; Chiu, F.-C. Selective localization of carbon nanotube and organoclay in biodegradable poly(butylene succinate)/polylactide blend-based nanocomposites with enhanced rigidity, toughness and electrical conductivity. *Compos. Part A Appl. Sci. Manuf.* **2018**, *114*, 30–39. [CrossRef]
171. Chen, Y.-F.; Tan, Y.-J.; Li, J.; Hao, Y.-B.; Shi, Y.-D.; Wang, M. Graphene oxide-assisted dispersion of multi-walled carbon nanotubes in biodegradable Poly(epsilon-caprolactone) for mechanical and electrically conductive enhancement. *Polym. Test.* **2018**, *65*, 387–397. [CrossRef]
172. Li, Y.; Li, N.; Ge, J.; Xue, Y.; Niu, W.; Chen, M.; Du, Y.; Ma, P.X.; Lei, B. Biodegradable thermal imaging-tracked ultralong nanowire-reinforced conductive nanocomposites elastomers with intrinsical efficient antibacterial and anticancer activity for enhanced biomedical application potential. *Biomaterials* **2019**, *201*, 68–76. [CrossRef] [PubMed]
173. Sun, J.; Kormakov, S.; Liu, Y.; Huang, Y.; Wu, D.; Yang, Z. Recent Progress in Metal-Based Nanoparticles Mediated Photodynamic Therapy. *Molecules* **2018**, *23*, 1704. [CrossRef] [PubMed]
174. Panda, S.; Behera, D.; Rath, P.; Bastia, T.K. Enhanced properties of UPE/ESOA partially bio-nanocomposites reinforced with chitosan functionalized graphene nanoplatelets: An innovative approach. *Bull. Mater. Sci.* **2018**, *41*, 12. [CrossRef]
175. Sim, S.; Andou, Y.; Bashid, H.A.A.; Lim, H.; Altarawneh, M.; Jiang, Z.T.; Eksiler, K.; Iikubo, S. Development of Organo-Dispersible Graphene Oxide via Pseudo-Surface Modification for Thermally Conductive Green Polymer Composites. *ACS Omega* **2018**, *3*, 18124–18131. [CrossRef]
176. Goodwin, D.G.; Boyer, I.; Devahif, T.; Gao, C.; Frank, B.P.; Lu, X.; Kuwama, L.; Gordon, T.B.; Wang, J.L.; Ranville, J.F.; et al. Biodegradation of Carbon Nanotube/Polymer Nanocomposites using a Monoculture. *Environ. Sci. Technol.* **2018**, *52*, 40–51. [CrossRef] [PubMed]
177. You, J.; Zhu, L.T.; Wang, Z.B.; Zong, L.; Li, M.J.; Wu, X.C.; Li, C.X. Liquid exfoliated chitin nanofibrils for re-dispersibility and hybridization of two-dimensional nanomaterials. *Chem. Eng. J.* **2018**, *344*, 498–505. [CrossRef]
178. Zare, E.N.; Lakouraj, M.M.; Mohseni, M. Biodegradable polypyrrole/dextrin conductive nanocomposite: Synthesis, characterization, antioxidant and antibacterial activity. *Synth. Met.* **2014**, *187*, 9–16. [CrossRef]
179. Du, Y.Z.; Ge, J.; Li, Y.N.; Ma, P.X.; Lei, B. Biomimetic elastomeric, conductive and biodegradable polycitrate-based nanocomposites for guiding myogenic differentiation and skeletal muscle regeneration. *Biomaterials* **2018**, *157*, 40–50. [CrossRef]
180. Mohan, R.; Subha, J.; Alam, J. Influence of Multiwalled Carbon Nanotubes on Biodegradable Poly(lactic acid) Nanocomposites for Electroactive Shape Memory Actuator. *Adv. Polym. Tech.* **2018**, *37*, 1–6. [CrossRef]
181. Lao, J.P.; Xie, H.A.; Shi, Z.Q.; Li, G.; Li, B.; Hu, G.H.; Yang, Q.L.; Xiong, C.X. Flexible Regenerated Cellulose/Boron Nitride Nanosheet High-Temperature Dielectric Nanocomposite Films with High Energy Density and Breakdown Strength. *ACS Sustain. Chem. Eng.* **2018**, *6*, 7151–7158. [CrossRef]
182. Pawde, S.; Deshmukh, K. Influence of γ irradiation on the properties of polyacrylonitrile films. *J. Appl. Polym. Sci.* **2008**, *110*, 2569–2578. [CrossRef]
183. Deshmukh, K.; Ahamed, M.B.; Pasha, S.K.; Deshmukh, R.R.; Bhagat, P.R. Highly dispersible graphene oxide reinforced polypyrrole/polyvinyl alcohol blend nanocomposites with high dielectric constant and low dielectric loss. *RSC Adv.* **2015**, *5*, 61933–61945. [CrossRef]

184. Kashi, S.; Gupta, R.K.; Baum, T.; Kao, N.; Bhattacharya, S.N. Dielectric properties and electromagnetic interference shielding effectiveness of graphene-based biodegradable nanocomposites. *Mater. Design.* **2016**, *109*, 68–78. [CrossRef]
185. Deshmukh, K.; Ahamed, M.B.; Deshmukh, R.R.; Pasha, S.K.K.; Sadasivuni, K.K.; Polu, A.R.; Ponnamma, D.; AlMaadeed, M.A.-A.; Chidambaram, K. Newly developed biodegradable polymer nanocomposites of cellulose acetate and Al_2O_3 nanoparticles with enhanced dielectric performance for embedded passive applications. *J. Mater. Sci.-Mater. Electron.* **2017**, *28*, 973–986. [CrossRef]
186. Zeng, X.; Deng, L.; Yao, Y.; Sun, R.; Xu, J.; Wong, C.-P. Flexible dielectric papers based on biodegradable cellulose nanofibers and carbon nanotubes for dielectric energy storage. *J. Mater. Chem. C* **2016**, *4*, 6037–6044. [CrossRef]
187. Choudhary, S. Characterization of amorphous silica nanofiller effect on the structural, morphological, optical, thermal, dielectric and electrical properties of PVA-PVP blend based polymer nanocomposites for their flexible nanodielectric applications. *J. Mater. Sci.-Mater. Electron.* **2018**, *29*, 10517–10534. [CrossRef]
188. Deshmukh, K.; Ahamed, M.B.; Sadasivuni, K.K.; Ponnamma, D.; AlMaadeed, M.A.A.; Deshmukh, R.R.; Pasha, S.K.K.; Polu, A.R.; Chidambaram, K. Fumed SiO_2 nanoparticle reinforced biopolymer blend nanocomposites with high dielectric constant and low dielectric loss for flexible organic electronics. *J. Appl. Polym. Sci.* **2017**, *134*, 1–11. [CrossRef]
189. Feili, D.; Schuettler, M.; Doerge, T.; Kammer, S.; Stieglitz, T. Encapsulation of organic field effect transistors for flexible biomedical microimplants. *Sens. Actuators A Phys.* **2005**, *120*, 101–109. [CrossRef]
190. Lee, Y.K.; Kim, J.; Kim, Y.; Kwak, J.W.; Yoon, Y.; Rogers, J.A. Room temperature electrochemical sintering of Zn microparticles and its use in printable conducting inks for bioresorbable electronics. *Adv. Mater.* **2017**, *29*, 1702665. [CrossRef]
191. Wu, G.; Lu, Y.; Teng, J.; Wang, J.; Nee, T. Preparation and characterization of pentacene-based organic thin-film transistors with PVA passivation layers. *Thin Solid Films* **2009**, *517*, 5318–5321. [CrossRef]
192. Xie, J.; Yang, Z.; Zhou, C.; Zhu, J.; Lee, R.J.; Teng, L. Nanotechnology for the delivery of phytochemicals in cancer therapy. *Biotechnol. Adv.* **2016**, *34*, 343–353. [CrossRef] [PubMed]
193. Sun, J.; Li, H.; Huang, Y.; Zheng, X.; Liu, Y.; Zhuang, J.; Wu, D. Simple and Affordable Way To Achieve Polymeric Superhydrophobic Surfaces with Biomimetic Hierarchical Roughness. *ACS Omega* **2019**, *4*, 2750–2757. [CrossRef]
194. Xie, J.; Zhang, K.; Wang, Z.; Zhao, Q.; Yang, Y.; Zhang, Y.; Ai, S.; Xu, J. Biodegradable poly(vinyl alcohol)-based nanocomposite film reinforced with organophilic layered double hydroxides with potential packaging application. *Iran. Polym. J.* **2017**, *26*, 811–819. [CrossRef]
195. Xie, J.; Wang, Z.; Zhao, Q.; Yang, Y.; Xu, J.; Waterhouse, G.I.N.; Zhang, K.; Li, S.; Jin, P.; Jin, G. Scale-Up Fabrication of Biodegradable Poly(butylene adipate-co-terephthalate)/Organophilic-Clay Nanocomposite Films for Potential Packaging Applications. *ACS Omega* **2018**, *3*, 1187–1196. [CrossRef]
196. Xie, J.; Zhang, K.; Wu, J.; Ren, G.; Chen, H.; Xu, J. Bio-nanocomposite films reinforced with organo-modified layered double hydroxides: Preparation, morphology and properties. *Appl. Clay. Sci.* **2016**, *126*, 72–80. [CrossRef]
197. Wang, S.; Jing, Y. Effects of formation and penetration properties of biodegradable montmorillonite/chitosan nanocomposite film on the barrier of package paper. *Appl. Clay. Sci.* **2017**, *138*, 74–80. [CrossRef]
198. Ren, P.G.; Liu, X.H.; Ren, F.; Zhong, G.J.; Ji, X.; Xu, L. Biodegradable graphene oxide nanosheets/poly-(butylene adipate-coterephthalate) nanocomposite film with enhanced gas and water vapor barrier properties. *Polym. Test.* **2017**, *58*, 173–180. [CrossRef]
199. Pawar, S.P.; Kumar, S.; Misra, A.; Deshmukh, S.; Chatterjee, K.; Bose, S. Enzymatically degradable EMI shielding materials derived from PCL based nanocomposites. *RSC Adv.* **2015**, *5*, 17716–17725. [CrossRef]
200. Gupta, T.K.; Singh, B.P.; Mathur, R.B.; Dhakate, S.R. Multi-walled carbon nanotube–graphene–polyaniline multiphase nanocomposite with superior electromagnetic shielding effectiveness. *Nanoscale* **2014**, *6*, 842–851. [CrossRef]
201. Kuang, T.R.; Chang, L.Q.; Chen, F.; Sheng, Y.; Fu, D.J.; Peng, X.F. Facile preparation of lightweight high-strength biodegradable polymer/multi-walled carbon nanotubes nanocomposite foams for electromagnetic interference shielding. *Carbon* **2016**, *105*, 305–313. [CrossRef]

202. Wang, G.L.; Zhao, G.Q.; Wang, S.; Zhang, L.; Park, C.B. Injection-molded microcellular PLA/graphite nanocomposites with dramatically enhanced mechanical and electrical properties for ultra-efficient EMI shielding applications. *J. Mater. Chem. C* **2018**, *6*, 6847–6859. [CrossRef]
203. Shih, Y.F.; Chen, L.S.; Jeng, R.J. Preparation and properties of biodegradable PBS/multi-walled carbon nanotube nanocomposites. *Polymer* **2008**, *49*, 4602–4611. [CrossRef]

© 2019 by the authors. Licensee MDPI, Basel, Switzerland. This article is an open access article distributed under the terms and conditions of the Creative Commons Attribution (CC BY) license (http://creativecommons.org/licenses/by/4.0/).

Review

Biodegradable Poly(Lactic Acid) Nanocomposites for Fused Deposition Modeling 3D Printing

Madison Bardot and Michael D. Schulz *

Department of Chemistry and Macromolecules Innovation Institute, Virginia Tech, Blacksburg, VA 24061, USA; mbardot2@vt.edu
* Correspondence: mdschulz@vt.edu; Tel.: +1-540-231-8244

Received: 30 November 2020; Accepted: 16 December 2020; Published: 21 December 2020

Abstract: 3D printing by fused deposition modelling (FDM) enables rapid prototyping and fabrication of parts with complex geometries. Unfortunately, most materials suitable for FDM 3D printing are non-degradable, petroleum-based polymers. The current ecological crisis caused by plastic waste has produced great interest in biodegradable materials for many applications, including 3D printing. Poly(lactic acid) (PLA), in particular, has been extensively investigated for FDM applications. However, most biodegradable polymers, including PLA, have insufficient mechanical properties for many applications. One approach to overcoming this challenge is to introduce additives that enhance the mechanical properties of PLA while maintaining FDM 3D printability. This review focuses on PLA-based nanocomposites with cellulose, metal-based nanoparticles, continuous fibers, carbon-based nanoparticles, or other additives. These additives impact both the physical properties and printability of the resulting nanocomposites. We also detail the optimal conditions for using these materials in FDM 3D printing. These approaches demonstrate the promise of developing nanocomposites that are both biodegradable and mechanically robust.

Keywords: 3D printing; poly(lactic acid) (PLA); additive manufacturing (AM); fused deposition modeling (FDM); cellulose; carbon nanoparticles

1. Introduction

Additive manufacturing (3D printing) enables rapid prototyping, convenient customization, and unique capabilities, while democratizing the manufacturing process in ways that are only just beginning to be leveraged on a large scale [1–3]. These burgeoning manufacturing trends, however, also intersect with growing concerns about the ecological impact of the materials used in manufacturing. As pollution from plastic waste grows worldwide, developing materials that are biodegradable and bio-renewable becomes increasingly important [4–6]. Unfortunately, most materials commonly used for 3D printing are neither.

One of the most common methods of 3D printing—fused deposition modeling (FDM), also known as fused filament fabrication—uses layer-by-layer addition of polymeric materials to form a completed piece. This addition is facilitated by computer-aided design, which instructs the printer where to add polymer [7]. FDM involves drawing a filament through a heated extrusion head, which deposits the molten polymer onto a bed where the 3D-printed part forms [7,8]. The FDM process requires specific parameters for drawability and processability that influence not only the filament production but also the layer deposition during printing [9]. For viable printing, the extruded material must have a low melting temperature and fast solidification time [7].

The printability and strength of printed parts also relies on good adhesion between layers and a homogeneous distribution of any additives [10]. Uniform distribution of additives ensures that agglomerates do not clog the printing apparatus or cause weak points in the printed material [11].

Compounding the complex nature of FDM 3D printing, various printing parameters—raster angle, raster width, layer thickness, build orientation, infill density and pattern, feed rate, and air gap—also have large effects on the quality and mechanical properties of FDM-printed parts [12–17].

Several polymers can be readily printed by FDM including acrylonitrile butadiene styrene (ABS), polycarbonate (PC), and poly(lactic) acid (PLA) [18]. Of these polymers, only PLA is biodegradable, and of the limited number of biodegradable polymers evaluated for 3D printing, PLA is by far the most common. However, its limited mechanical properties hinder its use in many applications, particularly in industrial settings. Additionally, PLA is insulating, which precludes its use in conducting parts [19]. To address each of these issues, various additives have been incorporated into PLA to increase its strength and conductivity. In this review we delve into different additives that enhance mechanical, thermal, or electrical properties while maintaining the biodegradability of the resulting PLA/additive nanocomposite. These nanocomposites exemplify the possibility of developing biodegradable materials while maintaining key physical properties required for industrial applications.

2. Poly(Lactic Acid) (PLA): 3D Printing Properties

PLA is popular for 3D printing due to its affordability, renewability (e.g., derived from corn or sugar cane), and biocompatibility [7,20,21]. PLA contains repeating lactic acid units as depicted in Figure 1, and can include both D- and L-stereoisomers, or be enantiomerically pure (e.g., PLLA contains only L stereocenters). PLA in conjunction with FDM produces 3D scaffolds that degrade via hydrolysis to lactic acid, a common biologically produced molecule [22,23]. PLA biodegradation depends on pH (degrading faster in highly acidic or basic media), temperature, autocatalytic behavior (catalysis by the lactic acid formed during degradation), and the degree to which water enters the matrix [23]. PLA also retains good mechanical strength while remaining processable through melt mixing, solution mixing, injection molding, and 3D printing [21]. However, several drawbacks limit its industrial use such as brittleness, poor thermal stability, low crystallinity, low elongation at break, poor impact strength, low heat-distortion temperature, and limited drawability [9,21,24]. These drawbacks, especially the slow crystallization, deter the replacement of fossil-based thermoplastics with PLA [25]. To increase the functionality of PLA, researchers have introduced additives such as cellulose, metals, carbon, continuous fibers, and others to modulate properties such as thermal conductivity, electrical conductivity, mechanical strength, viscosity, and degradation time [9,26].

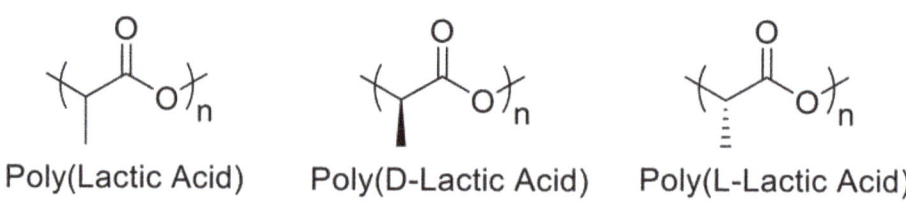

Figure 1. Poly(lactic acid) conformers.

FDM 3D printing parameters affect the printed materials and printing process. Manufacturing time and costs are directly linked to such parameters: production time (and therefore cost) decreases with increasing feed rate and layer thickness. Chacón et al. investigated the effects of these parameters on the properties of neat PLA FDM-printed tensile bars as depicted in Figure 2 [27]. Increasing layer thickness increased tensile and flexural strength for upright printed tensile bars. However, on-edge and flat printed tensile bars had only slight differences in the tensile and flexural strength [27,28]. Decreasing feed rate decreased tensile and flexural strength in upright samples but had limited effects on materials produced by on-edge and flat printing orientations. For all samples, the ductility decreased as layer thickness increased. Overall, on-edge orientation produced the best mechanical performance, ductility, and stiffness. Moreover, if an on-edge orientation is used, high layer thickness and low feed rate maximize ductility [27]. Tensile properties increased with infill density regardless of infill pattern, including concentric infill patterns (which produced the greatest tensile strength), grid patterns, and tri-hexagonal patterns [29,30].

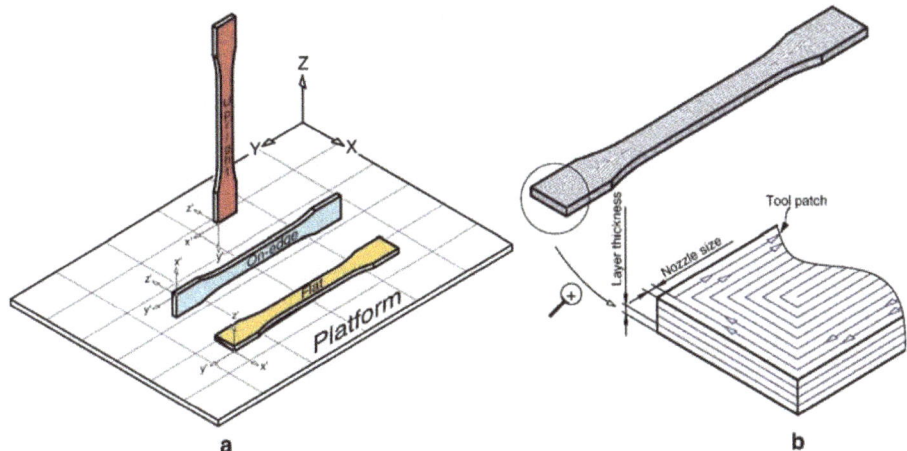

Figure 2. (**a**) Flat, on-edge, and up right printing orientation of tensile bars. (**b**) Processing parameters of tensile bars. Reproduced with permission from [27], Elseveir, 2017.

Thermal processing conditions also play a key role in sample material properties. For example, samples made with low build platform temperatures have increased mechanical properties, increased interfacial strength, larger crystal size, and lower crystallinity [31–33]. Additionally, post-printing annealing at 120 °C removes internal stresses, changes the crystallinity, removes the δ form (a more disorganized crystalline structure), and improves the storage modulus, especially in samples with large layer thickness [31]. The surface structure and friction can also be modulated by changing printing parameters [34].

Traditionally, PLA is sold as filaments for FDM with colorants and additives already incorporated. These additives can have profound effects on the properties of the resulting printed material. For example, Cicala et al. observed a marked difference in elasticity among various commercial PLA samples, demonstrating the effect of different additives on mechanical properties [35]. Importantly, Cicala et al. determined that polymers with high viscosity print with increased precision because of their resistance to flow after printing, which allows them to hold their shape and minimize voids between printed layers [35]. Cuiffo et al. investigated commercial PLA samples with calcium carbonate additives and found that the $CaCO_3$ concentrated in the voids of the 3D-printed materials after FDM printing [36]. Additionally, these PLA samples underwent minor chemical reactions during the FDM process, as shown by changes in the Fourier transform infrared (FTIR) stretches for the C=O, C–O–C, -CH_3, and -OH functional groups, and by deviations in the cold crystallization (T_{cc}) and melting (T_m) temperatures [36].

Neat PLA has a T_g of 55–65 °C [36,37] and T_m of 173–178 °C [36], which enables FDM printing. While the thermal and mechanical properties of PLA are appropriate for FDM 3D printing, they are often inappropriate for many applications, which require a different set of properties. To achieve such materials, while maintaining the renewability and degradability of PLA, new additives have been explored beyond those typical of commercial samples. These additives have resulted in a range of new materials that show promise for applications beyond the capabilities of neat PLA.

3. Cellulose-Based Additives

An increasingly attractive nanofiller for PLA-based materials is cellulose. Cellulose, a biodegradable material, is the most abundant natural polymer shown in Figure 3, and forms a fully biodegradable nanocomposite when incorporated into PLA [7]. Cellulose nanocrystals (CNCs), in particular, attract attention as an additive because of their strength, low weight, transparency, and biocompatibility [24]. While the addition of cellulose may increase mechanical properties, the crystallinity and hydrophobicity of CNCs hinder their incorporation into the PLA matrix. The crystallinity of CNCs, approximately 54–88%, accounts for their high tensile modulus (105–250 GPa), which is greater than that of both glass fiber and steel wire [7,20]. This crystallinity also allows cellulose to act as a nucleating agent in the PLA matrix, increasing PLA/CNC crystallinity and affecting mechanical properties [7].

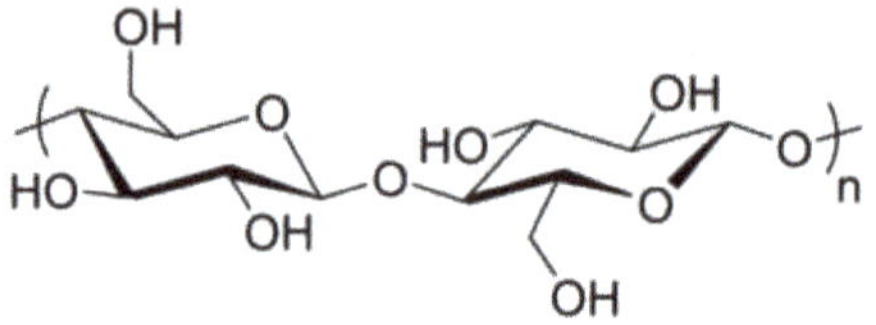

Figure 3. Cellulose chemical structure.

Kumar et al. evaluated the properties of 1, 2, 5, and 10 wt % PLA/CNC composites [7]. They indicate that the crystallinity, melting temperature, and glass transition temperatures change with the introduction of CNCs. The optimal incorporation, 1 wt % CNCs in PLA, had the highest cold crystallization and an increased elastic modulus of 4550 MPa compared to 3030 MPa for neat PLA FDM-printed materials [7]. Importantly, all samples, regardless of CNC incorporation, retain a constant and steady viscosity at the extruder shear rate, allowing for 3D printing without clogging the extrusion head. The yield strength of 1 wt % CNC also increased to 61.07 MPa compared to 51.4 MPa for neat PLA. However, there was a 2.8% drop in strain at break for 1 wt % PLA/CNC compared to neat PLA, indicating that the orientation of the CNCs allows for absorption of the load, but reduced strain [7].

Other forms of cellulose have also been incorporated into PLA to enhance mechanical properties [38]. For example, previous studies indicate that incorporating cellulose nanowhiskers (CNWs) silylated with (3-mercaptopropyl) trimethoxysilane into PLA nanocomposites increases the elongation at break 250.8% compared to neat PLA [39]. The silylation reaction functionalizes the hydroxyl groups on cellulose, increasing CNW compatibility with PLA. Functionalization also modifies the usable temperatures of the PLA/CNW nanocomposite. The addition of silane A-151 increased the T_{max} to 304.4 °C [24]. Importantly, the concentration of A-151 must be high enough, at least 8 wt %, to fully coat the CNWs and create an even surface. While silylation increased the compatibility of the PLA and CNWs, the tensile strength and tensile modulus decreased minimally with increasing silane concentration. However, the elongation at break increased significantly (from 12.3% to 213.8%) with increasing silane concentration. The stiffness also increased with silane addition. In general, the tensile strength and tensile modulus increase with silylation of CNWs; however, the thermal properties, including the glass transition temperature, crystallinity, melting temperature, and crystallization temperature, decrease after silylation.

Cellulose nanofibers (CNFs) have also been investigated in the context of PLA 3D printing. Interestingly, the method of 3D printing affects the mechanical properties of CNF-containing PLA nanocomposites [40]. Specifically, the strength and modulus of FDM-printed neat PLA is 49 and 41% lower than its compression molded counterparts. With the addition of CNFs, at just 1 wt %, the strength and modulus of 3D-printed PLA/CNFs increased by 84% and 63% compared to PLA, respectively [40]. Incorporating CNFs into PLA significantly decreased voids and facilitated nucleation and crystallization, leading to increased matrix crystallinity. This higher crystallinity and reduction of defects caused overall increased mechanical properties.

Another cellulose-based derivative, industrial hemp hurd (HH), is a lignocellulose byproduct of the cannabis industry. HH contains 18–24% hemicellulose, 21–24% lignin, and 48% cellulose [41]. Xiao et al. introduced HH into PLA in the presence of toughening agent poly(butylene adipate-co-terephthalate)

(PBAT) and interfacial modifier ethylene-methyl acrylate-glycidyl methacrylate (EGMA) terpolymer. The polymer matrices were all 87 wt % PLA and 13 wt % PBAT with 10–40 phr HH. Increasing HH concentration increased the complex viscosity, indicating a decrease in melt flow and 3D printability. The PLA/HH nanocomposites also showed increased shear-thinning behavior with increasing HH loading, which assists in 3D printability. Storage and loss moduli also increased with increasing HH content, especially at low frequencies, due to inhibition of chain mobility [41]. Contrastingly, the tanδ and angular frequency decreased with increasing HH loading. Interestingly, the addition of HH did not affect the thermal transitions such as T_g and T_m but did increase the cold crystallization temperature, indicating that increasing HH decreased crystallization enthalpy. With increased loading of HH the fracture surfaces showed less interfacial adhesion, indicating that HH decreased interfacial adhesion of polymer surfaces. HH also increased surface roughness and darkened the color of the composite. Overall, both the flexural strength and tensile strength decreased with increasing HH loading. Nevertheless, PLA/HH-30 had an impact strength comparable to commercial PLA FDM samples [41].

Poly(3-hydroxybutyrate) (PHB) has been incorporated into PLA/CNC nanocomposites through single-step blending with a crosslinking agent, dicumyl peroxide, which caused increased adhesion between the PLA and PHB phases and increased dispersion of CNCs. After 3D printing, the polymers align parallel to the extrusion direction indicating that FDM 3D printing changed the distribution and orientation of the CNCs. While the initial FDM printing of the nanocomposite caused some aggregation of CNCs, reprocessing those printed materials produced good dispersion of CNCs in the copolymer matrix [21].

PLA/PHB nanocomposites also facilitate introducing recycled cellulose in the form of pinewood fibers [42]. These PLA/PHB/wood composites are processable from 210 to 230 °C. This narrow window enables PLA/PHB melting without degradation of the wood nanofibers. This low print temperature not only makes the polymer desirable for home printing applications but also lowers the energy, and therefore cost, associated with printing. While CNCs improve mechanical properties, the PLA/PHB/wood nanocomposites show a 35% decrease in strength and 41% decrease in stiffness compared to neat PLA. The tensile strength was also low, not exceeding 21 MPa [42].

Adding plasticizers such as poly(ethylene glycol) (PEG) into PLA may increase ductility and toughness, but often decreases strength and stiffness [20]. These drawbacks may be mitigated by introducing nucleating agents, such as CNCs or CNFs, which increase the crystallinity of the PLA/PEG matrix and thus increase the strength and stiffness. Adding CNFs and CNCs gives significantly higher crystallinity than neat PLA or commercially available PLA/talc nanocomposites [20].

Solvent effects on CNC dispersion have recently been investigated in both amorphous and semi-crystalline PLA samples. Thermodynamic analysis identified dimethyl sulfoxide/tetrahydrofuran (DMSO/THF) as an optimal solvent system to incorporate CNCs into PLA because it can both dissolve the PLA and distribute the CNCs [43]. In semi-crystalline PLA nanocomposites, the CNCs distribute effectively but in amorphous nanocomposites the CNCs aggregate [43]. With the addition of CNCs, the storage modulus and complex viscosity increase. Significantly, residual solvent in the nanocomposite matrices caused dramatic decreases in complex viscosity (1 to 2 orders of magnitude) [43].

Cellulose in the form of wood fiber also increases the mechanical properties of PLA. Adding wood fiber to the PLA filament and subsequent FDM 3D printing produced materials with increased stiffness (18%), toughness (44%), thermal conductivity (23%), fracture strain (15%), strength (9%), and decreased density (10%) [44]. In FDM 3D printing, the printer typically prints line by line, with subsequent layers depositing directly on top of the previous layer. However, with the introduction of "isomixed" printing, a printing method where the printer prints crosshatched layers with each subsequent layer rotated by 45 degrees such that each new layer lies not perfectly on top but askew from each other as seen in Figure 4, there is an increase in properties. Isomixed microarchitectures of PLA/wood fiber nanocomposites show a 91% increase in stiffness and 48% increase in strength compared to neat PLA [44].

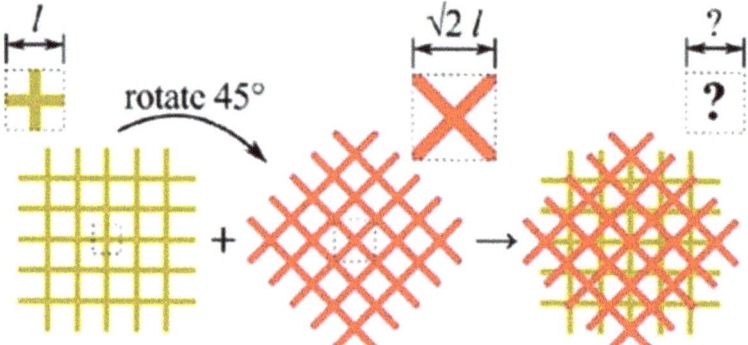

Figure 4. Depiction of rotation and crosshatched 3D printing pattern utilized to increase mechanical properties of 3D-printed materials. Reproduced with permission from [44], Wiley, 2020.

Cellulose, in the form of lignocellulosic fillers (LCFs), has also been incorporated into PLA/graphene nanoplatelet (GNP) nanocomposites to reduce the required GNP content and increase the mechanical properties of the materials. The introduction of both LCFs and GNPs allowed for good dispersion in the PLA matrix. Further, the introduction of these nanoparticles in combination increased nanocomposite stiffness up to 55%, elongation at break up to 35% and tensile strength up to 48% [45]. These results indicate that the introduction of GNPs and LCFs mitigate the reduced mechanical properties often found with natural fibers while strengthening the matrix. Varying the ratio of GNP to LCF modulated the degradation from 8% to 70% after just 900 h submersion in an alkaline environment and 1% to 30% in a neutral environment, indicating that this formulation can be tuned for many applications and degradation rates [45].

4. Metal-Based Additives

Metal additives can increase the mechanical and antibacterial properties of PLA nanocomposites. For example, Bayraktar et al. loaded silver nanowires into PLA, through solution mixing, as an antibacterial agent [46]. During processing, the suspended silver nanowires aligned in the direction of shear force homogeneously but broke down. Specifically, the shear force and heat required for 3D printing broke the nanowires into smaller particles but conserved the surface morphology. TGA showed that the nanowires influence the degradation of the PLA matrix: They increased the degradation temperature, increased crystallinity before printing, decreased T_g, decreased crystallinity after printing, and did not change T_m [46]. Interestingly, the concentration of nanowires decreased after printing, indicating that some silver nanowires stuck to the inside of the polymer extrusion nozzle. Overall, the addition of silver nanowires added a barrier to degradation of the PLA/Ag nanocomposite while adding an antibacterial property, killing 100% of both *S. aureus* and *E. coli* for all concentrations of silver nanowires studied [46]. More recent studies on introducing silver nanoparticles indicate that composites with no significant change in bulk properties can be formed with an addition of 0.01–5 wt % silver nanoparticles. At all loadings of silver nanoparticles studied, these new PLA/Ag nanocomposites showed antimicrobial properties against *E. coli*, *P. aeruginosa*, and *S. aureus* [47].

Many industrial applications cannot use FDM-printed parts because fractures occur between layers due to poor interfacial adhesion and low surface quality [48]. Therefore, modifying FDM PLA nanocomposites to increase this surface quality may increase industrial use of these nanocomposites. Incorporating aluminum into nanocomposites produces air-cooled heat exchangers with high thermal conductivity at a low cost. These PLA/Al composites, when laser polished, show increased surface quality, decreased surface roughness, increased storage modulus, decreased loss tangent, increased tensile strength, and increased Young's modulus [48]. These enhanced properties arise from the polymers becoming molten on the surface and reforming into a much more dense and interconnected structure during laser polishing.

Similarly, copper fiber can be incorporated into a PLA matrix. Copper fiber/PLA composites contain uneven surface morphologies after FDM 3D printing due to the layer-by-layer addition, which often leaves ridges in the surface of printed materials. To ameliorate the effects of this surface structure, laser polishing is employed to melt the polymer matrix at the surface, producing a smoother surface [49]. The surface roughness decreases over 90% after laser treatment to 0.87 µm Sα with a 5 W laser of 200 µm (ideal parameters) as seen in Figure 5. This polishing also significantly improved the glass transition, storage modulus, Young's modulus (34.2%), loss modulus, and tensile strength (52.98%) of Cu/PLA due to strong interfacial adhesion between PLA and Cu fibers after treatment [49].

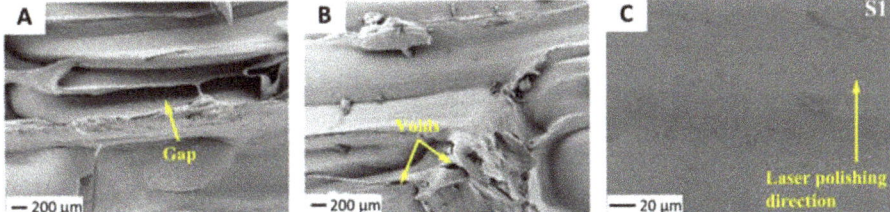

Figure 5. Poly(lactic acid) (PLA)/copper fiber surfaces (**A**,**B**) before laser treatment showing gaps, voids, and uneven surface morphology and (**C**) after 5 W laser treatment with a 175 µm beam. Reproduced with permission from [49], Wiley, 2020.

Hybrid FDM (HFDM) methods incorporate metal meshes into the matrix of PLA. In these methods, a tensile bar is printed, wire mesh is laid on top of that piece, and a top layer is then printed, encasing the copper mesh into the 3D-printed materials and forming a composite. These composites show increased resistance to tearing, fracture load, and flexural strength [50]. Interestingly, in three-point flexural tests neat PLA shows a clean break at the point where force is applied. In contrast, PLA/Cu HFDM materials experience striations (small cracks) throughout the matrix rather than a clean break at the point of force application [50].

Introducing wood, as particleboard wood flours (PWF), into PLA nanocomposites would create a fully biodegradable and strong nanocomposite while maintaining the aesthetic appearance of wood. However, introducing PWF into PLA does not produce a glossy appearance or antibacterial properties. Therefore, micrometer copper–zinc (mCu–Zn) alloys have been introduced [51]. First, PWF was treated with hydrogen peroxide under alkaline conditions to remove cellulose, which may degrade at high processing temperatures and cause the material to discolor or lose strength [51]. Maximum mechanical properties occurred when 5 wt % PWF was introduced into the PLA matrix. The addition of mCu–Zn increased the thermal properties of PLA/PWF nanocomposites and, with an introduction of 2 wt % mCu–Zn, and produced glossy, antibacterial nanocomposites [51].

Adding magnesium into PLA matrices can improve mechanical properties while remaining completely biodegradable and bioresorbable [52]. Adding magnesium also increases the PLA degradation rate, reducing the time it exists in the environment [53–56]. Importantly, homogenous magnesium distribution is not readily achieved, but increases with addition of vitamin E. This PLA/Mg/vitamin E filament enabled printing of surgical implant screws [52].

More recently magnesium has been incorporated into PLLA matrices in the form of MgO nanowhiskers. The functionalization of magnesium to magnesium oxide increased the dispersion of the nanoparticles and improved the mechanical properties [57,58]. These MgO nanowhisker-containing materials undergo in vivo degradation, which is modulated by the amount of MgO incorporated and the molecular weight of PLLA in the matrix. With the incorporation of MgO, an increase in bone cells occurs, likely due to the positive effect of the Mg^{2+} degradation product on osteogenesis [57]. This finding indicates that increased MgO content increased desirable properties for bone implants.

Magnesium oxide nanowhiskers modified with stearic acid improve strength and toughness of PLLA nanocomposites. The addition of stearic acid increases PLA compatibility, and while some aggregation of nanowhiskers in the PLLA nanocomposite occurs, no significant cavities resulted [59]. At a nanowhisker content of 1 wt %, the PLLA/MgO nanocomposites increase the tensile strength by 17% and Young's modulus by 78% compared to neat PLLA. The introduction of MgO nanocomposites also promoted nucleation, increasing crystallinity and decreasing the T_{cc} [59].

To further understand in vivo and in vitro degradation, the shape effect of magnesium oxide (nanowhiskers versus nanoparticles) was explored [60]. Introducing MgO in any form significantly accelerated degradation by accelerating water uptake into the polymer. After water uptake, the polymer first loses the nanocrystalline regions, followed by degradation of the crystalline regions [60]. Importantly, the addition of nanowhiskers increased the degradation rate more than the introduction of nanoparticles. In previous studies, modulating the MgO nanowhiskers content in PLLA nanocomposites modulated the degradation rate with an increase in degradation occurring with increased MgO nanowhisker content [61]. The introduction of MgO nanowhiskers also increased the cytocompatibility of the PLLA composites as seen through increased cell counts, consistent with previous studies on unfunctionalized magnesium [55,61].

To further compatibilize magnesium with the PLA matrix, the surface of Mg nanoparticles has been modified with adsorption dispersants. Polyethyleneimine (PEI) and cetyltrimethylammonium bromide (CTAB) were added in aqueous suspension and facilitated the ionic interactions of PLA and Mg nanoparticles. The introduction of PEI and CTAB enabled good dispersion in the PLA matrix as seen in Figure 6 [53]. In 3D-printed samples, PEI and PLA become covalently bonded through amide bonds, likely resulting from the temperatures required for printing. The addition of surface-functionalized Mg nanoparticles increased the Young's modulus and strength when compared to neat PLA. The matrix then improves the extrusion and printability of PLA by compatibilizing the magnesium and PLA and eliminating nozzle clogging [53].

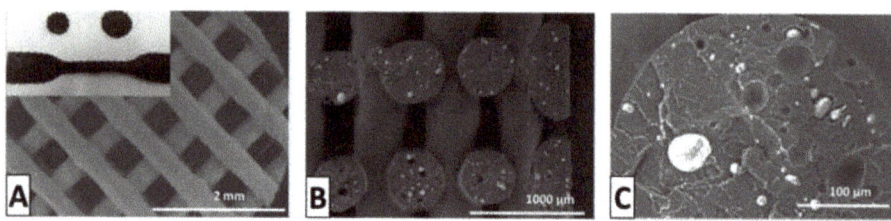

Figure 6. Compatibilized magnesium nanoparticle PLA filaments. (**A**) Fused deposition modelling (FDM) 3D-printed scaffold, (**B**) layer structure of scaffold, (**C**) cross section of individual layer showing magnesium dispersion. Reproduced with permission from [53], Elsevier, 2020.

5. Continuous Fibers

Recently, introducing continuous fibers into PLA matrices has been explored. These fibers made of carbon, flax, pineapple leaf, and others are strong uninterrupted materials, which can increase mechanical properties [62]. Adding these continuous fibers, if used in conjunction with FDM 3D printing, occurs in the extrusion head when the polymer matrix is molten. Figure 7 depicts one common method for entrapping continuous fibers where the molten polymer completely surrounds the fiber [63].

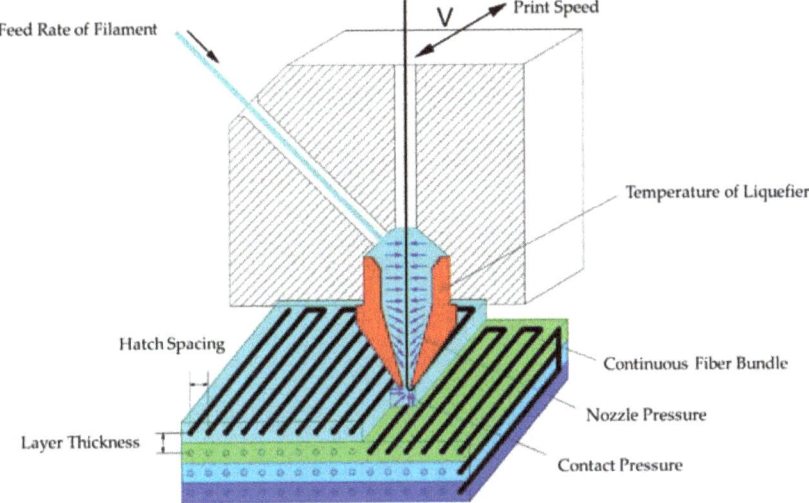

Figure 7. FDM 3D printing apparatus with continuous fibers. Reproduced with permission from [63], Elsevier, 2016.

Carbon fiber-reinforced (CFR) PLA matrices have the capacity for increased materials properties. Importantly, effectively impregnating the carbon fiber into the PLA matrix and forming the composite material requires a temperature of 200–230 °C. This temperature allows adequate flow of PLA while limiting the liquid-like and gravity-dependent flow properties when exiting the feed head [63]. The addition of 27% carbon fiber in CFR PLA composites increased the flexural strength of the composite to 335 MPa and flexural modulus to 30 GPa. Decreased layer thickness in FDM printed parts, and thus increased contact pressure and increased lamination between layers, caused increased flexural strength [63].

New methods for CFR PLA formation require optimization of printing parameters. The printing of continuous fibers requires that the carbon fiber remain at optimal tensions to reduce problems from wavy fibers in FDM 3D-printed materials [64]. For this reason, an "embedding on the component" method has been employed. This method allows the computer to control the PLA filament and carbon fiber filament speed when entering the extrusion head. This method, along with PVA pretreatment of carbon fibers, has increased the tensile strength by 35% and bending strength by 108% compared to neat PLA [64]. The failure methods of this composite, when exposed to external stress, are primarily delamination-induced matrix cracking and delamination. Importantly, the PLA/carbon fiber interface is well bonded with little delamination before cracking of the PLA matrix [64]. Interestingly, continuous carbon fiber-containing PLA composites printed with square-shaped nozzle outlets have increased fracture toughness and decreased bonded areas compared to those printed with a circular extrusion nozzle [65].

Further insights into the strengthening mechanism of CFR PLA composites came through studying the fiber-matrix bonding interface strength, relative fiber content, and failure form of FDM-printed materials through field emission scanning electron microscopy [66]. With increased layer height (from 0.2 to 0.4 mm) and extrusion width (0.86 to 1.5 mm), tensile properties decreased, largely due to the increased concentration of PLA in comparison to carbon fiber content. Tensile properties also decreased slightly with increased temperature (from 190 to 230 °C) and increased feed rate (50 to 400 mm/min) [66]. Interestingly, the main failure mode for continuous carbon fiber-reinforced PLA is fiber pull-out caused by interface failure, meaning that the carbon fiber releases from the PLA matrix and pulls through the polymer rather than fracturing with the PLA as depicted in Figure 8. Pull-out is directly affected by the cohesion of the carbon fiber/PLA interface, indicating that increased adhesion would increase tensile properties [66].

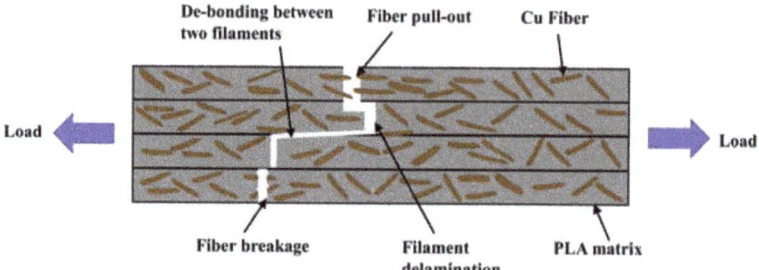

Figure 8. Depiction of fibers in FDM 3D-printed material and the different defects that occur with stress including fiber pull-out, filament delamination, debonding, and fiber breakage. Reproduced with permission from [49], Wiley, 2020.

Other continuous fibers also lend themselves to incorporation into polymer matrices. Flax fibers are among the strongest natural fibers and therefore a good option for introducing into PLA while maintaining biodegradability [67]. When coated with PLA, the twisted nature of the flax yarn causes imperfect impregnation and fiber-rich areas with minimal bonding to the PLA matrix. Even though imperfect binding occurred between the yarn and PLA matrix, the resulting composite exhibited good tensile behavior [67]. Further probing of slicing parameters, including layer height and number of layers, on FDM-printed parts determined that these parameters affect the properties of the continuous flax fiber/PLA composites [68]. Decreasing layer height from 0.6 to 0.2 mm increased stiffness and strength by over 210%, decreased porosity, and decreased layer thickness. Interestingly, layer thickness is determined, in part, by the width and compressibility of the flax yarn as the flax yarn must be within the PLA matrix. Increasing the number of layers (1 to 10) increased compaction ratio, tensile modulus (over 50%), and tensile strength (over 73%) [68]. For manufacturing, curved and bending 3D-printed parts, the introduction of continuous flax fibers increased the curved bending force by 39% and stiffness by 115% for FDM materials compared to the flat-slicing method of 3D printing (printing with supports where each layer is printed flat rather than printing onto an already curved area) [69].

Other natural fibers have also been incorporated into PLA matrices. Pineapple leaf fiber is currently used for wallpaper, textiles, and rope. This fiber has a diameter ranging from 0.1 to 0.5 mm, length ranging from 55 to 75 cm, tensile strength between 170 and 1627 MPa, and an elongation before break of 0.8 to 2.4% [70]. Interestingly, the pineapple leaf fibers do not require functionalization and can feed through a typical FDM setup with the PLA filament rather than the embedding method required for other continuous fibers depicted in Figure 8. Increasing the feed rate increased the tensile strength and decreased the printing time at 200 and 205 °C. At 210 °C increasing the feed rate decreased the tensile strength. Increasing the extrusion temperature from 200 to 210 °C increased the tensile strength from 85.30 MPa to 101.51 MPa at a feed rate of 15 mm/s [70].

The addition of continuous fibers brings into question the recyclability and biodegradability of the resulting new composites. Recycling the continuous carbon-impregnated PLA through melting and then introducing virgin PLA increased the mechanical properties and increased flexural strength by 25% compared to the original 3D-printed material, due to the improved interfacial properties compared to the first printing process. Since the continuous fiber does not break though tensile testing and does not pull out, this recycling process is a non-downgrade recycling process, the first for continuous fibers [71]. With good material recovery, 100% carbon fiber and 73% PLA, this recycling process represents an exciting step forward toward recyclable, high-quality PLA composites. While this process is by no means low energy (67.7 MJ/kg for recycling and 66 MJ/kg for remanufacturing), it is a promising approach towards recyclable materials [71].

Fiberglass, another widely used industrial continuous fiber, is used in everything from boat hulls to wind turbine blades. Recently, these turbine blades have been broken down, after decommissioning of the turbine, to collect the fiberglass fibers inside. These fibers are then embedded in PLA matrices for FDM 3D printing [72]. Addition of 5 wt % recycled fiberglass in FDM-printed materials increased the elastic modulus 16% and strength 10% compared to neat PLA. The Young's modulus also increased 8%, to 3.35 GPa [72]. This increase in mechanical properties indicates that the PLA/fiberglass composites are a viable option for repurposing fiberglass materials.

6. Carbon-Based Additives

Carbon-based additives can be incorporated into PLA to increase thermal/electrical conductivity and mechanical properties [73]. The most common carbon-based additives include carbon nanotubes (CNTs), multiwalled carbon nanotubes (MWCNTs), and graphene nanoplatelets (GNPs) [8,74]. Such modifications, however, increase the risk that the additive will reduce or eliminate the biodegradability of the PLA matrix. Another concern, especially with carbon-based additives, is that the degradation byproducts contain hazardous particles. However, a study by Ivanov et al. indicates that PLA/carbon nanocomposites biodegrade into safe byproducts under standard degradation conditions [11,75–77].

Adding CNTs—cylindrical sheets of graphene—into PLA affects the mechanical and conductive properties of the material. Including CNTs in any concentration reduces the melting temperature of PLA and inhibits crystallization, indicating that adding CNTs adversely affects FDM printing. While CNTs may hinder the printing process, the addition of only 6 wt % CNT increased the tensile strength of PLA/CNT 64% and flexural strength 29% [78]. The electrical resistivity of these nanocomposites also increased to between 10^{12} Ω/sq and 10^2 Ω/sq with CNT content from 0 to 8 wt % [78]. To achieve the best electrical conductivity in the PLA/CNT nanocomposites studied, a high extrusion nozzle temperature, large layer thickness, and small filling velocity were employed [78]. In a related study, increasing the CNT content was found to increase the degree of crystallinity of PLA/CNT nanocomposites, likely due to the small defects on CNT surfaces, which become nucleating sites. The addition of 3 wt % CNTs produced a material with a surface resistivity of 10^5 Ω/m^2, indicating that these printed materials have resistance similar to the human body [79]. This resistance proves that the CNTs are well ordered after printing and that these PLA/CNT nanocomposites may be usable as thermal resistance plastics [79].

MWCNTs—CNTs with multiple concentric layers—and GNPs (graphene sheets) have been introduced into PLA through solution blending and melt mixing. In both melt-mixed and solution-blended samples, increasing the GNP content caused aggregation. In mixed GNP/MWCNT materials, this aggregation likely results from the MWCNTs adsorbing onto the surface of GNPs. Agglomerates of approximately 50 nm appear even at the lowest studied concentration of GNPs/MWCNTs in solution-blended nanocomposites. However, the melt-mixed nanocomposites of PLA/1.5 wt % GNP/4.5 wt % MWCNT show no aggregation [74]. Formation of the PLA/GNP/MWCNT nanocomposites into filaments is a necessity for FDM printing and thus the nanocomposite must retain good rheological and mechanical properties during extrusion into a filament. The viscosity of nanocomposites doubled in the 6 wt % materials compared to neat PLA. To achieve electrical conductivity, PLA nanocomposites require 1.5–3 wt % and 3–6 wt % for MWCNTs and GNPs, respectively. Interestingly, the electrical conductivity was greater in melt-mixed nanocomposites compared to their solution-blended counterparts because of the increased surface contact area of nanoparticles when introduced via melt-mixing. Overall, the melt-mixed polymers gave better rheological properties, higher electrical conductivity, stronger electromagnetic shielding, and were easier to process than those produced by solution blending [74]. Further examination of the thermal and electrical conductivity of GNP and MWCNT nanocomposites determined that the addition of 12% GNPs increased thermal conductivity by 263%, compared to 99% with MWCNTs and 190% with a bifiller system [8]. Improved thermal properties were also seen with MWCNT, GNP, and a bifiller composition of up to 4.54 S/m, 6.57 S/m, and 0.95 S/m respectively [8].

Batakliev et al. found that better mechanical properties occur with the addition of 6 wt % MWCNTs or GNPs compared to 1.5, 3, 8, and 12 wt % [80]. Initially, the Young's modulus of GNP/PLA nanocomposites remains largely unchanged for all GNP-loaded samples due to their high degree of exfoliation. Compared to neat PLA, the tested nanocomposites showed a 20–50% increase in elasticity and 30–50% increase in hardness resulting from the good dispersion of GNPs and MWCNTs [80]. When compared to neat PLA, the best mechanical properties in this study come from bifiller systems with equal concentrations, specifically a 12 wt % total concentration (6 wt % GNP/ 6 wt % MWCNT) [80].

Introducing carbon-based additives also enables printing electrically conductive parts. To conduct electricity the nanocomposite must pass the percolation threshold, the transition from a conductive to insulating polymer, while remaining viable for 3D printing [11]. MWCNT-containing systems did not show aggregation at 6 wt %; however, GNP systems at the same concentration showed agglomeration, likely due to π–π and Van der Waals interactions between the graphene sheets. Bifiller PLA/GNP/MWCNT materials have increased dispersion, which increased the electrical conductivity 7–8 decades compared to neat PLA to a value of 8.4×10^{-3} S/m. Importantly, all samples over 3 wt % total monofiller or bifiller concentration pass the percolation threshold, allowing for electrical conductivity [11]. If heat is not dissipated in electrically conductive materials, the lifetime of the printed materials diminishes. Thermal conductivity depends on several factors including heating temperature, polymer crystallinity, and orientation of macromolecules [11]. In these polymer nanocomposites, the thermal conductivity increased linearly with increasing filler content, while thermal diffusivity also increased [11].

While previous nanocomposites containing MWCNTs and GNPs showed enhanced mechanical properties, the agglomeration of nanoparticles causes issues such as nozzle clogging or inconsistent mechanical properties throughout the matrix [81]. Oxidizing the MWCNT surface introduces hydroxyl groups, leading to better incorporation into the PLA matrix. The low flow index and high initial viscosity, even at a low loading of 1.5 wt %, indicate that the oxidized MWCNT-containing nanocomposites show high dispersion. Interestingly, MWCNTs with higher specific surface area, 250–300 m^2/g compared to 110 m^2/g, have about two times better dispersion. In contrast, GNP nanoparticles show low dispersion and cause 12% (aspect ratio 500) and 25% (aspect ratio 240) reductions of strength and elongation [81]. In contrast, the strength and elongation for oxidized MWCNT samples only slightly decreased (at 12 wt % loading) [81]. The thermal conductivity of PLA/GNP nanocomposites increased with the addition of GNPs more than with the introduction of oxidized MWCNTs [81]. This data indicates that oxidation of MWCNTs increases dispersion while decreasing the mechanical properties of the nanocomposite.

The creep behavior—the tendency to deform under stress—of 3D-printed PLA nanocomposites impacts their viability and long-term strength. Bustillos et al. studied PLA/graphene 3D-printed materials that contained a good dispersion of GNPs, porous structure, and a well-defined pore-size gradient as depicted in Figure 9 [82]. The PLA/GNP printed materials showed a higher cooling and nucleation rate compared to neat PLA due to the introduction of GNP nanoparticles with higher thermal conductivity. The fast cooling causes thermal strain mismatch between the GNPs and PLA matrix, causing poor interlayer and intralayer bonding. The printed material and filament also show a lower T_g compared to neat PLA; however, the 3D-printed material had a 12% higher T_g than its corresponding filament, indicating the 3D printing process increases the interactions between graphene and PLA. GNP-containing samples also showed a higher nucleation rate and a corresponding decrease in the T_{cc}. Crystallinity also decreased about 8% in the 3D-printed PLA/GNP material, which was attributed to the rapid cooling and therefore increased amorphization [82]. The GNP-reinforced samples also showed an increase in hardness (18%), elastic modulus (11%), and resistance to displacement (25%). The permanent deformation also decreased: After elongation, neat PLA samples only recover 25% compared to the 43% recovery of the PLA/GNP nanocomposite. Overall, adding GNPs to the PLA matrix caused a 20.5% decrease in creep displacement, 14% increase in wear resistance, and overall increase of mechanical properties [82].

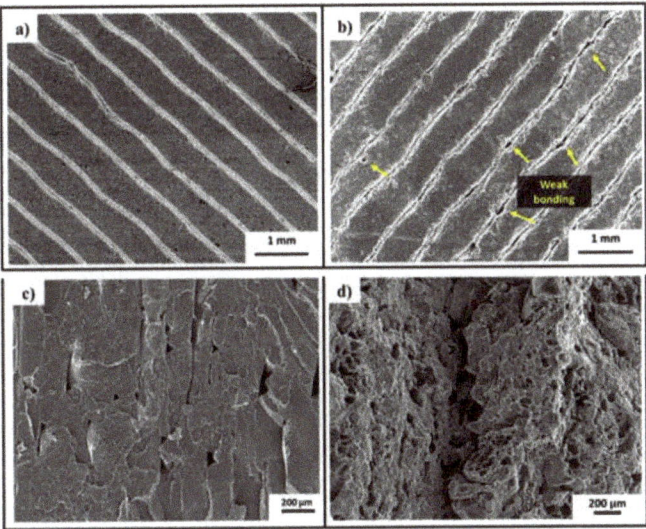

Figure 9. Integration of graphene nanoparticles into FDM 3D-printed PLA. (**a**) Neat PLA after FDM 3D printing. (**b**) PLA/graphene nanocomposite after FDM 3D printing. (**c**) PLA and (**d**) PLA/graphene fractured cross-sectional microstructure. Reproduced with permission from [82], Wiley, 2018.

Incorporating additional chemicals, such as L-arginine, into PLA/GNP matrices decreases phase separation and increases GNP dispersion in the matrix. For example, at 2 wt % L-Arg/GNP loadings, limited agglomeration occurs; however, increasing GNP concentration increases agglomeration. 3D-printed materials produced from solution-mixed filaments of these composites [83]. The fracture surfaces of the 3D-printed structures occur in the interlayer space, indicating good adhesion and interlayer homogeneity. The elongation at break of these PLA nanocomposites increased from 8.3% to 14.8%, indicating that the introduction of GNPs decreased the brittleness of the polymer and transformed it into a more ductile material by transferring stress and preventing cracking. Both the flexural strength and tensile strength also increase with increased L-Arg [83]. The enhanced mechanical properties are presumably caused by the polar groups on the GNP surface; however, at GNP concentrations greater than 2 wt %, increased concentration decreases mechanical properties. When GNP content is held constant, the matrices containing L-Arg have higher T_g, stronger interfacial binding, and lower T_m. The crystallinity of 2 wt % PLA/L-Arg/GNP samples also increased to 16.74%, almost six times that of neat PLA. At the optimal loading of 2 wt %, the degradation temperature increased by 60 °C [83].

The printing parameters, including infill and layer thickness, also affect mechanical properties of PLA/GNP FDM-printed materials. Tensile strength and flexural strength increased with increasing layer thickness and infill to a maximum of 33.7 and 60.9 MPa, respectively, with a layer thickness of 0.27 mm and infill of 78% [84]. The impact energy reached a maximum value of 70.9 J/m with an infill of 50% and layer thickness of 0.20 mm [84]. Interestingly, the impact energy decreased with increasing infill and layer thickness until the midpoint of each factor after which it increased.

Modification of MWCNTs and graphene through oxygen plasma etching (OPE) increases oxidation and exfoliation on their surface, which leads to better incorporation and increased in vivo viability. Introducing these OPE MWCNTs into PLA decreased mechanical properties, increased cell adhesion, and increased cell viability [85]. While the introduction of graphene oxide and MWCNT oxides decreased the elongation-at-break and tensile strength compared to neat PLA, the mechanical performance remained sufficient for bone implants. Importantly, the introduction of graphene oxide and MWCNT oxide nanoparticles significantly increased the osteoblast ALP cell activity in vitro and significantly increased bone cell activity in vivo [85].

While the addition of carbon nanoparticles into PLA matrices has increased the mechanical properties significantly in many cases, a key barrier remains to FDM-printed part acceptance in industrial applications: fractures at the layer interface. Interlayer cracking accounts for a large portion of breaks in FDM parts and therefore is one of the key reasons that FDM parts are not used industrially. Microwave irradiation can strengthen the interlayer adhesion after printing by melting the layers together, increasing the fracture strength by 275% [86].

7. Other Additives

Additives for PLA reinforcement are not limited to the materials previously described. Other additives including nanoclay, silica, other polymers, and various biodegradable bioderived powders have been incorporated into PLA to form biodegradable and well-dispersed PLA nanocomposites.

Nanoclay has been added to PLA to enhance its properties. These samples, PLA 4032D (T_m 170 °C, 1.5% D-isomer) and PLA 2003D (T_m 150 °C, 4% D-isomer), were processed at varying nozzle temperatures to determine the effect of nozzle temperature on clay-containing nanocomposites at different crystallinities [9]. Adding nanoclay increased the storage modulus of all samples, indicating the effectiveness of the nanoclay in strengthening the composite. Specifically, dynamic mechanical analysis (DMA) showed an increase in the storage modulus at 35 °C of 8 and 23% for PLA 4032 and PLA 2003 nanoclay composites. Nanoclay also acted as a nucleating agent, as shown by differential scanning calorimetry (DSC), and increased the thermal stability of composites. All printed samples were largely amorphous. Printing temperatures, when increased, increased the transparency of the PLA/clay nanocomposites. The increased nozzle temperature also increased brittleness of 3D-printed parts.

Adding silica, collected as sand from local beaches, leads to enhanced thermostatics, which increases handling and quality performance of thermoplastic polymers. Silica, one of the most abundant inorganic ceramics, is a good choice for incorporation into polymer matrices due to its low cost and good mechanical properties [87]. Incorporating silica during recycling not only increases the tensile strength to 121.03 MPa (with a 10 wt % addition) but also increases toughness, yield stress, and ductility [87]. The improved mechanical properties promote recycling of PLA while retaining their biodegradability.

Lyu et al. took an alternative approach to toughen the PLA matrix [88]: incorporating poly (butylene-adipate-co-terephthalate) (PBAT) into PLA-g-GMA (PLA grafted with glycidyl methacrylate) and PLA systems. PLA-g-GMA was added as a compatibilizer to stabilize the interface between the PLA and PBAT. All samples containing PLA-g-GMA compatibilizer showed an island phase morphology indicating increased homogeneity and surface compatibility. Samples without PLA-g-GMA compatibilizer had low viscosity with easily seen crystalline sections, indicating that without compatibilizer PBAT and PLA are completely incompatible and, therefore, PLA cannot hinder the PBAT crystallization. With the addition of only 10 wt % compatibilizer the viscosity increased and no PBAT crystalline region were visible.

The introduction of secondary polymers, such as polyhydroxybutyrate (PHB), enables mechanical properties to be tuned without reducing biocompatibility or biodegradability. PHB, a biocompatible, biodegradable, and sustainable polyester, is compatible with PLA [89]. Importantly, the amount of water absorbed into the polymer matrix from the environment affects 3D printability by affecting the viscosity. Also, if parts are dried after printing at 80 °C for 60 min, the materials become much more stable than their non-dried counterparts. Interestingly, printing shape also affects long term stability and material properties [89].

Polyhedral oligomeric silsesquioxane (POSS), another biocompatible and non-toxic polymer nanoparticle, also increases the flexural strength (22%), flexural modulus (9%), and fracture toughness (117%) of 3D-printed PLA/POSS nanocomposites compared to neat PLA [90]. PLA/POSS nanocomposites were formed through melt-mixing of pre-dried samples and then extrusion and FDM 3D printing, resulting in a good distribution of nanoparticles in the PLA matrix. Importantly, the addition of POSS allowed the samples to remain in a phosphate buffered saline solution at 37 °C for 120 days with little mechanical or physical deterioration, suggesting their potential use for longer-term medical implants [90].

Eggshell powder, functionalized with calcium phenylphosphonate, provides nucleation sites when incorporated into PLA matrices. This addition of eggshells, in combination with PEG plasticizer, occurs through melt blending [91]. Incorporating both functionalized eggshell and PEG into the PLA matrix produces a significantly increased nucleation and crystallization rate [91]. Further investigation into the addition of eggshell into PLA showed that eggshells modified with CaO-oleic acid show good dispersion and higher impact strength and elongation at break compared with unmodified or propionic acid modified eggshell composites [92]. These materials also show wider melt-processing windows than neat PLA [92]. Further analysis indicated that the incorporation of eggshell into PLA increases the Young's modulus, thermal resistance, and plasticity [93].

8. Conclusions

PLA is an important biodegradable polymer produced from some of the most renewable feedstocks available. While PLA is useful in FDM 3D printing, its drawbacks—brittleness, poor thermal stability, low crystallization, low elongation at break, poor impact strength, low heat distortion temperature, and limited drawability—reduce its prevalence as an industrial material. Therefore, additives have been incorporated into PLA to form nanocomposites with enhanced mechanical, electrical, or thermal properties.

Cellulose is commonly incorporated into PLA matrices to enhance the mechanical properties while maintaining complete biodegradability. Key findings include:

- Silylation of cellulose increases the cellulose–PLA compatibility and results in increased tensile strength and tensile modulus but decreased glass transition temperature, crystallinity, melting temperature, and crystallization temperature [24].
- Adding cellulose decreases the processable temperatures due to cellulose thermal degradation [24].
- The incorporation of CNFs into PLA decreases voids, increases crystallinity, and decreases defects during the FDM 3D printing process [40].
- When aggregation occurs with cellulose-based additives, recycling and reprinting that material may lead to good dispersion [21].
- Printing in an "isomixed" orientation produces materials that are stronger and have increased stiffness [44].

Metal-based additives have also been explored because of their ability to increase strength, promote bone cell growth, and prevent bacterial growth. The incorporation of various metals, including silver, aluminum, copper, and magnesium, has increased mechanical and antibacterial properties. Specifically:

- At all concentrations, PLA/Ag nanowire nanocomposites show antimicrobial properties against *E. coli*, *P. aeruginosa*, and *S. aureus* [47].
- PLA/Al nanocomposites show increased surface quality, decreased surface roughness, increased storage modulus, decreased loss tangent, increased tensile strength, and increased Young's modulus after laser polishing [48].
- The introduction of mCu-Zn into PLA/wood flour nanocomposites increases the aesthetic appearance by creating a glossy finish while also increasing the antibacterial properties [51].
- The distribution of magnesium nanoparticles increased with the addition of vitamin E, addition of stearic acid, the oxidation of the magnesium to magnesium oxide, or the functionalization of magnesium with PEI and CTAB [53].
- The Mg^{2+} ions released during PLA/MgO degradation increased bone cell growth [57].

Continuous fibers, another PLA additive, are long strands of microscale fibers that allow for increased mechanical strength. While these fibers have incredible strength, they suffer from a lack of compatibility with the PLA matrix; however, many advances in continuous fibers have occurred, specifically:

- Carbon fiber-reinforced PLA shows increased flexural strength, flexural modulus, delamination between layers, and contact pressure, but decreased layer thickness [63].
- The failure modes of carbon fiber/PLA nanocomposites are delamination, delamination-induced matrix cracking, or pull out [64].
- Carbon fiber-reinforced PLA has greater fracture toughness when extruded through square shaped extruder nozzles rather than circular [65].
- Tensile properties decrease with increased layer height and extrusion width [66].
- Carbon fiber-reinforced PLA matrices are not only partially biodegradable but are almost fully recyclable [71].

Carbon-based additives—including carbon nanotubes (CNTs), multiwalled carbon nanotubes (MWCNTs), and graphene nanoplatelets (GNPs)—increase the thermal, electrical, and mechanical properties of PLA nanocomposites. Key findings include:

- Increased CNT content increases degree of crystallinity in PLA/CNT nanocomposites [79].
- Nanocomposites of PLA/1.5% GNP/4.5% MWCNT show no aggregation and are, therefore, viable for FDM 3D printing [74].
- GNP addition increases thermal conductivity more than adding the same concentration of MWCNTs [80].
- Oxidizing MWCNT surfaces enables better incorporation into the PLA matrix with limited agglomeration [81].
- Introducing GNPs decreases creep, decreases wear resistance, and increases overall mechanical properties of PLA [82].
- L-arginine acts as a compatibilizer for GNPs and PLA, increasing the dispersion of GNPs in the PLA matrix [83].
- Oxygen plasma etching of MWCNTs increases the PLA/MWCNT nanocomposite cell adhesion in vivo, while maintaining adequate mechanical properties for bone implants [85].

While incorporating additives into PLA matrices has improved the mechanical properties of PLA nanocomposites, more advances are needed before PLA nanocomposites overtake petroleum-based materials in terms of ease of use and mechanical robustness. Current PLA composites often lack good dispersion and do not accommodate large loadings of nanomaterials in the PLA matrix. This problem has been partially addressed by functionalizing nanoparticles, which enables better surface contact area.

Moving forward, this field would benefit from further exploration of functionalized nanoparticles, specifically introducing functionalities that would stabilize the nanoparticle/PLA interface. Additionally, new additives and new combinations of additives should also be explored. Ultimately, these composites have the potential to replace environmentally harmful, non-degradable materials in a variety of applications, thereby offering a viable approach to a more environmentally responsible future.

Author Contributions: Writing—Original draft preparation, M.B.; Writing—Review and editing, M.D.S. Both authors have read and agreed to the published version of the manuscript.

Funding: This research received no external funding.

Acknowledgments: We would like to acknowledge the Open Access Subvention Fund at Virginia Tech for funding support.

Conflicts of Interest: The authors declare no conflict of interest.

References

1. Popescu, D. FDM process parameters influence over the mechanical properties of polymer specimens: A review. *Polym. Test.* **2018**, *69*, 157–166.
2. Gao, W. The status, challenges, and future of additive manufacturing in engineering. *Comput. Aided Des.* **2015**, *69*, 65–89.
3. Surange, V.G.; Gharat, P.V. 3D printing process using fused deposition modelling (FDM). *Int. Res. J. Eng. Technol.* **2016**, *3*, 1403–1406.
4. Chae, Y.; An, Y.-J. Current research trends on plastic pollution and ecological impacts on the soil ecosystem: A review. *Environ. Pollut.* **2018**, *240*, 387–395. [PubMed]
5. Sharma, S.; Chatterjee, S. Microplastic pollution, a threat to marine ecosystem and human health: A short review. *Environ. Sci. Pollut. Res.* **2017**, *24*, 21530–21547.
6. Li, W.C.; Tse, H.F.; Fok, L. Plastic waste in the marine environment: A review of sources, occurrence and effects. *Sci. Total Environ.* **2016**, *566–567*, 333–349. [PubMed]
7. Dinesh Kumar, S.; Venkadeshwaran, K.; Aravindan, M.K. Fused deposition modelling of PLA reinforced with cellulose nano-crystals. *Mater. Today Proc.* **2020**, *33*, 868–875.
8. Spinelli, G. Nanocarbon/Poly(Lactic) Acid for 3D Printing: Effect of Fillers Content on Electromagnetic and Thermal Properties. *Materials* **2019**, *12*, 2369.
9. Coppola, B. 3D Printing of PLA/clay Nanocomposites: Influence of Printing Temperature on Printed Samples Properties. *Materials* **2018**, *11*, 1947.
10. Levenhagen, N.P.; Dadmun, M.D. Interlayer diffusion of surface segregating additives to improve the isotropy of fused deposition modeling products. *Polymer* **2018**, *152*, 35–41.
11. Ivanov, E. PLA/Graphene/MWCNT composites with improved electrical and thermal properties suitable for FDM 3D printing applications. *Appl. Sci.* **2019**, *9*, 1209.
12. Caminero, M.A. Impact damage resistance of 3D printed continuous fibre reinforced thermoplastic composites using fused deposition modelling. *Compos. Part B Eng.* **2018**, *148*, 93–103.
13. Rajpurohit Shilpesh, R.; Dave Harshit, K. Effect of process parameters on tensile strength of FDM printed PLA part. *Rapid Prototyp. J.* **2018**, *24*, 1317–1324.
14. El Magri, A. Mechanical properties of CF-reinforced PLA parts manufactured by fused deposition modeling. *J. Compos. Mater.* **2019**. [CrossRef]
15. Valerga, A.P. Influence of PLA Filament Conditions on Characteristics of FDM Parts. *Materials* **2018**, *11*, 1322.
16. Rodríguez-Panes, A.; Claver, J.; Camacho, A.M. The Influence of Manufacturing Parameters on the Mechanical Behaviour of PLA and ABS Pieces Manufactured by FDM: A Comparative Analysis. *Materials* **2018**, *11*, 1333.
17. Naveed, N. Investigate the effects of process parameters on material properties and microstructural changes of 3D-printed specimens using fused deposition modelling (FDM). *Mater. Technol.* **2020**, 1–14. [CrossRef]
18. Huang, B. Optimizing 3D printing performance of acrylonitrile-butadiene-styrene composites with cellulose nanocrystals/silica nanohybrids. *Polym. Int.* **2019**, *68*, 1351–1360.
19. Farah, S.; Anderson, D.G.; Langer, R. Physical and mechanical properties of PLA, and their functions in widespread applications—A comprehensive review. *Adv. Drug Deliv. Rev.* **2016**, *107*, 367–392.
20. Clarkson, C.M. Crystallization kinetics and morphology of small concentrations of cellulose nanofibrils (CNFs) and cellulose nanocrystals (CNCs) melt-compounded into poly(lactic acid) (PLA) with plasticizer. *Polymer* **2020**, *187*, 122101.
21. Frone, A.N. Morpho-Structural, Thermal and Mechanical Properties of PLA/PHB/Cellulose Biodegradable Nanocomposites Obtained by Compression Molding, Extrusion, and 3D Printing. *Nanomaterials* **2019**, *10*, 51. [CrossRef] [PubMed]
22. Wilson, J.A. Magnesium catalyzed polymerization of end functionalized poly (propylene maleate) and poly (propylene fumarate) for 3D printing of bioactive scaffolds. *J. Am. Chem. Soc.* **2018**, *140*, 277–284. [CrossRef] [PubMed]
23. Elsawy, M.A. Hydrolytic degradation of polylactic acid (PLA) and its composites. *Renew. Sustain. Energy Rev.* **2017**, *79*, 1346–1352. [CrossRef]
24. Qian, S. Improved properties of PLA biocomposites toughened with bamboo cellulose nanowhiskers through silane modification. *J. Mater. Sci.* **2018**, *53*, 10920–10932. [CrossRef]

25. Adesina, O.T. Mechanical property prediction of SPS processed GNP/PLA polymer nanocomposite using artificial neural network. *Cogent Eng.* **2020**, *7*, 1720894. [CrossRef]
26. Melocchi, A. Hot-melt extruded filaments based on pharmaceutical grade polymers for 3D printing by fused deposition modeling. *Int. J. Pharm.* **2016**, *509*, 255–263. [CrossRef]
27. Chacón, J.M. Additive manufacturing of PLA structures using fused deposition modelling: Effect of process parameters on mechanical properties and their optimal selection. *Mater. Des.* **2017**, *124*, 143–157. [CrossRef]
28. Yao, T. Tensile failure strength and separation angle of FDM 3D printing PLA material: Experimental and theoretical analyses. *Compos. Part B Eng.* **2020**, *188*, 107894. [CrossRef]
29. Rismalia, M. Infill pattern and density effects on the tensile properties of 3D printed PLA material. *J. Phys. Conf. Ser.* **2019**, *1402*, 044041. [CrossRef]
30. Vigneshwaran, K.; Venkateshwaran, N. Statistical analysis of mechanical properties of wood-PLA composites prepared via additive manufacturing. *Int. J. Polym. Anal. Charact.* **2019**, *24*, 584–596. [CrossRef]
31. Liao, Y. Effect of Porosity and Crystallinity on 3D Printed PLA Properties. *Polymers* **2019**, *11*, 1487. [CrossRef] [PubMed]
32. Wang, L.; Gardner, D.J. Contribution of printing parameters to the interfacial strength of polylactic acid (PLA) in material extrusion additive manufacturing. *Prog. Addit. Manuf.* **2018**, *3*, 165–171. [CrossRef]
33. Wang, L.; Gramlich, W.M.; Gardner, D.J. Improving the impact strength of Poly(lactic acid) (PLA) in fused layer modeling (FLM). *Polymer* **2017**, *114*, 242–248. [CrossRef]
34. Aziz, R.; Ul Haq, M.I.; Raina, A. Effect of surface texturing on friction behaviour of 3D printed polylactic acid (PLA). *Polym. Test.* **2020**, *85*, 106434. [CrossRef]
35. Cicala, G. Polylactide (PLA) filaments a biobased solution for additive manufacturing: Correlating rheology and thermomechanical properties with printing quality. *Materials* **2018**, *11*, 1191. [CrossRef]
36. Cuiffo, M.A. Impact of the Fused Deposition (FDM) Printing Process on Polylactic Acid (PLA) Chemistry and Structure. *Appl. Sci.* **2017**, *7*, 579. [CrossRef]
37. Patanwala, H.S. The microstructure and mechanical properties of 3D printed carbon nanotube-polylactic acid composites. *Polym. Compos.* **2018**, *39*, E1060–E1071. [CrossRef]
38. Long, H. Mechanical and thermal properties of bamboo fiber reinforced polypropylene/polylactic acid composites for 3D printing. *Polym. Eng. Sci.* **2019**, *59*, E247–E260. [CrossRef]
39. Qian, S.; Zhang, H.; Sheng, K. Cellulose nanowhiskers from moso bamboo residues: Extraction and characterization. *BioResources* **2017**, *12*, 419–433. [CrossRef]
40. Ambone, T.; Torris, A.; Shanmuganathan, K. Enhancing the mechanical properties of 3D printed polylactic acid using nanocellulose. *Polym. Eng. Sci.* **2020**, *60*, 1842–1855. [CrossRef]
41. Xiao, X. Polylactide/hemp hurd biocomposites as sustainable 3D printing feedstock. *Compos. Sci. Technol.* **2019**, *184*, 107887. [CrossRef]
42. Guessasma, S.; Belhabib, S.; Nouri, H. Microstructure and Mechanical Performance of 3D Printed Wood-PLA/PHA Using Fused Deposition Modelling: Effect of Printing Temperature. *Polymers* **2019**, *11*, 1778. [CrossRef] [PubMed]
43. Mohammadi, M. CNC dispersion in PLA and PBAT using two solvents: Morphological and rheological properties. *Cellulose* **2020**, *27*, 9877–9892. [CrossRef]
44. Estakhrianhaghighi, E. 3D-Printed Wood-Fiber Reinforced Architected Cellular Composites. *Adv. Eng. Mater.* **2020**, *20*, 2000565. [CrossRef]
45. Scaffaro, R. Lignocellulosic fillers and graphene nanoplatelets as hybrid reinforcement for polylactic acid: Effect on mechanical properties and degradability. *Compos. Sci. Technol.* **2020**, *190*, 108008. [CrossRef]
46. Bayraktar, I. 3D printed antibacterial silver nanowire/polylactide nanocomposites. *Compos. Part B Eng.* **2019**, *172*, 671–678. [CrossRef]
47. Podstawczyk, D. Preparation of antimicrobial 3D printing filament: In situ thermal formation of silver nanoparticles during the material extrusion. *Polym. Compos.* **2020**, *41*, 4692–4705. [CrossRef]
48. Chen, L. Laser polishing of Cu/PLA composite parts fabricated by fused deposition modeling: Analysis of surface finish and mechanical properties. *Polym. Compos.* **2020**, *41*, 1356–1368. [CrossRef]
49. Chen, L.; Zhang, X. Modification the surface quality and mechanical properties by laser polishing of Al/PLA part manufactured by fused deposition modeling. *Appl. Surf. Sci.* **2019**, *492*, 765–775. [CrossRef]
50. Butt, J. Hybrid Manufacturing and Mechanical Characterization of Cu/PLA Composites. *Arab. J. Sci. Eng.* **2020**, *45*, 9339–9356. [CrossRef]

51. Yang, F. Micrometer Copper-Zinc Alloy Particles-Reinforced Wood Plastic Composites with High Gloss and Antibacterial Properties for 3D Printing. *Polymers* **2020**, *12*, 621. [CrossRef] [PubMed]
52. Antoniac, I. Magnesium filled polylactic acid (PLA) material for filament based 3D printing. *Materials* **2019**, *12*, 719. [CrossRef] [PubMed]
53. Ferrández-Montero, A. Study of the matrix-filler interface in PLA/Mg composites manufactured by Material Extrusion using a colloidal feedstock. *Addit. Manuf.* **2020**, *33*, 101142. [CrossRef]
54. Cifuentes, S.C. In vitro degradation of biodegradable polylactic acid/magnesium composites: Relevance of Mg particle shape. *Acta Biomater.* **2016**, *32*, 348–357. [CrossRef]
55. Zhao, C. Development of PLA/Mg composite for orthopedic implant: Tunable degradation and enhanced mineralization. *Compos. Sci. Technol.* **2017**, *147*, 8–15. [CrossRef]
56. Li, X. In vitro degradation kinetics of pure PLA and Mg/PLA composite: Effects of immersion temperature and compression stress. *Acta Biomater.* **2017**, *48*, 468–478. [CrossRef]
57. Liang, H. Fabrication, Crystalline Behavior, Mechanical Property and In-Vivo Degradation of Poly(l–lactide) (PLLA)–Magnesium Oxide Whiskers (MgO) Nano Composites Prepared by In-Situ Polymerization. *Polymers* **2019**, *11*, 1123. [CrossRef]
58. Zhao, Y. Effects of modifying agents on surface modifications of magnesium oxide whiskers. *Appl. Surf. Sci.* **2016**, *388*, 370–375. [CrossRef]
59. Zhao, Y. Effects of MgO whiskers on mechanical properties and crystallization behavior of PLLA/MgO composites. *Mater. Des.* **2016**, *89*, 573–581. [CrossRef]
60. Zhao, Y. Effects of Magnesium Oxide (MgO) Shapes on In Vitro and In Vivo Degradation Behaviors of PLA/MgO Composites in Long Term. *Polymers* **2020**, *12*, 1074. [CrossRef]
61. Zhao, Y. The Degradation Properties of MgO Whiskers/PLLA Composite In Vitro. *Int. J. Mol. Sci.* **2018**, *19*, 2740. [CrossRef] [PubMed]
62. Scaffaro, R.; Lopresti, F.; Botta, L. PLA based biocomposites reinforced with Posidonia oceanica leaves. *Compos. Part B Eng.* **2018**, *139*, 1–11. [CrossRef]
63. Tian, X. Interface and performance of 3D printed continuous carbon fiber reinforced PLA composites. *Compos. Part A Appl. Sci. Manuf.* **2016**, *88*, 198–205. [CrossRef]
64. Heidari-Rarani, M.; Rafiee-Afarani, M.; Zahedi, A.M. Mechanical characterization of FDM 3D printing of continuous carbon fiber reinforced PLA composites. *Compos. Part B Eng.* **2019**, *175*, 107147. [CrossRef]
65. Papon, E.A.; Haque, A. Fracture toughness of additively manufactured carbon fiber reinforced composites. *Addit. Manuf.* **2019**, *26*, 41–52. [CrossRef]
66. Dou, H. Effect of Process Parameters on Tensile Mechanical Properties of 3D Printing Continuous Carbon Fiber-Reinforced PLA Composites. *Materials* **2020**, *13*, 3850. [CrossRef]
67. Le Duigou, A. 3D printing of continuous flax fibre reinforced biocomposites for structural applications. *Mater. Des.* **2019**, *180*, 107884. [CrossRef]
68. Le Duigou, A. Tailoring the mechanical properties of 3D-printed continuous flax/PLA biocomposites by controlling the slicing parameters. *Compos. Part B Eng.* **2020**, *203*, 108474. [CrossRef]
69. Zhang, H. Three-Dimensional Printing of Continuous Flax Fiber-Reinforced Thermoplastic Composites by Five-Axis Machine. *Materials* **2020**, *13*, 1678. [CrossRef]
70. Suteja, J. Properties investigation of 3D printed continuous pineapple leaf fiber-reinforced PLA composite. *J. Thermoplast. Compos. Mater.* **2020**. [CrossRef]
71. Tian, X. Recycling and remanufacturing of 3D printed continuous carbon fiber reinforced PLA composites. *J. Clean. Prod.* **2017**, *142*, 1609–1618. [CrossRef]
72. Rahimizadeh, A. Recycling of fiberglass wind turbine blades into reinforced filaments for use in Additive Manufacturing. *Compos. Part B Eng.* **2019**, *175*, 107101. [CrossRef]
73. Pentek, A. The Effect of Printing Parameters on Electrical Conductivity and Mechanical Properties of PLA and ABS Based Carbon Composites in Additive Manufacturing of Upper Limb Prosthetics. *Crystals* **2020**, *10*, 398. [CrossRef]
74. Spinelli, G. Morphological, Rheological and Electromagnetic Properties of Nanocarbon/Poly(lactic) Acid for 3D Printing: Solution Blending vs. Melt Mixing. *Materials* **2018**, *11*, 2256. [CrossRef] [PubMed]
75. Kotsilkov, S.; Ivanov, E.; Vitanov, N.K. Release of Graphene and Carbon Nanotubes from Biodegradable Poly(Lactic Acid) Films during Degradation and Combustion: Risk Associated with the End-of-Life of Nanocomposite Food Packaging Materials. *Materials* **2018**, *11*, 2346. [CrossRef]

76. Velichkova, H. Influence of polymer swelling and dissolution into food simulants on the release of graphene nanoplates and carbon nanotubes from poly(lactic) acid and polypropylene composite films. *J. Appl. Polym. Sci.* **2017**, *134*, 45469. [CrossRef]
77. Velichkova, H. Release of carbon nanoparticles of different size and shape from nanocomposite poly(lactic) acid film into food simulants. *Food Addit. Contam. Part A* **2017**, *34*, 1072–1085. [CrossRef]
78. Yang, L. Effects of carbon nanotube on the thermal, mechanical, and electrical properties of PLA/CNT printed parts in the FDM process. *Synth. Met.* **2019**, *253*, 122–130. [CrossRef]
79. Zhou, X. Additive manufacturing of CNTs/PLA composites and the correlation between microstructure and functional properties. *J. Mater. Sci. Technol.* **2021**, *60*, 27–34. [CrossRef]
80. Batakliev, T. Nanoindentation analysis of 3D printed poly (lactic acid)-based composites reinforced with graphene and multiwall carbon nanotubes. *J. Appl. Polym. Sci.* **2019**, *136*, 47260. [CrossRef]
81. Kotsilkova, R. Essential Nanostructure Parameters to Govern Reinforcement and Functionality of Poly(lactic) Acid Nanocomposites with Graphene and Carbon Nanotubes for 3D Printing Application. *Polymers* **2020**, *12*, 1208. [CrossRef] [PubMed]
82. Bustillos, J. Integration of graphene in poly(lactic) acid by 3D printing to develop creep and wear-resistant hierarchical nanocomposites. *Polym. Compos.* **2018**, *39*, 3877–3888. [CrossRef]
83. Wang, Y. 3D printing biocompatible l-Arg/GNPs/PLA nanocomposites with enhanced mechanical property and thermal stability. *J. Mater. Sci.* **2020**, *55*, 5064–5078. [CrossRef]
84. Camargo, J.C. Mechanical properties of PLA-graphene filament for FDM 3D printing. *Int. J. Adv. Manuf. Technol.* **2019**, *103*, 2423–2443. [CrossRef]
85. Silva, E. PDLLA honeycomb-like scaffolds with a high loading of superhydrophilic graphene/multi-walled carbon nanotubes promote osteoblast in vitro functions and guided in vivo bone regeneration. *Mater. Sci. Eng. C* **2017**, *73*, 31–39. [CrossRef] [PubMed]
86. Sweeney, C.B. Welding of 3D-printed carbon nanotube–polymer composites by locally induced microwave heating. *Sci. Adv.* **2017**, *3*, e1700262. [CrossRef]
87. Ahmed, W.; Siraj, S.; Al-Marzouqi, A.H. 3D Printing PLA Waste to Produce Ceramic Based Particulate Reinforced Composite Using Abundant Silica-Sand: Mechanical Properties Characterization. *Polymers* **2020**, *12*, 2579. [CrossRef]
88. Lyu, Y. Manipulating phase structure of biodegradable PLA/PBAT system: Effects on dynamic rheological responses and 3D printing. *Compos. Sci. Technol.* **2020**, *200*, 108399. [CrossRef]
89. Findrik Balogová, A. Determination of geometrical and viscoelastic properties of PLA/PHB samples made by additive manufacturing for urethral substitution. *J. Biotechnol.* **2018**, *284*, 123–130. [CrossRef]
90. Meyva-Zeybek, Y.; Kaynak, C. A comparative study for the behavior of 3D-printed and compression molded PLA/POSS nanocomposites. *J. Appl. Polym. Sci.* **2020**. [CrossRef]
91. Li, Y. Crystallization behaviors of poly(lactic acid) composites fabricated using functionalized eggshell powder and poly(ethylene glycol). *Thermochim. Acta* **2018**, *663*, 67–76. [CrossRef]
92. Threepopnatkul, P. Preparation and Properties of Polylactide Reinforced with Eggshell Modified with Different Fatty Acids. *Key Eng. Mater.* **2019**, *824*, 16–22. [CrossRef]
93. Urtekin, G.; Hazer, S.; Aytac, A. Effect of eggshell and intumescent flame retardant on the thermal and mechanical properties of plasticised PLA. *Plast. Rubber Compos.* **2020**, 1–10. [CrossRef]

Publisher's Note: MDPI stays neutral with regard to jurisdictional claims in published maps and institutional affiliations.

© 2020 by the authors. Licensee MDPI, Basel, Switzerland. This article is an open access article distributed under the terms and conditions of the Creative Commons Attribution (CC BY) license (http://creativecommons.org/licenses/by/4.0/).

MDPI
St. Alban-Anlage 66
4052 Basel
Switzerland
Tel. +41 61 683 77 34
Fax +41 61 302 89 18
www.mdpi.com

Nanomaterials Editorial Office
E-mail: nanomaterials@mdpi.com
www.mdpi.com/journal/nanomaterials